NATURAL CLIMATE VARIABILITY
AND GLOBAL WARMING

NATURAL CLIMATE VARIABILITY AND GLOBAL WARMING: A HOLOCENE PERSPECTIVE

EDITED BY

RICHARD W. BATTARBEE
AND
HEATHER A. BINNEY

ENVIRONMENTAL CHANGE RESEARCH CENTRE,
UNIVERSITY COLLEGE LONDON

A John Wiley & Sons, Ltd., Publication

This edition first published 2008, © 2008 by Blackwell Publishing Ltd

Blackwell Publishing was acquired by John Wiley & Sons in February 2007. Blackwell's publishing program has been merged with Wiley's global Scientific, Technical and Medical business to form Wiley-Blackwell.

Registered office: John Wiley & Sons Ltd, The Atrium, Southern Gate, Chichester, West Sussex, PO19 8SQ, UK

Editorial offices: 9600 Garsington Road, Oxford, OX4 2DQ, UK
The Atrium, Southern Gate, Chichester, West Sussex, PO19 8SQ, UK
111 River Street, Hoboken, NJ 07030-5774, USA

For details of our global editorial offices, for customer services and for information about how to apply for permission to reuse the copyright material in this book please see our website at www.wiley.com/wiley-blackwell

Library of Congress Cataloguing-in-Publication Data

Natural climate variability and global warming : a Holocene
perspective / edited by R.W. Battarbee and H.A. Binney.
      p. cm.
   Includes bibliographical references and index.
   ISBN 978-1-4051-5905-0 (hardcover : alk. paper)
   1. Paleoclimatology—Holocene.   2. Climatic changes—Research.
3. Global warming—Research.   I. Battarbee, R.W.   II. Binney, H.A.
(Heather A.)

   QC884.2.C5N38 2008
   551.609′01—dc22

                                                2007038976

ISBN: 978-1-4051-5905-0

A catalogue record for this book is available from the British Library.

Set in 10/12.5pt Minion
by Graphicraft Limited, Hong Kong
Printed and bound in Singapore
by Fabulous Printers Pte Ltd

1   2008

Contents

Contributors

Carin Andersson Bjerknes Centre for Climate Research, Allégaten 55, 5007 Bergen, Norway

Richard W. Battarbee Environmental Change Research Centre, University College London, 26 Bedford Way, London WC1E 6BT, UK

Jürg Beer Swiss Federal Institute of Aquatic Science and Technology (EAWAG), CH-8600 Dübendorf, Switzerland

H. John B. Birks Department of Biology, and Bjerknes Centre for Climate Research, University of Bergen, N-5007 Bergen, Norway, and Environmental Change Research Centre, University College London, London WC1E 6BT, UK

Raymond S. Bradley Climate System Research Center, Department of Geosciences, University of Massachusetts Morrill Science Center, 611 North Pleasant Street, Amherst, MA 01003-9297, USA

Dan J. Charman School of Geography, University of Plymouth, Plymouth, Devon PL4 8AA, UK

Martin Claussen Max-Planck-Institute for Meteorology and Meteorological Institute, University Hamburg, Bundesstr. 53, D-20146 Hamburg, Germany

Michel Crucifix Institut d'Astronomie et de Géophysique G. Lemaître, Université Catholique de Louvain, 2 Chemin du Cyclotron, B-1348 Louvain-la-Neuve, Belgium

Bas van Geel Institute for Biodiversity and Ecosystem Dynamics, Universiteit van Amsterdam, Kruislaan 318, 1098 SM Amsterdam, The Netherlands

Hugues Goosse Université Catholique de Louvain, Institut d'Astronomie et de Géophysique G. Lemaître, Chemin du Cyclotron, 2, B-1348 Louvain-la-Neuve, Belgium

Eystein Jansen Bjerknes Centre for Climate Research, Allégaten 55, 5007 Bergen, Norway and Department of Earth Science, University of Bergen, Norway

Michael E. Mann Department of Meteorology and Earth and Environmental Systems Institute (EESI), Pennsylvania State University, University Park, PA 16802-5013, USA

Matthias Moros Bjerknes Centre for Climate Research, Allégaten 55, 5007 Bergen, Norway and Institut für Ostseeforschung, Warnemunde, Germany

Kerim H. Nisancioglu Bjerknes Centre for Climate Research, Allégaten 55, 5007 Bergen, Norway

Birgitte F. Nyland Bjerknes Centre for Climate Research, Allégaten 55, 5007 Bergen, Norway and Department of Earth Science, University of Bergen, Norway

Frank Oldfield Department of Geography, University of Liverpool, Liverpool L69 7ZT, UK

Hans Renssen Faculty of Earth and Life Sciences, Vrije Universiteit Amsterdam, De Boelelaan 1085, 1081 HV Amsterdam, The Netherlands

Richard J. Telford Bjerknes Centre for Climate Research, Allégaten 55, 5007 Bergen, Norway and Department of Biology, University of Bergen, Norway

Dirk Verschuren Limnology Unit, Department of Biology, Ghent University, K.L. Ledeganckstraat 35, B-9000 Gent, Belgium

Abbreviations, acronyms, and terminology

AMO	Atlantic Multi-decadal Oscillation
AO	Arctic Oscillation (also known as Northern Annular Mode)
AOGCM	Atmosphere–Ocean General Circulation Model
AUPs	abrupt, unprecedented and persistent climate anomalies
CCM	Community Climate Model
CCSM	Community Climate System Model
CFR	climate field reconstruction
CLIMAP	Climate: Mapping, Analysis, and Prediction
CLIMBER	CLIMate and BiosphERe Group – a climate model of the Climate System Department, Potsdam Institute for Climate Impact Research
CLIO	coupled large-scale ice–ocean
COHMAP	Co-operative Holocene Mapping Project
CPS	composite plus scale
CSM	climate space model
DATUN	data assimilation through upscaling and nudging
DPSIR	drivers–pressure–state–impact–response
EBM	Energy Balance Model
ECBILT	ECBilt is a coupled atmosphere–ocean–sea-ice model containing simplified parameterizations of the subgrid-scale physical processes (an intermediate-complexity climate model)
ECHO-G	ECHO-G is a global coupled atmosphere–ocean climate model consisting of two-component models, the atmospheric component (ECHAM4) and the oceanic component (HOPE-G)
ELLDB	European Lake-level Database
EMICS	Earth models of intermediate complexity
ENSO	El Niño–Southern Oscillation
EOF	empirical orthogonal function
EPICA	European Programme for Ice Coring in Antarctica
ERIK	GCM-based experiment run for the past 1000 years with the ECHO-G model (the "Erik" simulation)
GCM	General Circulation Model
GDD	growing degree days
GFDL	Geophysical Fluid Dynamics Laboratory

GISS	Goddard Institute for Space Studies
GKSS	GKSS Research Centre, Hamburg, Germany
IGBP	International Geosphere–Biosphere Program
IPCC	Intergovernmental Panel on Climate Change
IPSL	Institut Pierre et Simon Laplace [climate model of]
ITCZ	Intertropical Convergence Zone
HOLIVAR	European Science Programme on "Holocene Climate Variability"
LGM	Last Glacial Maximum
LIA	Little Ice Age
LLN-2D	Louvain-la-Neuve [two-dimensional climate model of]
MCA	Medieval Climate Anomaly
MIS	Marine Isotope Stage
MIT-EMIC	Michigan Institute of Technology – Earth System Model of Intermediate Complexity
MoBidiC	Modèle Bidimensionnel du Climat
MODIS	Moderate Resolution Imaging Spectroradiometer (NASA/EOS instrument)
MWP	Medieval Warm Period
NCAR	National Center for Atmospheric Research
NOA	North Atlantic Oscillation
PAGES	Past Global Changes
PANASH	Paleoclimate and Environments of the Northern and Southern Hemispheres
PCI	Polar Circulation Index
PDF	probability density function
PDO	Pacific Decadal Oscillation
PEP	Pole–Equator–Pole
PMIP	Paleoclimate Modeling Intercomparison Project
POLLANDCAL	POLlen-LANDscape CALibration
RegEM	regularized expectation maximization
SNR	signal to noise ratio
SOI	Southern Oscillation Index
SSI	spectral solar irradiance
TIMS	thermal ionization mass spectrometry
TSI	total solar irradiance
YD	Younger Dryas

Dating conventions

Unless otherwise stated, the following conventions for dates are used in the text and figures:

BP	Before Present (Present = 1950) when associated with a radiocarbon date
cal. years BP	years before present for radiocarbon dates after calibration to calendar years using a standard function
years BP	years before present where mixed chronologies are shown on one figure (e.g. uranium series and radiocarbon dates)
years AD	calendar age
ka	thousand years ago
kyr	thousand year time span

Holocene climate variability and global warming

Richard W. Battarbee

Keywords

Holocene, natural climate variability, global warming

Introduction

This book addresses one of the key questions facing climate scientists today: how important is natural variability in explaining global warming? The book aims to place the past few decades of warming in the context of longer term climate variability and considers the causes of such variability on different time-scales through the Holocene, the period of Earth history covering approximately the past 11 500 years that has elapsed since the last major Ice Age. In particular it reviews the evidence for past climate change based on the analysis of data from naturally occurring climate archives (such as tree rings, peat bogs, corals, and lake and marine sediments) and describes progress being made in developing the climate models needed to simulate and explain past climate variability. It also considers how people in the past have changed the environment and responded to climate change.

Over the past decade it has become increasingly clear that there is now a human contribution to global warming (IPCC 2007). Antarctic ice-core records (e.g. Petit 1999; EPICA Community Members 2004) show that greenhouse-gas concentrations are already higher than at any time in the past 750 000 years, temperatures in the Northern Hemisphere are now on average probably higher than the previous 1000 years (Mann *et al.* 1998) and climate models can only simulate temperatures accurately over the past 150 years if greenhouse gases are included as a forcing mechanism (Stott *et al.* 2001).

Evidence is also accumulating to suggest that changes in natural ecosystems that can be unambiguously attributed to rising temperatures are also occurring. In particular most mountain glaciers across the world are receding (Oerlemans 2005) and unprecedented changes in the ecology of remote arctic lake ecosystems have been recorded by lake sediments (Smol *et al.* 2005).

The evidence for human impact on the climate system is thought now to be so compelling that Crutzen has argued that the recent period of Earth history dating from the late 18th century increase in atmospheric CO_2 should be given a new geologic name, the Anthropocene (Crutzen and Stoermer 2000). Indeed Ruddiman has even argued that human activity may have affected atmospheric greenhouse-gas concentrations much earlier in the Holocene as a result of deforestation and land-cover change associated with early agriculture (Ruddiman 2003).

Yet despite the strength of the evidence for human-induced change, climate-change sceptics still remain, arguing that the role of natural variability is being underestimated. It can indeed be maintained that recent changes in climate, exemplified by ice-cover loss on lakes (Magnuson *et al.* 2000) or earlier spring flowering (Menzel *et al.* 2006) are still within the long-term natural range of the climate system, if viewed on centennial time-scales. In Europe, for example, historians can point to the more northerly cultivation of vines in Medieval and Roman times and in Africa major periods of very low lake-levels in previous centuries are well documented (e.g. Verschuren 2004).

This debate, about the relative importance of natural variability and pollutant greenhouse gases in explaining recent warming, is therefore still very much alive. In this book we consider this issue in a Holocene perspective. We present evidence for climate change on different time-scales using both paleoclimate reconstructions and modeling, and we include the results of recent research from both high- and low-latitude environments.

Preview

The opening two chapters by John Birks and Frank Oldfield, respectively, provide a comprehensive introductory context for the chapters that follow. John Birks traces Holocene research back to its roots in the early 19th century and describes early debates, principally in Scandinavia, about the interpretation of plant remains preserved in peat bogs and their relevance to climate change. His account takes in the development of pollen analysis and radiocarbon dating, the use of transfer functions in an attempt to quantify past climate reconstruction from proxy records and the pioneering work of COHMAP (Cooperative Holocene Mapping Project) in paleoclimate modeling. He highlights the principal debates and developments in Holocene climate change research that have taken place in recent years and points specifically to the importance of understanding the spatial as well as temporal component of natural climate variability.

Frank Oldfield's chapter is concerned with the role of people in the Holocene. He stresses the need to take into account a much longer history of interactions between human activity and climate change than simply the very recent past. Using data from many different regions he shows that people, especially in the Old World, have had a major impact on land-use and land-cover over many millennia. He argues that the extent of land-cover change may have been sufficient to modify local and regional climate and that these changes in turn were responsible for causing alterations in the hydrologic cycle and in soil erosion. The evidence for land-cover change presented is not inconsistent with Ruddiman's claim (see above) that early agriculture may have been the cause of increased atmospheric greenhouse gas concentrations over the past 8000 years and Oldfield stresses the need for further paleoecological research to test this hypothesis. Finally he reviews evidence for the interaction between climate change and human society in the past and calls for a more balanced dialogue between the physical and social science communities in debating this issue, calling on the need to develop models that couple biophysical and social systems and that acknowledge the adaptive nature of human society.

Chapter 4 by Michel Crucifix is divided into two main sections. The first describes the principles of climate modeling. He stresses the difficulty of modeling an inherently complex and chaotic system and the need for models of different kinds: conceptual, comprehensive and intermediate. He points out the importance of specifying initial conditions and boundary conditions, describes some of the problems of parameterization and the different ways in which equilibrium or transient experiments are conducted. He also indicates how paleodata are used by modelers, not only for direct comparison of output, but also, using data assimilation techniques, for providing improved model parameterization. The second section of the chapter is concerned with the results of two model applications. The first asks the question "how long will the Holocene last?", a question relevant to the debate opened by Ruddiman (see above) about the role of human activity in the early Holocene in increasing atmospheric concentrations of greenhouse gases. Output from models of intermediate complexity do not rule out the Ruddiman hypothesis in scenarios where early Holocene CO_2 concentrations are allowed to fall below 240 ppmv (parts per million by volume). On the other hand projections forward from the present-day suggest that glacial conditions are not now likely to return for approximately 50 000 years.

The final section of Crucifix's chapter focuses on ocean stability through the Holocene. He argues that regional ocean instabilities, such as sudden coolings related to the reduction in deep-ocean convection, could have occurred throughout the Holocene in the North Atlantic by convective feedback related to interactions with sea-ice and atmosphere dynamics.

Eystein Jansen and colleagues review data from the North Atlantic region that indicate the Holocene "climate optimum" is recorded in many but not all marine sediment cores and that different proxies from the same core have different patterns. The differences are attributed to the seasonality of the insolation forcing and the relationship of the proxy to surface ocean stratification. The results indicate that the thermal maximum is mainly caused by orbital forcing, enhanced by sea-ice albedo feedbacks. The data also show that on shorter century to millennial time-scales variability is an important aspect of the marine climate in the high-latitude Atlantic Ocean, possibly increasing after the end of the thermal maximum. There is little evidence for stationary cyclicity in this variability and the authors conclude that the variability may be a response to long time-scale dynamics of the climate system and not necessarily to a specific external forcing factor.

Jürg Beer and Bas van Geel give an overview of the mechanisms causing natural climate change on decadal to millennial time-scales focusing especially on solar forcing. They argue that although the change in total solar irradiance over the course of an 11-year Schwabe cycle is quite small, the variability of the solar radiation is strongly wavelength dependent and that large changes in the spectral solar irradiance strongly influence photochemistry in the upper atmosphere, and in particular the ozone concentration, which may cause shifts in the tropospheric circulation systems and therefore climate. As direct measurements of solar irradiance are lacking prior to the advent of satellite technologies, evidence for centennial-scale change needs to be derived from proxy records, especially from ^{10}Be from ice cores and ^{14}C from tree rings.

In the second half of their chapter Beer and van Geel argue that there are a rapidly growing number of examples of Holocene climate change that point to the Sun as a major forcing factor. They present the well-known 850 BC event, equivalent to the Sub-boreal–Sub-atlantic transition in the original Blytt and Sernander scheme for the Holocene as a good example. This event is associated with increased peat bog growth and lake-level increase in north-west Europe and with changing husbandry and agricultural practices in south-central Siberia and central Africa. The beginning of this event is coincident with a significant increase in the atmospheric production of ^{14}C. By extension they argue that as the amplification mechanisms for changing solar activity are not well understood, and therefore cannot yet be sufficiently quantified in climate models, solar forcing of climate change may be more important than has been suggested to date. They argue that if the Little Ice Age and the subsequent warming were mainly driven by changes in solar activity this component of natural forcing may well play an important role in estimating future trends in climate.

In Chapter 7, Hugues Goosse, Michael Mann, and Hans Renssen present a strong defence of the "hockey-stick" curve of Northern Hemisphere temperature trends for the past 1000 years, pointing out that since the first curve was presented (Mann *et al.* 1998) there are now several additional independent analyses covering the same period. All are essentially in agreement in showing anomalously high temperatures over the past few decades. Goosse *et al.* also show from data–model comparisons how natural (especially volcanic) forcing could explain many features of pre-19th climate variability, including the regional patterns of change associated with the North Atlantic Oscillation (NAO) and El Niño.

Goosse *et al.* present simulations of the past 1000 years that explore the separate and combined role of internal and forced variability. They present model results that show how temperature differences between regions could be due to spatial responses to particular forcings and/or to internal variability. They also show how progress could be made in data–model comparisons by using paleoproxy data to select the best realization in an ensemble. In this way a climate reconstruction could be derived that was consistent with the paleorecord, model physics, and the forcings.

Dirk Verschuren and Dan Charman stress the difficulty of relating past hydrologic variability on decadal to century time-scales to external forcing. Using proxy evidence from Europe and Africa, however, they argue that a number of periods of cooler and wetter conditions inferred from peatland and lake-level changes in Europe correspond to periods of reduced solar activity. In Africa there is evidence for substantial spatial variation across the continent, but some evidence, especially in eastern Equatorial Africa, for an inverse relationship between solar activity and moisture.

Martin Claussen presents evidence that rapid climate change, capable of affecting early civilizations, occurred in North Africa during the Holocene, and that the climate at 5500 years BP was especially unstable. Earth system models are now capable of simulating rapid swings between arid and wet phases in the past but may not yet be able to reliably predict future transitions.

Finally, Ray Bradley provides a perspective on Holocene climate change and presents an array of evidence to demonstrate the relevance of understanding past climate in order to provide insights for the future. He stresses the importance of reconstructing the history of climate forcing and the need to understand the causes and consequences of rapid changes especially AUPs (abrupt, unprecedented and persistent climate anomalies), for which there are many examples, but mainly droughts, in the paleorecord.

Acknowledgments

The chapters in this book are based on the keynote lectures delivered at the University College London (UCL) Open Science Meeting in June 2006. That meeting was financed by the European Science Foundation (ESF), with co-funding from the International Geosphere–Biosphere Programme and Past Global Changes (IGBP–PAGES), who sponsored bursaries for young scientists from Developing Countries. I also thank UCL who provided conference facilities and Heather Binney and Mike Hughes who were the principal organizers. The production of this book has further benefited from the help of many expert reviewers, from Cathy Jenks who carried out the technical editing and from Heather Binney, my co-editor, without whom little would have been possible.

Finally I would like to thank all those who have contributed to the success of HOLIVAR (Holocene Climate Variability) over the past few years: to the Steering Committee, the organizers of the workshops and training courses, the tutors on the training courses, the participants in the workshops and training courses, the UCL-based administrators, Andrew McGovern, Heather Binney and Cath Rose, and the support team in the ESF, especially Joanne Goetz.

References

Crutzen P.J. & Stoermer E.F. (2000) The "Anthropocene". *Global Change News-letter*, **41**, 12–13.

EPICA Community Members (2004) Eight glacial cycles from an Antarctic ice core. *Nature*, **429**, 623–628.

IPCC (2007) *Climate Change 2007 – The Physical Science Basis Working Group I.* Contribution to the Fourth Assessment Report of the Intergovernmental Panel on Climate Change, Cambridge University Press.

Magnuson J.J., Robertson D.M., Benson B.J., *et al.* (2000) Historical trends in lake and river ice cover in the Northern Hemisphere. *Science*, **289**, 1743–1746.

Mann M.E., Bradley R.S. & Hughes M.K. (1998) Global-scale temperature patterns and climate forcing over the past six centuries. *Nature*, **392**, 779–787.

Menzel A., Sparks T.H., Estrella N., *et al.* (2006) European phenological response to climate change matches the warming pattern. *Global Change Biology*, **12**, 1969–1976.

Oerlemans J. (2005) Extracting a climate signal from 169 glacier records. *Science*, **308**, 675–677.

Petit J.R., Jouzel J., Raynaud D., *et al.* (1999) Climate and atmospheric history of the past 420 000 years from the Vostok ice core, Antarctica. *Nature*, **399**, 429–436.

Ruddiman W.F. (2003) The anthropogenic greenhouse era began thousands of years ago. *Climatic Change*, **61**, 261–293.

Smol J.P., Wolfe A.P., Birks H.J.B., *et al.* (2005) Climate-driven regime shifts in the biological communities of arctic lakes. *Proceedings of the National Academy of Sciences*, **102**, 4397–4402.

Stott P.A., Tett S.F.B., Jones G.S., Ingram W.J. & Mitchell J.F.B. (2001) Attribution of twentieth century temperature change to natural and anthropogenic causes. *Climate Dynamics*, **17**, 1–21.

Verschuren D. (2004) Decadal and century-scale climate variability in tropical Africa during the past 2000 years. In: *Past Climate Variability through Europe and Africa* (Eds R.W. Battarbee, F. Gasse & C.E. Stickley), pp. 139–158. Springer-Verlag, Berlin.

2 Holocene climate research – progress, paradigms, and problems

H. John B. Birks

Keywords

Peat stratigraphy, pollen analysis, climate history, radiocarbon dating, COHMAP, transfer functions, paleolimnology, paleohydrology, ice-cores, Anthropocene, glacier history, PEP transects, human activities

Introduction

Everyone is fascinated by climate and by history. Both can be studied over a wide range of spatial and temporal scales, ranging from small areas and single years to continents and thousands of years. The reconstruction of how climate has varied in time and space over the past 11 500 years of the Holocene has been a major challenge for natural scientists for the past 200 years. Such reconstructions were probably motivated initially by natural curiosity about the past and the attractive and even seductive idea of "secrets of the past". Today such reconstructions are vitally important in the current debate about recent climate change and global warming, because reconstructions of past climate provide a powerful means of assessing the magnitude and rate of natural climate variability over the perspective of the past 11 500 years at a range of spatial and temporal scales from small areas to continents and from single years to centuries and millennia.

There has been enormous progress in reconstructing the history of Holocene climate since the pioneering investigations by Dau (1829) in Denmark. Progress in science, like climate and almost all other phenomena in the natural world, varies continuously in time and space. Historians of science (e.g. Kuhn 1970) suggest that the temporal continuum of gradual directional scientific progress is interrupted by abrupt changes, so-called paradigm shifts. Science is thought to progress by the gradual accumulation of observations and data within a basic agreed intellectual framework (Kuhn's "normal science") until a "revolution" occurs and the basic research framework or paradigm of the old conceptual structure is overturned and a new research framework or paradigm rapidly develops and becomes the new "normal" science. The spatial patterns of scientific progress are less explored by science historians, but Crane's (1972) "invisible college" effect is clearly important. In this, major researchers and their laboratories develop as nuclei of scientific influence through their methodologies, publications, presentations, research students, and visiting researchers. Visitors assimilate the methods and concepts of the center they have visited, and transfer the learning and experience to their own laboratory and research group. Patterns of scientific progress can often only be explained, in part at least, by the "invisible college" effect and by considering who was where when. For example, many developments in Quaternary pollen analysis in the past 50 years are only explicable in terms of the "invisible college" effect (Birks 2005).

Holocene climate research has witnessed several major paradigm shifts in the past 200 years as a result not only of new ideas and conceptual breakthroughs, but also because of new and improved techniques, studies of different climate proxies, increased scientific rigor, improved project design, increased quantification, greater attention to detail, and investigations in different geographic areas and climate regimes (e.g. low latitudes, high latitudes, high altitudes, arid areas). The effects of "invisible colleges" and the inevitable geographic concentration of research effort were clearly important in the development of Holocene climate research, especially in the early pioneering stages.

This chapter provides a historical overview of progress and associated paradigm shifts in Holocene climate research. My review is inevitably incomplete for several reasons. First, it reflects my geographic parochialism as there is an inevitable bias towards areas where I have had some direct research interest and experience. Second, it reflects a methodologic bias with an inevitable bias towards terrestrial proxies and Holocene terrestrial paleoecology of which I have had some direct experience. Third, the recent primary literature is so vast that I have mainly cited recent reviews and books rather than original research papers to keep the bibliography a manageable size. Fourth, I pay greatest attention to the early pioneering studies and research stages and the early publications than to the later stages, several of which are being actively pursued today by many researchers. This concern with the pioneering studies is because there is an increasing tendency in these days of the Internet and electronic sources, such as Wikipedia, for the early primary literature and the early pioneering researchers to be forgotten, or at least largely ignored. I apologize for any major omissions in the geographic areas, methodologic developments, research studies, and literature discussed here, and for any unevenness in my accounts of the different stages in Holocene climate research.

Progress and paradigm shifts

Pioneering Holocene climate research was, I suspect, motivated by natural curiosity about our past. Since about 1985 and increasing concerns about global warming, however, Holocene climate research has acquired a key role in providing unique information about natural climate variability since the last glaciation (Chambers and Brain 2002). In this chapter, I identify 14 major stages or paradigm shifts in the development of Holocene climate research, all of which were due to or were associated with major methodologic developments, conceptual advances, increased scientific rigor, greater attention to detail, and/or investigations in different geographic areas. The first five stages form a unidirectional succession of paradigms from 1829 to about 1988, the time of COHMAP (Co-operative Holocene Mapping Project) and the first use of climate models to simulate past Holocene climate. Since then, the research activities in Holocene climate research have become so diverse, with many new techniques and research approaches, that progress and associated paradigm shifts are occurring rapidly and in parallel.

Peat stratigraphy, megafossils, and macrofossils

The impressive occurrence of large fossil trunks and stumps (megafossils) of pine trees preserved in peat bogs in north-west Europe (Figure 2.1a and b) naturally attracted attention from naturalists as early as the late 18th century (e.g. Tait, 1794), and raised problems for many scientists who assumed that the environment did not change greatly. For example, in Scotland, Maxwell (1915) suggested that "one of the greatest enigmas of natural science is presented in the remains of pine forest buried under a dismal treeless expanse on the Moor of Rannoch, and on the Highland hills up to and beyond 2000 feet altitude" (Figure 2.1a).

The first scientific study of peat and pine stumps was probably by Heinrich Dau (1790–1831) in Denmark. Dau (1829) recognized and described several different types of peat bog, the occurrence of pine trunks in peat, and the stratigraphic differences in peat color and peat type (fresh, pale unhumified peat and dark, humified peat – see Figure 2.1c). Dau interpreted the occurrence of pine megafossils as reflecting a phase in his hypothetical forest history of Denmark. Sadly, Dau died two years after the publication of his monograph and he was not able to test his forest-history hypothesis. The occurrence of buried pine trees in Danish peat bogs attracted so much public attention in the 1830s that the Danish Academy of Sciences offered a prize for "solving the problem" about how did pine trees once grow on Danish bogs and what caused the extinction of pine as a native tree in Denmark (Iversen 1973). The Danish zoologist and geologist Japetus Steenstrup (1813–1897; Figure 2.2) won the prize and he proposed (1841) that there had been four periods in Danish forest history – the aspen, pine, oak, and alder periods. Steenstrup emphasized the importance of plant and animal remains preserved in peat bogs as the best available means of investigating past environmental changes, including climate. He tentatively suggested that during the Danish post-glacial there had been changes in moisture and possibly temperature, thereby explaining the observed changes in peat stratigraphy and the occurrence of tree remains in peats. Japetus Steenstrup can thus be regarded as one of the fathers of Holocene paleoecology and climate research.

(c)

Figure 2.1 (a) Fossil pine stumps on Rannoch Moor, western Scotland. These stumps are about 4000 years old (Birks 1975). (b) Fossil pine stump at Cooran Lane, Galloway, south-west Scotland. This stump is about 6000 years old (Birks 1975). (c) The "Grenzhorizont" or major recurrence surface at Chat Moss, Lancashire, north-west England. There is a conspicuous change from dark, humified peat to pale, less decomposed peat. This transition is radiocarbon-dated to about 2600 years ago. (Photographs: John Birks.)

Christian Vaupell (1821–1862; Figure 2.3) continued investigations on pine megafossils in Denmark. In addition to providing an ecologic explanation for the causes of long-term forest succession, he examined the width of the tree-rings of the buried pines. He concluded (1857) that the pines had not grown in cold conditions, as Steenstrup had proposed in his forest history, but that the pines had grown under warm but dry conditions. Vaupell suggested therefore that there had been changes not only in moisture (to explain the changes in peat stratigraphy) but also in temperature (to explain the changes in tree composition and growth) during the Holocene. The Swedish paleobotanist A.G. Nathorst (1850–1921; Figure 2.4) investigated plant macrofossils in the clays underlying the peat in southern Sweden (Nathorst, 1870) and subsequently elsewhere in Europe (Nathorst, 1892). He was the first to find remains of arctic plants such as *Dryas octopetala*, *Salix polaris*, and *S. reticulata*, plants that today are confined to high

Figure 2.2 Japetus Steenstrup (1813–1897), a Danish scientist who pioneered the study of plant and animal remains preserved in peat as a means of reconstructing past environmental change and who provided one of the first explanations for how pine trees could once grow on Danish peat bogs. (With permission from the Royal Library, Denmark.)

latitudes and/or high altitudes. With these discoveries Nathorst transformed peat-stratigraphic and macrofossil research and early ideas about climate change (Holmboe 1921). Temperature change immediately became recognized as the major factor influencing the history of plants and animals since the last Ice Age as a result of Nathorst's macrofossil studies. The final acceptance of the Ice Age theory

Figure 2.3 Christian Vaupell (1821–1862), a Danish scientist who pioneered the use of tree-rings in fossil pine stumps as a means of inferring past conditions of temperature and moisture. (Picture from Iversen 1973. Copyright: Geological Survey of Denmark and Greenland.)

Figure 2.4 Alfred Gabriel Nathorst (1850–1921), a Swedish paleobotanist who studied plant macrofossils in late-glacial and Holocene deposits in Scandinavia and elsewhere in Europe. His discovery of the remains of arctic–alpine plants in the clays below Holocene peats highlighted the large changes in temperature that occurred at the onset of the Holocene. (Photographer: Thor G. Halle, 1916. The Swedish Museum of Natural History.)

by geologists at the same time also contributed greatly to climate change occupying a key role in Quaternary research (Iversen 1973).

The idea that there had also been changes in temperature during the Holocene was investigated in detail in Sweden in the early 20th century. Sweden is ideal for such studies as there are marked south–north gradients in summer warmth today and hence well-defined northern geographic limits of many thermophilous plants and animals. Gunnar Andersson (1865–1928) discovered plant macrofossils (seeds, fruits, nuts, leaves, etc.) preserved in peat bogs. Finds of fossil nuts of *Corylus avellana* (hazel) well north of the present range of hazel suggested that the climate had once been warmer than today (Figure 2.5). The present-day northern limit of hazel coincides closely with the mean July temperature isotherm of 12°C today, whereas fossil nuts occurred as far north as the present-day July isotherm of

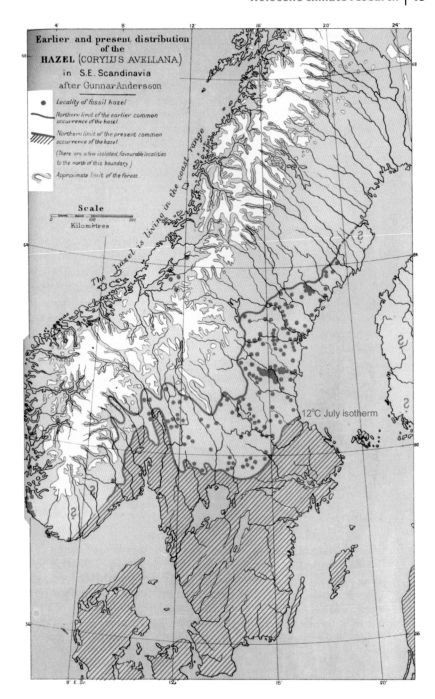

Figure 2.5 The present-day distribution (red cross-hatching) of *Corylus avellana* (hazel) in Denmark and Sweden in relation to the mean July temperature isotherm of 12°C and the distribution of fossil hazel nuts (red dots) of early- or mid-Holocene age. (From Wright 1936. Reproduced with permission of Palgrave Macmillan.)

9.5°C, suggesting a change in July temperature of 2–2.5°C. On the basis of these and related investigations, Andersson (1902, 1909) emphasized changes in summer temperature and presented the idea of one long early- to mid-Holocene period with a "higher-than-today" temperature comparable with the modern

Figure 2.6 Axel Blytt (1843–1898), a Norwegian botanist who proposed a series of alternating wet and dry periods based on peat-stratigraphic changes and used these periods to explain the history of the Norwegian flora. (Photograph by Carl Størmer with permission from the Norwegian Museum of Science and Technology.)

concept of the Holocene thermal maximum. Support for this idea came from discoveries of macrofossils of *Cladium mariscus* (saw-sedge), *Carex pseudocyperus* (cyperus sedge), *Trapa natans* (water-chestnut), *Emys orbicularis* (European pond tortoise), and other warmth-demanding plants and animals well north of their present northern limits in Europe.

In Norway, Axel Blytt (1843–1898; Figure 2.6), perhaps influenced by the work of Dau (1829), Steenstrup (1841), and Vaupell (1857), interpreted tree layers in peat bogs and changes from dark, humified to pale, fresh peat (Figure 2.1c) as evidence for alterations between dry (continental = Boreal) and wet (oceanic = Atlantic) periods (Blytt, 1876). He proposed an elaborate theory for the immigration of the Norwegian flora and its various floristic elements during these oceanic and continental periods. Blytt assumed that the floristic elements of the present-day flora of Norway (e.g. boreal, atlantic) had immigrated during successive climatic periods, with the arctic element first and the sub-atlantic element last (Mangerud *et al.* 1974). In light of the present-day distributions of these elements, Blytt proposed that the boreal and sub-boreal elements had immigrated during periods of continental climate, whereas the atlantic and sub-atlantic elements had immigrated during periods of oceanic climate. In this early work Blytt used the terms boreal and atlantic to refer to floristic elements and not to phases within the Holocene. He widened their use in 1893 after Sernander's early publications.

The Swedish botanist Rutger Sernander (1866–1944; Figure 2.7), who had strong interests in both climate history and plant geography (Fries 1950), combined the Swedish ideas of summer-temperature changes with Blytt's (1881) moisture changes to propose the famous Blytt–Sernander four periods of post-glacial time (Sernander, 1893, 1894, 1908, 1909, 1910 – see Table 2.1). Sernander (1889) initially used the concepts of Blytt's "atlantiska period" and "sub-boreala period"

Figure 2.7 Rutger Sernander (1866–1944), a Swedish botanist and Quaternary geologist who expanded Blytt's ideas into the Blytt–Sernander scheme for Holocene climate history (see Table 2.1) that became the major paradigm for Holocene climate until about 1960. (From unknown Internet source that is no longer on line.)

Table 2.1 The Blytt–Sernander division of the Holocene as proposed by Sernander (1890, 1894). The Pre-Boreal was added by Fægri (1940) for the earliest Holocene with a cool–sub-arctic climate

Period	Inferred climate	Approximate age (years BP)
Sub-Atlantic	Cool, wet	0–2500
Sub-Boreal	Warm, dry	2500–5000
Atlantic	Warmest, wet	5000–8000
Boreal	Warm, dry	8000–10 000
Pre-Boreal	Cool, sub-arctic	>10 000

to link Blytt's floristic terms atlantic, sub-boreal, etc. to Blytt's descriptions of peat stratigraphy in Norway and Blytt's associated climatic interpretations (Mangerud *et al.* 1974). Later Sernander (1890) used all Blytt's floristic terms for successive time periods, characterized by a different climate based on Blytt's peat investigations. Blytt (1893) then applied his own floristic terms to his peat-stratigraphic layers, thereby introducing the sub-atlantic peat-bed, the sub-boreal stump layer, the atlantic peat-bed, etc. (Mangerud *et al.* 1974).

The Blytt–Sernander paradigm of Holocene climate change, in the form that later became widespread, was probably fully established by Sernander (1894) in his doctoral thesis at the University of Uppsala (Mangerud 1982). In this remarkable monograph, Sernander first discussed the basis for dating and correlation of geologic sequences. He regarded the stratigraphic sequences based on plant macrofossils

Table 2.2 Sernander's (1894) correlation scheme for the Holocene (modified from Mangerud 1982)

Sea-level phase	Peat bogs in southern Scandinavia (Steenstrup, Nathorst, *et al.*)	Gotland vegetation history		Period	Correlations
Littorina	Spruce zone	Spruce		Sub-Atlantic Sub-Boreal	Littorina transgression maximum
	Oak zone	Oak		Atlantic	
Ancylus	Pine zone	Pine	*Cladium* *Carex* *pseudocyperus* *Iris pseudacorus*	Boreal Sub-Arctic	Ancylus Lake transgression maximum
	Aspen zone	Pine	Northern *salix* *Betula nana* *Dryas*		
Polar Sea	Dryas zone	*Dryas* with *Salix polaris*		Arctic	

and megafossils described by Steenstrup (1841) as the basis for Holocene climate change in Scandinavia. However, in contrast to other researchers at that time, Sernander doubted that vegetational changes in, for example, Denmark could be synchronous with changes in, for example, north-central Sweden because plant spread over several hundred kilometers must have required a long time. Sernander thus sought other methods to try to determine the age of these changes. In terms of climate history, Sernander accepted Blytt's observations and interpretations and wrote (Sernander, 1894, p. 71 – translated from Swedish)

> "We introduce therefore to the peat bog-sequences in Gotland the terms of Blytt, from the top downwards: sub-atlantic, sub-boreal, atlantic, and boreal beds, and we will also use these names for the time periods corresponding to the beds."

An approximate chronology for these periods was proposed based on correlations with archaeology, sea-level changes (Littorina and Ancylus transgressions), and the Swedish varve chronology of Gerard de Geer (1858–1943) by Sernander (1894) (see Table 2.2). The change from the Sub-Boreal to the Sub-Atlantic was thought by Sernander to be an abrupt climate change, even a catastrophe, the Fimbulwinter of the Sagas (Iversen 1973).

The 1910 International Geologic Congress in Stockholm was, according to Knut Fægri (personal communication 1995), a major event in the dissemination and acceptance of the Blytt–Sernander classification scheme. Interestingly there was close co-operation at this time between Sernander and Lennart von Post (the founder of pollen analysis – see below) and they organized a major excursion to Swedish peat bogs (von Post and Sernander 1910). After 1910, the Blytt–Sernander scheme was widely used in Scandinavia for many decades without any major change in its meaning or interpretation (Mangerud *et al.* 1974; Mangerud 1982). Fægri (1940) introduced the term Pre-Boreal as a unit between the Boreal and the Younger Dryas of the late-glacial (Table 2.1).

A major conflict and acrimonious debate ensued for about 20 years between the "grand old men" (Lundqvist 1965) of Swedish post-glacial climate research, namely Gunnar Andersson and Rutger Sernander. A rapid polarization of ideas developed between the Blytt–Sernander scheme with its alternating dry and wet periods and rapid climate change and Andersson's more uniform, gradually rising temperature curve, a thermal maximum, and subsequent decrease. The debate was partly one of scientific personalities (Fries 1950; Danielsen *et al.* 2000) and partly one about research techniques (Iversen 1973). Andersson's approach was based exclusively on macrofossils whereas the Blytt–Sernander scheme was based on peat stratigraphy and megafossils. The conflict between the two Swedish schools was not satisfactorily resolved until another Swede, Lennart von Post (1946), proposed that post-glacial climate history involved both broad-scale temperature changes (Andersson) and finer-scale precipitation changes (Blytt–Sernander). This new paradigm did not develop until the development of pollen analysis, which will be discussed in the next section.

As part of the Andersson–Sernander debate, Samuelsson (1916) analyzed the northern limit of hazel in Sweden in considerable detail and showed that summer temperature was far from uniform along the limit today. He showed that a lower summer temperature could be compensated for by a longer growing season, and he modeled the climatic demands of hazel in terms of both summer temperature and the length of the growing season. He proposed that both summer and winter temperatures, and hence the length of the growing season, may have changed during the Holocene. This idea was followed up in detail by Iversen (1944) using pollen analysis and by Hintikka (1963) in plant geography.

During the early part of the 20th century, the Blytt–Sernander scheme became the dominant paradigm for Holocene climate history and many peat and mega-fossil stratigraphies in Scotland, central Europe, and the Alps were interpreted in terms of the Blytt and Sernander model (e.g. Samuelsson 1910; Gams and Nordhagen 1923). As peat stratigraphies were examined in more detail, several recurrence surfaces (changes from dark, humified to fresh, unhumified peat – Figure 2.1c) were identified by Granlund (1932), suggesting several shifts in mois-ture during the late Holocene. Granlund proposed at least two "Sub-Atlantic" climate phases during Sernander's Sub-Boreal period and after Sernander's Fimbulwinter the climate changed twice over to the Sub-Boreal type and then again to the Sub-Atlantic. Von Post (1946) summarized Granlund's (1932) modification of the Blytt–Sernander paradigm as follows:

"Granlund's work showed that the post-glacial warm period did not end abruptly with a climatic catastrophe about 500 BC, as supposed by Sernander; but that instead of this event we must suppose a gradually advancing climatic deterioration spread over at least 4000 years. This process was, however, not continuous. Just in the same way as, during autumn, summer gradually changes to winter during repeated alternations between days of summer warmth and wintry cold, so the climate falls and rises from the warm period to the present time, through the bronze age, iron age, and our so-called historic time, swaying between fimbul winters and phases which were echoes of the glorious climate of the stone age."

In other words, the accepted paradigm by the 1930s based on peat stratigraphy and megafossils was of alternating wet and dry phases but perhaps with increased variability and several short-lived climate shifts in the late Holocene and with a mid-Holocene warm period.

Independently of the Scandinavian botanists and peat stratigraphers, the paradigm of a Holocene climate optimum or thermal maximum had already been presented by Thomas F. Jamieson (1829–1913; Figure 2.8) in his study of the fossil molluscan fauna of mid-Holocene estuarine clays in central Scotland (Wright 1936). Jamieson found species that do not occur as far north as Scotland today. The Irish naturalist Robert Lloyd Praeger (1865–1963) reached similar conclusions in his studies of estuarine clays and raised beaches in Northern Ireland, as did W.C. Brøgger (1900/01) in the Oslo Fjord (Wright 1936). Mitchell (1976) discusses whether W.B. Wright's (1936) dedication in his second edition of *The Quaternary Ice Ages* to "R. Lloyd Praeger, Discoverer of the Climatic Optimum" is appropriate, given Jamieson's pioneer work in 1865 that Praeger cites frequently in his Irish studies of 1887 and 1892. Perhaps credit for the discovery of the climatic optimum should be given to T.F. Jamieson in 1865, and not to Praeger, Brøgger, or Andersson.

In retrospect, the advances made by Dau, Steenstrup, Vaupell, Andersson, Jamieson, Praeger, Blytt, Sernander, Brøgger, and Granlund in reconstructing

Figure 2.8 Thomas F. Jamieson (1829–1913), a Scottish geologist who discovered the mid-Holocene climatic optimum or thermal maximum in his investigations of Mollusca preserved in estuarine clays in central Scotland. (Picture from www.fettes.com/(Artist anknown.))

Holocene climate change were very remarkable, given the very limited range of techniques then available. A flavor of the excitement of these early investigations based around peat stratigraphy and macrofossils can be obtained from Godwin (1978, 1981) and Mitchell's (1990) autobiographies. It was not until the development of pollen analysis that the different views of Andersson and Sernander could be satisfactorily reconciled and the Holocene climate history reconstructed for other parts of the world where peat stratigraphy is not always possible. I will return to modern detailed studies of peat stratigraphy later when I discuss paleolimnology, paleohydrology, and evidence for hydrologic changes during the Holocene.

The study of peat stratigraphy in Scandinavia was largely ignored after the development of pollen analysis as a tool for reconstructing Holocene climate change. It was not until the mid-1970s that peat stratigraphy was revived in Scandinavia using new techniques for estimating peat humification, detecting changes in bog moisture using testate amoebae, and establishing detailed chronologies by radiocarbon dating (e.g. Aaby and Tauber 1975; Aaby 1976). These new studies suggested consistent shifts from dry to moist conditions with a periodicity of 260 years over the past 5500 years.

The study of plant macrofossils and megafossils in north-west Europe was similarly ignored for about 70 years. Hilary Birks (1975) re-examined buried pine stumps (Figure 2.1) in Scottish blanket bogs using pollen analysis, peat stratigraphy, plant macrofossils, and radiocarbon dating and showed that the temporal patterns of pine stumps were considerably more complex than predicted from the application of the Blytt–Sernander scheme to Scottish peats by Samuelsson (1910). In recent decades Kullman (e.g. 1998a,b 2000, 2002) has studied tree megafossils and macrofossils in several areas of the Swedish mountains and has made spectacular discoveries of *Picea* and *Pinus* remains in the early Holocene and the late-glacial, of thermophilous trees in the early Holocene, and of *Larix sibirica* in the Swedish Mountains in the early Holocene. *Larix sibirica* today is found 1000 km to the east in Russia. The finds of its cones and wood (Kullman 1998a) suggest major changes in its distribution since the early Holocene and its extinction from Fennoscandia. These studies and others (see Birks 2005; Birks *et al.* 2005) emphasize the importance of integrated studies involving detailed plant macrofossil and megafossil analyses, pollen analysis, lithostratigraphic investigations, and radiocarbon dating if the complex patterns of tree spreading and climate change in the early Holocene are to be unraveled.

Pollen analysis

The Swedish geologist Lennart von Post (1884–1951; Figure 2.9) presented in 1916 the technique of pollen analysis at the 16th Scandinavian meeting of natural scientists in Christiania (now Oslo) and demonstrated its potential as a technique for relative dating and for reconstructing past vegetation and past climate. Although the original idea of pollen analysis can be attributed to the Swedish botanist Gustaf Lagerheim (1860–1926), it was von Post who had the vision of

Figure 2.9 Lennart von Post (1884–1951), a Swedish geologist who developed the technique of Quaternary pollen analysis as a tool for dating and for reconstructing past climate. Picture from (http://en.wikipedia.org/wiki/ Lennart_von_Post (accessed 30 October 2006).)

using pollen analysis as a dating technique and as a paleoclimatic tool (Fries 1950; Lundqvist 1965; Manten 1967a,b). In 1916 von Post proposed that in contrast to megafossils and macrofossils in peat, pollen could give a continuous record of vegetational and hence climatic change. He demonstrated strong pollen-stratigraphic similarities within southern and central Sweden and strong pollen-stratigraphic differences between the two regions (Manten 1967b; von Post 1967). He later proposed (von Post 1946) the concept of regional parallelism, namely that the same overriding climate change will be reflected in different ways by different taxa in different geographic and climatic regions. Based on the concept of regional parallelism, pollen analysis was used prior to about 1960 as a means of "determining geologic time" (von Post 1946) by assuming that different pollen zones (e.g. Jessen's zones I–IX, Godwin's zones I–VIII) in an area were synchronous. Pollen analysis provided the major means of dating botanic, zoologic, and archaeologic material in relation to the regional pollen zonation. Attempts were naturally made to correlate the pollen zonation schemes with the Blytt–Sernander climate periods (Iversen 1973; Fægri 1981; Mangerud 1982). Pollen stratigraphy soon became the only approach used to detect past climate changes and to identify the Blytt–Sernander periods. These periods almost became synonyms for pollen zones.

In a series of publications based on pollen-stratigraphic investigations in southern and central Sweden, von Post (e.g. 1920, 1924, 1933) emphasized climate as the principal cause of the inferred vegetation changes and interpreted the pollen record in a straightforward way and proposed that summer temperatures in the mid-Holocene were about 2°C higher than present-day. Even by 1933, von Post emphasized the need for a synthesis of paleoclimatic records and for improved chronologies. He wrote (von Post 1933, p. 58 – translated from Swedish)

"At some future date, as and when the time proves ripe, the variety of time-scales for the history of climate development might be fashioned into a unified whole. The system so finalized, naturally enough, should not only comprise a succession of periods. It ought to be based, firmly and squarely, on a time-scale of years, centuries, and millennia."

In the 1920s–1940s, many investigators made pollen-analytic studies in Europe, North America, South America, New Zealand, and China. Although these studies are, by modern standards, rather rudimentary, they lay the foundation for von Post's (1946) major synthesis on "The Prospect for Pollen Analysis in the Study of Earth's Climatic History", along the ideas he suggested in 1933. This global synthesis was the lecture von Post delivered when he was awarded the Vega Medal from the Swedish Anthropological and Geographic Society in 1944.

In this far-ranging and forward-looking paper, von Post emphasized that pollen analysis is "the most complete and most realistic register of climatic fluctuations throughout the past which we now have at our disposal". He proposed, based on pollen diagrams from Europe, New Zealand, Tierra del Fuego, North America, Hawaii, and China, that there is a consistent three-fold division of Holocene pollen stratigraphies. The first he interpreted as a period of increasing warmth, the second as a period of culminating warmth, and the third as a period of decreasing warmth. He characterized the dominant elements of the middle period as mediocratic, and of the earlier and later periods as protocratic and terminocratic, respectively. Von Post attempted to show that similar patterns of changing warmth had affected not only many parts of Europe, but many widely spaced regions on Earth. In this synthesis, von Post asked "now is it conceivable that anything other than climatic change could have brought about this general and, within certain regions, fundamental transformation of the forest distribution of our part of the world?" Besides discussing these broad-scale climatic trends, von Post also considered "minor paleoclimatic fluctuations" and suggested that the reader must be "wondering about these 'ripples'". He presented a Holocene pollen diagram from a bog near Stockholm and examined fluctuations in tree pollen concentrations (pollen per gram of peat) and the *Betula* (birch) pollen percentage curve. Using an early form of smoothing and spectral decomposition, von Post identified two long-term trends and three cycles with period lengths of 1700, 800–900, and 200–400 years within the birch pollen curve. The chronology was very crude and was based on peat-stratigraphic changes and the assumed ages of Granlund's peat recurrence surfaces.

Although von Post (1946) does not openly criticize the Blytt–Sernander scheme (von Post was after all a strong supporter of Rutger Sernander in the Sernander–Andersson feud and they had published together in 1910), von Post appears discretely to have modified his general views towards Andersson's idea of a mid-Holocene warmth. Von Post recognized that there were fluctuations in climate that correspond to the Blytt and Sernander phases but these were small compared with the major long-term climatic changes proposed by Andersson. This view, obliquely presented by von Post (1946), represents the paradigm of

Holocene climate change that had developed by the 1940s, despite the widespread application of the Blytt–Sernander scheme and its terminology (see also Hafsten 1970).

In his 1946 paper, von Post widened his discussion of Holocene climate change to the global scale and discussed "registrations of climate by the ocean surface", "the post-glacial changes in the world's system of circulation", and "theories" of climate change. He identified areas on Earth where pollen-analytic studies should be made (e.g. western North America, north-east Australia, Japan, Chile, western Africa, southern Africa, India, and Malaysia) to determine Holocene changes in the Earth's circulation systems. Under theories of climate change, von Post wrote (p. 212):

"In what I have said I have hesitated in my speech, when dealing with the nature of climatic changes, in such a manner as to astonish my audience, and about the causes, which are, nevertheless, the kernel of the problem, I have kept silent. Both these things I have done intentionally. For it might be indisputable that there was a wave of warmth in Europe, which determined the general course of post-glacial development, very likely this has in fact been the case in North America and Tierra del Fuego, but in the South Island of New Zealand and in China it is only more copious precipitation that is clearly indicated. About the driving force at the back of the climatic fluctuations of higher and lower order, which we may define or surmise, we will stumble in deep ignorance, in spite of all speculations about it."

He warns about the potential dangers of uncritical comparisons between data and models, when

"attempts are made to fit a course of development into a chronological system which is borrowed from a range of outside phenomena. If the facts appear to agree approximately then the theory becomes positively dangerous, for it may tempt us to wishful thinking which obscures our vision of the empirical realities. This has happened to a lamentable extent in dealing with Quaternary climatic history. We must, of course, at first define objectively the course of events that are to be explained, and neither for the post-glacial climate development nor for the climatic revolutions of the older Quaternary phases have we got the basis for a sound interpretation."

Despite his global view of vegetational and climatic development, von Post (1946) showed remarkable caution in relating observations to the theories about climate change then current (e.g. variations in solar radiation). Von Post mentions Milutin Milankovitch (1879–1958) and his ideas of orbital forcing to explain glacials and interglacials, but suggested "once the secret of the cause of the post-glacial waves of climate is exposed, then the riddle of the Quaternary ice ages will no doubt be solved." In fact, it was the solving of the riddle of the Quaternary ice ages first (Imbrie and Imbrie 1979) that contributed to our understanding of Holocene climate change, rather than the reverse order predicted by von Post (1946).

Although von Post (1946) never mentioned the discovery by Johs. Iversen, Knut Fægri, Franz Firbas, and others in the late 1930s and 1940s (Iversen 1973; Fægri

1981; Lang 1994) that pollen analysis can detect the early impact of prehistoric people on vegetation, von Post was very aware of the potential role of climate change on human societies and societal change. He wrote (pp. 215–217) that:

"It is axiomatic that such considerable revolutions (climatic change is meant here – my addition) in the very heart of natural conditions, which we find registered in pollen diagrams from different parts of the world, must have set their trace on mankind's destiny from the disappearance of the very latest land ice and right on till material culture reached stages in which people could, at any rate in some measure, become independent of natural conditions of life. But until this took place, both the changes in climate themselves and their geographic results must often have brought about weal or woe for considerable groups of mankind. Particularly at the limits of human cultivation, such as Greenland or Iceland, or the extensive plains of Asia, Africa, and America, which hovered between, on the one side, the barrenness of a desert, and on the other, the more or less good pastures and possibilities of cultivation in the grass steppes and forest land, it is inconceivable that changes in the natural conditions should not have had sweeping results on both the distribution and standards of living of mankind. Certainly cultural, social, political, and folk-psychological factors have many times been the driving force both of migrations of peoples or the flowering or fading of former cultures. None the less the conclusion seems inescapable that changes in natural factors must be considered if we wish to reach a true understanding of these phenomena, indeed that these changes may have occasionally been fundamental. In dealing with the problems of cultural history the reconstruction of the history of nature should be considered a necessity."

This is a theme that is explored in depth by Oldfield (2005, this volume).

Von Post (1946) foresaw the need for major international collaboration in establishing the patterns of Holocene climate change by writing a "plan of work of the extent here sketched can scarcely be carried out without organized international collaboration. At the head of this must stand a scientist with experience, foresight, mental acuity and activity, and he must be young, for the task is a life's work."

Von Post's Vega lecture was published 60 years ago and his original paper on pollen analysis was published 90 years ago. It is little wonder that the late Ed Deevey (1967) wrote "von Post's simple idea that a series of changes in pollen proportions in accumulating peat was a four-dimensional look at vegetation, must rank with the double-helix as one of the most productive suggestions of modern times."

Major advances in pollen-analytic methodology and interpretation were made in the 1940s–1960s by the Danish botanist Johs. Iversen (1904–1971; Figure 2.10a), the Norwegian botanist Knut Fægri (1909–2001; Figure 2.10b), and others (see, for example, Lang's (1994) review). In terms of Holocene climate research, one of the most important developments was Iversen's (1944) use of *Viscum album* (mistletoe), *Hedera helix* (ivy), and *Ilex aquifolium* (holly) as "indicator species" of summer and winter temperatures. Iversen studied in detail the occurrence, growth, flowering, and fruit production of these three shrubs at or near their

Figure 2.10 (a) Johs. Iversen (1904–1971), a Danish botanist (left) and (b) Knut Fægri (1909–2001), a Norwegian botanist. Together they made major advances in pollen-analytic methodology and interpretation. (The picture of Iversen is from Iversen (1973; copyright: Geological Survey of Denmark and Greenland). The picture of Fægri was taken by Jan Berge on the occasion of Fægri's 90th birthday.)

northern and/or eastern limits in Denmark in relation to the mean temperatures of the warmest and coldest months. He delimited "thermal limits" for these species. All three have distinctive pollen. From fossil pollen occurrences in Holocene sediments in southern Scandinavia, Iversen inferred that mid-Holocene summer temperatures may have been 2–2.5°C warmer than today and that winter temperatures may have been 0.5–1.5°C warmer than today, thereby building on Samuelsson's (1916) idea that both summer and winter temperatures may have changed in the Holocene. Iversen's (1944) brilliant analysis (see also Fægri 1950 and Hintikka 1963) of the climatic sensitivity of *Ilex* has been elegantly confirmed by Walther *et al.* (2005), who show shifts in its northern margin in the past 50 years in response to recent climate changes. These shifts are entirely predictable, given Iversen's thermal limit and recent climate changes.

In terms of direct relevance to Holocene climate research, the other major development in pollen analysis in the 1940s–1960s was the demonstration of the impact of prehistoric people on vegetation through forest clearance and agriculture. This demonstration was a result of improved pollen identifications (e.g. cereals, agricultural weeds), fine resolution analyses, identification and counting of charcoal particles, and research collaboration between pollen analysts and archaeologists (e.g. Iversen 1973). Although the interpretation of the observed pollen-stratigraphic changes in terms of human activity is not in doubt, recent concerns about the changing extent of forest and cleared land during the Holocene and the role of humans in Holocene climate change (Oldfield 2005, this volume) have stimulated research into the relationship between landscape cover and pollen deposition (e.g. Sugita *et al.* 1999). In general, pollen-analytic data underestimate the extent of cleared land because of the huge differences in pollen representation between forest trees and cleared-land herbs and shrubs (Sugita *et al.* 1999).

The paradigm of Holocene climate change based on pollen analysis and earlier macrofossil and megafossil analysis at about 1950 was that proposed by von Post (1946), namely long-term regional-scale or even global changes in temperature and precipitation at broad millennial scales and more local-scale shifts in moisture and possibly in temperature at centennial scales. Changes in both summer and winter temperature had occurred and there may have been periods of rapid climate change in the late Holocene. There was also evidence for the effects of prehistoric people on vegetation, and interest naturally centered on whether the extent of human activity and resulting land-use change were related to climate change. At that time there were no means of testing the implicit assumption that pollen-stratigraphic changes and hence inferred climate changes were synchronous over large areas. The next major stage in Holocene research dawned with the development of radiocarbon dating and absolute chronologies.

Radiocarbon dating and absolute chronologies

The development of radiocarbon dating by Willard F. Libby (1908–1980) in the early 1950s provided a means of deriving an absolute chronology (in theory at least!) for events in the Holocene. Harry Godwin (1901–1985) was one of the first Holocene paleoecologists, along with Eric Willis, Donald Walker, and others, to take advantage of radiocarbon dating to provide an absolute chronology for pollen-zone boundaries, peat-stratigraphic changes, and other events in the Holocene (Godwin 1960) and to discuss the implications of radiocarbon dating for Holocene climate research (Godwin 1966). A fascinating account of its early applications in Holocene research is given by Godwin (1981). As a result of a concerted dating program, Smith and Pilcher (1973) showed that major pollen-stratigraphic changes and pollen-zone boundaries were not synchronous, even within an area as small as the British Isles. A similar picture emerges at broader scales such as Europe (Huntley and Birks 1983; Berglund *et al.* 1996) or eastern North America (Webb 1988), raising serious doubts about the assumed synchroneity implicit in the von Post paradigm of Holocene climate history.

As more and more radiocarbon-dated pollen-stratigraphic data became available in Europe and eastern North America, attempts were made to display the spatial and temporal patterns of variation in the abundance of major pollen types, along the lines pioneered by Szafer (1935) in Poland. By mapping pollen values at selected time intervals from as many localities as possible, so-called iso-pollen maps (Szafer 1935) could be constructed. Such maps have been compiled at a range of spatial scales, ranging from the European continental scale (Huntley and Birks 1983) to single countries (e.g. Ralska-Jasiewiczowa *et al.* 2004). An alternative mapping approach, so-called iso-chrone maps, has also been developed (e.g. Moe 1970; Davis 1976; Birks 1989) to display the spatial patterns of the ages at which different pollen types have their first consistent occurrence or expansion, thereby illustrating the patterns of range expansion and contraction. Results from these early studies and from more recent, more detailed syntheses (e.g. Giesecke and Bennett 2004) suggest unexpectedly complex patterns of tree spreading, with trees apparently spreading at amazingly fast rates from a range of presumed source areas. The results raise many questions about the interpretation of pollen-analytic data, about what factors may have caused major pollen-stratigraphic changes (Giesecke 2005; Tinner and Lotter 2006), about how trees spread over large areas so quickly (McLachlan *et al.* 2005), and whether tree assemblages existed in the past that appear to have no modern analog today (Jackson and Williams 2004).

An important development in Holocene climate research involved a combination of pollen analysis and radiocarbon dating to study pollen-stratigraphic changes in space and time. This development took advantage of natural climate gradients and major vegetational ecotones (e.g. the prairie–forest ecotone, arctic tundra–forest ecotone) to study Holocene palynologic and hence vegetational changes in areas of potentially high sensitivity to climate change. A classic example of such a study was McAndrews' (1966) transect of sites near Itasca in north-west Minnesota. The transect ran from mixed coniferous–deciduous forest in the east, through deciduous forest and oak savanna, to prairie in the west and paralleled a major climatic gradient in precipitation today. McAndrews (1966) was able to show major changes in vegetation and infer major shifts in precipitation and soil-moisture deficit during the mid-Holocene "prairie period" after major vegetational and temperature changes at the onset of the Holocene. This research design of sites along major climate gradients has also been elegantly applied in studies in northern Fennoscandia using not only pollen percentages but also pollen concentrations and accumulation rates to detect Holocene changes in the northern extent of pine and birch, presumably in response to regional climate changes (Hyvärinen 1975; Seppä 1996).

By the early 1970s, Holocene researchers had accumulated large amounts of pollen-stratigraphic and other biologic data, often with radiocarbon dates providing an independent chronology. Climate reconstructions based on these data were primarily qualitative and based on indicator species or a comparative approach where modern and fossil assemblages were compared visually. Climate reconstructions were often molded into the paradigms of Holocene climate change derived from von Post (1946) or from the original Blytt and Sernander scheme. The next major paradigm shift in Holocene climate research was the development of transfer functions and the quantitative reconstruction of past climate.

Transfer functions and the quantitative reconstruction of past climate

A major paradigm shift in Quaternary paleoclimate research occurred in the early 1970s. John Imbrie (awarded the Vega Medal in 1999, 45 years after von Post) and the late Nilva Kipp (1925–1989) revolutionized marine paleoceanography by developing transfer functions to reconstruct quantitatively summer and winter sea-surface temperatures and salinity from fossil foraminiferal assemblages preserved in a deep-sea core covering several glacial and interglacial stages (Imbrie and Kipp 1971). The idea of transfer or calibration functions was also developed by Hal Fritts and colleagues for deriving quantitative estimates of past climate from tree rings in the western USA (Fritts *et al.* 1971; Fritts 1976) and by Tom Webb,

Reid Bryson, and associates for the quantitative reconstruction of past climate from fossil pollen assemblages in the Great Lakes region of North America (Webb and Bryson 1972).

The basic idea of transfer functions is very simple. If modern biologic assemblages (e.g. foraminifers in an ocean sediment core top) and modern environmental (e.g. climate) data are available for a wide range of sites and environmental conditions today, it is possible to model the relationship between the modern assemblages and the contemporary environment by some form of regression or calibration (inverse regression) procedure, to derive modern transfer functions. These functions summarize mathematically the relationships between modern biota and modern environment. If the assumption is made that the transfer functions are invariant in space and time, they can be applied to fossil assemblages to derive quantitative estimates of the past environment at the time the fossil assemblages were deposited.

Although there are now many numerical procedures for deriving, applying, and evaluating transfer functions (e.g. Birks 1995), the basic principles and assumptions presented by Imbrie and Kipp (1971) remain the same. Transfer functions are now widely used in Holocene climate research to derive quantitative estimates of several climatic variables (e.g. summer, winter, and annual temperatures, length of growing season, annual precipitation, ratio of actual to potential evapotranspiration) and climate-related variables (e.g. lake-water temperature and salinity, sea-surface temperatures and salinity, bog moisture) from a wide range of fossil assemblages (e.g. pollen, cladocera, chironomids, diatoms, foraminifers, radiolarians, dinoflagellate cysts, chrysophytes, coccolithophorids, testate amoebae). Several of the chapters in Mackay *et al.* (2003) give examples of the application of transfer functions in Holocene climate research.

Transfer functions played an important role in the CLIMAP Project Members (1976) (Climate: Mapping, Analysis, and Prediction) project that attempted to reconstruct the climate of the last ice-age Earth, primarily from paleoceanographic data. This project also involved the compilation and synthesis of a range of biologic and geologic paleoclimate data relating to 18 000 years ago and the first attempts at using climate models to simulate an 18 000-year climate (e.g. Williams *et al.* 1974; Gates 1976; Manabe and Hahn 1977).

Transfer functions and CLIMAP rapidly provided new paradigms for paleoclimate research, namely transfer functions for quantitative reconstructions of past climate and the use of climate models to simulate past climates for comparison with geologic and biologic "proxy-climate" evidence. This new paradigm led directly to COHMAP (Co-operative Holocene Mapping Project) and major developments in paleoclimate modeling.

COHMAP and paleoclimate modeling

The international COHMAP project represented a major paradigm shift in Holocene climate research with its close integration of paleoclimate "proxy-data" scientists with paleoclimate modelers. Although COHMAP started life in 1977 as Climates of the Holocene – Mapping based on Pollen Data, it quickly became global in its research area and in the climate proxies it considered. It changed its name to Co-operative Holocene Mapping Project. It was masterminded by John Kutzbach, Tom Webb, Herb Wright, and others (Wright *et al.* 1993). The basic idea of COHMAP was to simulate past climates at 18 000, 15 000, 12 000, 9000, 6000, 3000, and 0 years ago using the Community Climate Model (CCM) and to compare the model simulations with the available paleoclimate data, particularly terrestrial pollen, lake-levels, and pack-rat midden data and marine plankton data (COHMAP Members 1988; Wright *et al.* 1993).

Prior to COHMAP, paleoceanographic studies (Hays *et al.* 1976) had established that orbitally induced variations in insolation as proposed by James Croll (1821–1890) and Milutin Milankovitch are the pacemakers of the ice ages (Imbrie and Imbrie 1979). Variations in insolation are the result of the 22 000-year precession cycle, the 40 000-year tilt cycle, and the 100 000-year eccentricity cycle in the geometric relationships between the Earth and the Sun. COHMAP investigated in detail the role of such orbital variations in insolation on Holocene climate history. The 22 000-year precession cycle controls the time of year when the Earth–Sun distance is at a maximum or minimum and hence determines seasonality. The 40 000-year tilt affects the latitudinal distribution of solar radiation (COHMAP Members 1988).

Input to COHMAP CCM simulations of past climate at 3000-year time-slices were external orbitally determined insolation conditions and internal surface boundary conditions of mountain and ice-sheet orography, atmospheric trace-gas concentrations, sea-surface temperatures, sea-ice limits, snow cover, albedo, and effective soil moisture (COHMAP Members 1988). The boundary conditions used in the COHMAP simulations are summarized in Figure 2.11. These conditions provided a major new paradigm for Holocene climate research at the global spatial scale and at broad millennial temporal scales. The seasonal and latitudinal distributions of solar radiation at 18 000 years ago, the time of the Last Glacial Maximum, were similar to conditions today. The atmospheric conditions simulated for glacial times must therefore largely be a result of the very different internal surface boundary conditions (COHMAP Members 1988). Between 15 000 and 9000 years ago climate seasonality increased in the Northern Hemisphere and decreased in the Southern Hemisphere due to the precession and tilt cycles. As a result, continental ice sheets and sea-ice retreated and the oceans warmed. By about 9000 years ago, solar radiation over the Northern Hemisphere was, on average, 8 percent higher in July and 8 percent lower in January than today. After 9000 years ago, the extremes in solar radiation decreased towards modern values (COHMAP Members 1988).

COHMAP's model–data comparisons showed that orbitally induced changes in insolation explain, in broad terms, the history of Holocene climate and landscape development. Clearly climatic events of short duration in the late-glacial or Holocene were not considered in the modeling experiments or data syntheses because COHMAP considered 3000-year time-slices only (Wright *et al.* 1993).

Figure 2.11 Boundary conditions for the COHMAP simulations of past climate for the past 18 000 years. External forcing is shown for Northern Hemisphere solar radiation in June–August (S_{JJA}) and December–February (S_{DJF}) as percentage difference from present-day radiation. Internal boundary conditions include land ice (Ice) as percentage of ice-volume at 18 000 years ago, global mean annual sea-surface temperatures (SST) as deviations from present-day values (degrees K), excess glacial-age aerosol (arbitrary scale), and atmospheric CO_2 concentration (parts per million by volume). In the COHMAP simulations only the 18 000-year model used a lowered CO_2 concentration (200 ppm). All the other simulations used a CO_2 concentration of 330 ppm. Aerosol loadings for the 18 000- and 15 000-year simulations were not used. The pre-Holocene part is shaded in blue. (Modified from COHMAP Members (1988). Reprinted with permission from AAAS.)

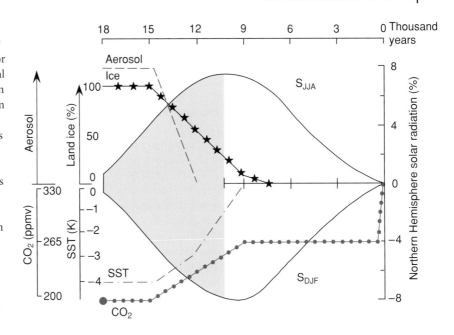

COHMAP's modeling experiment results showed how orbitally induced changes in insolation and changes in surface boundary conditions affect regional climates and hence the observed broad-scale changes in vegetation, marine plankton, and hydrology (COHMAP Members 1988). Variations in insolation changed seasonality at low- and mid-latitudes. By enhancing the thermal contrast between oceans and land, increased seasonality resulted in strong summer monsoons from 12 000 to 6000 years ago in the Northern Hemisphere tropics and sub-tropics and warm and dry summers in the continental interiors of northern mid-latitudes (COHMAP Members 1988). The model results also showed how internal surface boundary conditions (land and sea-ice extent, sea-surface temperatures) could have influenced atmospheric circulation patterns and thus patterns in temperature, precipitation, and wind. The climate of North America in the early Holocene was unlike any today and this may have led to biotic assemblages without modern analogs. The insolation maximum occurred before 9000 years ago and resulted in high temperatures south-west of the Laurentide ice sheet, whereas in eastern North America, the slow retreat of the ice sheet there delayed the summer thermal maximum to 6000 years ago (COHMAP Members 1988).

COHMAP was a major turning point in Holocene climate research for several reasons. First, it used what were then state-of-the-art climate models to simulate paleoclimate under specified boundary conditions. Second, it resulted in detailed compilations and syntheses of paleoclimate proxy data that were then used in data–model comparisons. Third, it considered the global climate system as a whole and revealed the strong regional interconnections between different components of the Earth's climate system. Fourth, it revealed the remarkable spatial and temporal variation in circulation patterns and climate at 3000-year intervals during the

Holocene. This variation helps explain the complex spatial and temporal patterns of tree spreading and of no-analog biotic assemblages in the early Holocene (Jackson and Overpeck 2000; Jackson and Williams 2004).

COHMAP Members (1988) recognized the implications of their work for understanding not only past climate but also future climates. They concluded their 1988 paper by suggesting:

> "Climate has influenced human activities. Two great cultural developments emerged at about the time of major environmental change between 12 and 10 ka – the earliest appearance of agriculture in the Old World, and the cultural changes accompanying extinction of the Pleistocene megafauna in the New World. Both developments may have been caused at least indirectly by the types of climatic change examined here. Now we may be faced with the reverse: human modification of climate and of related aspects of the physical environment. Application of a well-tested climate model is at present the only method for predicting these climatic and environmental changes and can help in planning a response to them. COHMAP research is contributing to the testing of these models and to the understanding of past climates." (COHMAP Members 1988, p. 1051)

Many of the researchers involved in COHMAP then attempted a unified global synthesis of paleovegetation data at the level of biomes for 6000 radiocarbon years ago, the so-called BIOME 6000 project (Prentice and Webb 1998). Such a synthesis was designed to be comparable to other global syntheses of sea-surface temperatures and ice sheets by CLIMAP Project Members (1976) and of lake-level changes by Street-Perrott and Harrison (1985). The BIOME 6000 data synthesis aimed to provide basic paleodata that could be used to evaluate simulations from coupled climate–biosphere models and to assess the extent of biogeophysical (vegetation–atmosphere) feedbacks within the global climate system (Anon 1994).

Climate models and paleoclimate and climate–biosphere modeling have made enormous advances since the pioneering CLIMAP, COHMAP, and BIOME projects with the development and application of box models, energy balance models, earth system models of intermediate complexity, and general circulation models. These advances and their contribution to our understanding of Holocene climate history are reviewed by Valdes (2003).

Paleolimnology and paleohydrology

Although lake sediments and their contained fossils have been a source of paleoenvironmental evidence since the 1940s (e.g. Wright 1966; Digerfeldt 1972), it was not until the late 1980s that the full potential of the paleolimnologic record began to be exploited. Paleolimnology is primarily concerned with the reconstruction of lake history from the sediments accumulating in the lake and from the fossils (e.g. diatoms, chironomids, cladocerans, ostracods) preserved in the sediments (Battarbee 2000; Smol 2002). Many advances have been made in paleolimnology in the past 25 years, including inorganic and organic sediment geochemistry, sediment paleomagnetism, stable isotope (^{13}C, ^{15}N, ^{18}O, D) analyses of sediments and their contained fossils, and the use of transfer functions to derive quantitative reconstructions of past lake conditions (e.g. salinity, pH, lake-water temperature) (Smol 2002). Several of these advances were made in direct response to the need to develop a quantitative paleolimnology to address critical environmental issues associated with "acid-rain" research in north-west Europe and eastern North America.

In terms of Holocene paleoclimate research (Verschuren and Charman, this volume), major contributions have come from paleolimnologic studies in low-latitude areas in Africa, South America, and south-east Asia and in the Great Plains of North America. These studies, involving a wide range of paleoenvironmental proxies such as diatoms, ostracods, stable isotopes, and geochemical ratios (e.g. Sr/Ca, Mg/Ca of ostracod carapaces), have provided unique evidence for major changes in lake-levels, salinity, and hydrology (Oldfield 2005). Many studies indicate a high degree of hydrologic variability, but a general tendency towards high lake-levels and wetter conditions during the early Holocene, followed by varying degrees of desiccation in the mid- and late Holocene (Oldfield 2005). Paleolimnologic evidence from northern and central Africa, for example, indicates extensive lakes and generally high lake-levels until about 6000 years ago (Gasse 2000), followed by irregular, often pulsed declines in lake-levels (Oldfield 2005).

Fine-resolution paleolimnologic studies also suggest high-frequency changes superimposed on the overall long-term changes at the orbital time-scales of COHMAP (Wright *et al.* 1993). For example, drought cycles detected by detailed paleolimnologic studies (e.g. Laird *et al.* 1998a,b) in the Great Plains of North America may be a complex climatic response to millennial-scale cooling events in the North Atlantic region and solar variation at several temporal scales (Fritz 2003).

Besides providing evidence of extremely low lake-levels, droughts, and drought-cycles (e.g. Digerfeldt 1988; Digerfeldt *et al.* 1992; Battarbee 2000; Fritz 2003), lake sediments can also provide records of other extreme events such as floods and debris flows (e.g. Nesje *et al.* 2001; Sletten *et al.* 2003; Bøe *et al.* 2006). Detailed peat-stratigraphic studies involving fine-resolution analyses of plant macrofossils, testate amoebae, and quantitative reconstructions of bog moisture (e.g. Barber and Charman 2003; Booth and Jackson 2003; Booth *et al.* 2005 2006; Charman *et al.* 2006; Hughes *et al.* 2006; Verschuren and Charman, this volume) also provide valuable information about hydrologic changes, especially in the late Holocene.

The paradigm of Holocene climate history that has emerged, and is continuing to emerge, from detailed paleolimnologic and peat-stratigraphic studies is that there have been major changes in regional and local hydrology at a wide range of temporal scales, especially in low-latitude regions (Oldfield 2005). In addition, there appears to be considerable spatial variation in changes in Holocene hydrologic patterns (Verschuren and Charman, this volume). There have also been a range of extreme events identified from lake sediments, as well as evidence for Holocene temperature changes in high-latitude areas (Oldfield 2005).

Ice-core research

A major breakthrough in Quaternary paleoclimatology came in the late 1960s with the drilling, extraction, and analysis of the Camp Century ice-core from north-west Greenland (Dansgaard *et al.* 1969, 1971). There are now many polar and alpine ice-cores that have been analyzed in detail for a wide range of paleoclimatic proxies. These include stable isotopes, borehole temperatures, and melt (proxies for paleotemperatures and humidity), ^{10}Be (solar activity), CO_2, CH_4, and N_2O content (atmospheric composition), conductivity, acidity, and sulfate (volcanic activity), microparticle content, trace elements, and electrical conductivity (tropospheric turbidity), mineral dust and particle size and concentration (wind speeds), major ions (atmospheric circulation), salt content (sea-ice extent, marine storminess), and seasonal signals and ^{10}Be (net accumulation rates) (Bradley 1999; Fisher and Koerner 2003). Alley (2000) and Bowen (2005) give fascinating and exciting accounts of work on Greenland ice-cores and on low-latitude alpine ice-cores, respectively. Fischer *et al.* (2006) reviews current ice-core projects worldwide.

The most important contribution from ice-core research to understanding Holocene climate history was the demonstration that there are fundamental differences between the climate of the Holocene and of the preceding 100 000 years. When viewed at this broad time-scale, Holocene records from Greenland ice-cores indicate that the Holocene generally has been a period of overall relative stability with small fluctuations in many paleoclimate proxies, in contrast to the preceding 100 000 years where there were rapid changes between two or more climatic modes, so-called Dansgaard–Oeschger events or oscillations. There appears to have been 24 such oscillations between 15 000 and 60 000 years ago with an amplitude of warming and cooling during each event of about 10°C in central Greenland (Bradley 1999; Alley 2000; Oldfield 2005).

When viewed in the context of the Holocene only, the major paleoclimatic contributions from ice-core research are (i) providing clear evidence for unprecedented rapid climate variations in the Greenland and Ellesmere Island ice-cores at centennial and even decadal scales (e.g. O'Brian *et al.* 1995), (ii) quantitative measurements of greenhouse-gas concentrations (CO_2, CH_4, etc.) from air bubbles enclosed in Antarctic ice (Raynaud *et al.* 2003), and (iii) providing near continuous records of ice-accumulation rates, volcanic activity, and dust deposition from both polar and low-latitude alpine areas. Fisher and Koerner (2003) discuss in detail Holocene ice-core research and its major findings, whereas Cecil *et al.* (2004) review alpine ice-core research from mid- and low-latitudes for the past 500 years.

The discovery of the abrupt "8.2 ka event" in a range of Greenland ice-core proxies (Alley *et al.* 1997) rekindled interest in Holocene climate instability and in the underlying mechanisms for rapid climate changes. The event of about 200 years duration appears to represent a major climatic instability event with a 5°C magnitude in Greenland and a considerable geographic extent in the North Atlantic regions and possibly elsewhere in the Northern Hemisphere (Alley and Ágústsdóttir 2005; Rohling and Pälike 2005). The event does not appear to be unique in the early Holocene as there appear to be other rapid but less extreme events at, for example, 9200 years ago and between 10 200 and 10 400 years ago, 10 800 and 10 900 years ago, and 11 300 and 11 500 years ago, the so-called Pre-Boreal oscillation (e.g. Schwander *et al.* 2000; Björck *et al.* 2001). The most likely cause for the 8.2 ka event and other early Holocene rapid climate events is the discharge of large amounts of glacial meltwater from glacial lakes Agassiz and Ojibway via the Hudson Strait into the north-west Atlantic (Clarke *et al.* 2003). This freshwater pulse reduced surface salinity in the north-west Atlantic, thereby reducing the formation of intermediate water in the Labrador Sea and of North Atlantic deep water. This in turn may have led to a reduction in the northward transport of heat associated with meridional overturning circulation in the North Atlantic (Clark *et al.* 2001; Labeyrie *et al.* 2003; Oldfield 2005). Considerable attention is being paid by Holocene researchers to the 8.2 ka event and other early Holocene abrupt events because they show that a sufficiently large freshwater pulse can rapidly disrupt ocean circulation and climate over a large area when the Earth is in an interglacial climate mode (Alley *et al.* 2003; Oldfield 2005; Schmidt and LeGrande 2005).

In terms of paradigm shifts, Holocene ice-core research has led to the addition of episodic meltwater input into the North Atlantic in the early Holocene to about 8000 years ago as a major forcing function to the COHMAP internal boundary conditions (e.g. Rensen *et al.* 2001; Rind *et al.* 2001).

Air bubbles preserved in the Antarctic Dome and Vostock ice cores provide a means for investigating the concentrations of atmospheric trace-gases (CO_2, CH_4, N_2O) in the Holocene (Raynaud *et al.* 2003). These indicate an increase of CO_2 concentrations about 8000 years ago and an increase in methane concentrations about 5000 years ago. These discoveries have led to the so-called Anthropocene hypothesis proposed by Ruddiman (2003) and Ruddiman and Thomson (2001).

CO_2 and CH_4 changes and the Anthropocene hypothesis

In a series of stimulating publications, Ruddiman interprets the increase in atmospheric CO_2 and methane at 8000 and 5000 years ago in ice-core records, respectively, as largely the result of human activity (Ruddiman and Thomson 2001; Ruddiman 2003, 2005a,b) Ruddiman's hypothesis of early anthropogenic influences on Holocene climate consists of three parts (Oldfield 2005). First, humans reversed a natural decrease in atmospheric CO_2 concentrations 8000 years ago through the early development of agriculture and associated forest clearance and carbon release to the atmosphere. Second, humans reversed a natural methane decrease about 5000 years ago by the development of intensive rice cultivation and flood-irrigated regions in Asia. Third, these human activities and associated changes in CO_2 and methane atmospheric concentrations caused an anthropogenic warming sufficient to counter a natural cooling and avoided the onset of a new glaciation in the past several thousand years (Ruddiman 2005a). Ruddiman (2003, 2005b) presents many arguments to support this hypothesis. He emphasizes that the orbital configurations to which CO_2 and methane concentrations have been linked in the past would have led to declines in both CO_2 and methane during the past 5000 years of the Holocene. He also emphasizes that none of the previous interglacials in the Antarctic ice-cores show increases in CO_2 or methane similar to the Holocene. He argues, using a wide range of paleoenvironmental records, that fluctuations in CO_2 in the past 2000 years are correlated with human demographic changes including plagues that result in forest regrowth, increased carbon sequestration, and reduced atmospheric CO_2 concentrations (Oldfield 2005). Oldfield (this volume) presents an up-to-date review of the evidence being presented from several sources for and against Ruddiman's hypothesis.

A global synthesis of fire-history records for the Holocene for different geographic regions (Carcaillet *et al.* 2002) shows that global fire indices parallel the increase in atmospheric CO_2 concentrations, suggesting that biomass burning may have been a major cause for the increase in CO_2 concentrations from 8000 years ago. Studies on the loss of storage of organic carbon in Chinese soils as a result of cultivation (Wu *et al.* 2003) similarly support Ruddiman's main hypothesis.

Joos *et al.* (2004) tested Ruddiman's hypothesis concerning CO_2 changes by driving the carbon component of a carbon-cycle climate model with simulated climate from two global climate models for the past 21 000 years. They conclude that the Holocene ice-core record of CO_2 is well reproduced (within a few ppm) by the model results and by the processes incorporated in the models without invoking any anthropogenic influences suggested by Ruddiman. Joos *et al.* (2004) argue that the early Holocene CO_2 rise was a result of several processes, including terrestrial carbon uptake and release, coral-reef build-up, and sea-surface temperature changes. They further claim that the level of terrestrial carbon release suggested by Ruddiman's hypothesis is not compatible with the ice-core record of $\delta^{13}C$ change in the pre-industrial times of the Holocene.

Ruddiman's hypotheses about CO_2 changes and associated climate responses have stimulated much interest and further work (e.g. Claussen *et al.* 2005; Crucifix *et al.* 2005; Ruddiman *et al.* 2005) and are already making a major impact on research and thinking (Mason 2004; Oldfield 2005, this volume). The hypotheses challenge current ideas on the spatial extent of deforestation and land abandonment in the Holocene as deduced from pollen-analytic data and on estimating carbon sequestration and release as a result of human activity. Attempts at quantitative modeling of the extent of forested land-area required to be cleared to be registered in pollen-stratigraphic data (Sugita *et al.* 1997, 1999) suggest that traditional interpretations by pollen analysts of the areas involved may be a gross underestimate.

As Oldfield (this volume) notes, Ruddiman's interpretation of the rise in atmospheric methane concentrations about 5000 years ago (see also Ruddiman and Thompson 2001) has not received as much criticism or attention as Ruddiman's hypothesis about changes in atmospheric CO_2 concentrations. A recent attempt at identifying possible causes of changes in atmospheric methane concentrations in the mid-Holocene by MacDonald *et al.* (2006) has involved a synthesis of over 1500 radiocarbon-dates from circum-arctic peatlands. This synthesis shows that the mid-Holocene rise in atmospheric methane concentrations cannot be explained by any major expansion of northern peatlands in the mid-Holocene. It adds further credibility to Ruddiman's hypothesis of an anthropogenic explanation for the mid-Holocene rise in methane concentrations.

Ruddiman's hypotheses are a major potential paradigm shift in Holocene climate research. As Oldfield (2005) concludes "Ruddiman has opened up a range of possibilities that will take a long time to evaluate fully."

Marine-core research

The study of marine cores and their contained fossil assemblages and the stable-isotope analyses of foraminifers revolutionized our understanding of Quaternary climate history with the demonstration that cyclical variations in the Earth's orbit around the Sun are the pacemakers of glacial–interglacial cycles (Hays *et al.* 1976; Imbrie and Imbrie 1979). In contrast, the contributions of marine research to Holocene climate history have been relatively small, compared with the contributions from terrestrial, limnologic, and peat records and from ice-core records. Jansen *et al.* (2004) provide an overview of Holocene climate variability from a marine perspective.

Perhaps the most influential Holocene marine studies have concerned the occurrence of ice-rafted debris (hematite-stained grains, Icelandic volcanic glass shards, and detrital carbonate) in cores from the North Atlantic (Bond *et al.* 1997, 2001), and the suggestion that this debris occurs with periodicities of 1500 years, related to unknown internal oscillations of the climate system (Bond *et al.* 1997) or to variations in solar output throughout the Holocene (Bond *et al.* 2001). Similar 1500-year periodicities have been reported from, for example, sediment grain-size data from a North Atlantic core, considered to be a proxy for the speed of deep-water flow (Bianchi and McCave 1999). In addition to a 1500–1600-year periodicity in sediment color data from a North Atlantic core (a possible proxy for North Atlantic Deep Water circulation), Chapman and Shackleton (2000) report 550- and 1000-year periodicities. The existence of millennial-scale periodicities during the Holocene has been questioned by, for example, Schulz and Paul (2000) and Risebrobakken *et al.* (2003). The detection of periodicities from geologic core data is fraught with difficulties – sampling resolution, irregular sample distribution in time, time control, and the reliability and inherent errors in the resulting age–depth models. Time control is particularly difficult in marine studies due to spatial and possible temporal variations in the marine reservoir effect and the effects of these variations on radiocarbon datings in the Holocene.

Multi-proxy high-resolution analyses of rapidly sedimented marine cores from the North Atlantic (e.g. Birks and Koç 2002; Andersson *et al.* 2003; Risebrobakken *et al.* 2003; Andersen *et al.* 2004; Moros *et al.* 2004) involving reconstructions of sea-surface temperatures from fossil assemblages and alkenone unsaturation ratios, stable-isotope analyses, and ice-rafted debris suggest four major climatic phases in the Holocene. There is an initial early Holocene thermal maximum that lasted to about 6700 years ago, followed by a distinct cooler phase associated with increased ice rafting between 6500 and 3700 years ago. There was then a transition to generally warmer but relatively unstable conditions between 3700 and 2000 years ago. Then there was a striking decline in sea-surface temperature between 2000 and 500 years ago. Although the dominant forcing factor for the early Holocene climate history was Northern Hemisphere summer insolation via strong seasonality at high northern latitudes, the trigger for the onset of relatively unstable climatic conditions beginning about 3700 years ago is less clear. Moros *et al.* (2004) suggest that this change may have been triggered by a late Holocene winter insolation increase at high northern latitudes and/or interhemispheric changes in orbital forcing. The late Holocene Neoglaciation trend beginning about 2000 years ago, which is a feature of many terrestrial records in northern Europe, may have been driven not only by a gradual decrease in orbitally forced summer temperatures but also by an increase in snow precipitation at high northern latitudes during generally milder winters. Such high-resolution multi-proxy marine studies highlight the role of a range of forcing functions and close ocean–terrestrial links. Jansen (this volume) reviews North Atlantic climate and ocean variability at different temporal scales.

In terms of paradigms in Holocene climate research, marine studies clearly show that climate change is frequent in the Holocene but whether it follows regular periodicities is presently unresolved.

Glacier history

Glaciers are a striking feature of many alpine landscapes globally, and glacial moraines provided early evidence for recent glacial fluctuations in response to climate change. Denton and Karlén (1973) dated glacial termini in the St Elias Mountains on the Yukon/Alaska border and in the mountains of Swedish Lapland and showed repeated intervals of glacial expansion and retreat in both areas during the Holocene. They suggested that major glacial expansions had occurred in the "Little Ice Age" of the past several centuries, and at 1050–1250, 2300–3000, 4900–5800, and 8000 years ago. They proposed a recurrence of major glacial activity about every 2500 years and hypothesized that Holocene glacial and climatic fluctuations were caused by varying solar activity because of close correlations with short-term atmospheric ^{14}C variations (see also Karlén and Kuylenstierna 1996). The glacial advances and retreats identified by Denton and Karlén (1973) are now termed "rapid climate changes" by Mayewski *et al.* (2004) in their survey of some globally distributed paleoclimate records and are now dated to 8000–9000, 5000–6000, 3800–4200, 2500–3500, 1000–1200, and 150–600 years ago by tuning to the high-resolution GISP2 ice-core record.

In some areas it is difficult to reconstruct reliably the Holocene history of alpine glaciers because moraines that pre-date the "Little Ice Age" of the 18th–19th centuries have often been overrun by "Little Ice Age" or Neoglacial ice advances.

Karlén (1976), working in the mountains of northern Sweden, showed the potential of sediments in lakes downstream of glaciers to provide a continuous record of Holocene glacial activity. By measuring the organic content (and hence the inorganic content) and quantifying changes in sediment color by X-radiography, Karlén (1976) reconstructed glacial activity for much of the Holocene in the drainage basin of the study lake. This approach of using proglacial lakes to reconstruct glacial activity and equilibrium-line altitudes (ELA) has been greatly developed by Nesje and Dahl (2000, 2003) (see, for example, Dahl *et al.* 2003; Bakke *et al.* 2005a,b).

Glacial ELAs are a curvilinear relationship between mean annual winter precipitation and mean ablation-season (summer) temperature (Nesje and Dahl 2000, 2003). If glacial ELA can be reconstructed using proglacial lake sediments and terrestrial geologic evidence (Dahl *et al.* 2003) and summer temperatures are reconstructed using biologic proxies (e.g. pollen) and modern organism–climate transfer functions (e.g. Bjune *et al.* 2005), changes in winter precipitation during the Holocene can be inferred (e.g. Bakke *et al.* 2005b; Bjune *et al.* 2005). Nesje *et al.* (2005) provide a recent synthesis of Holocene glacial history and reconstructed variations in winter temperature in southern Norway.

Although different glaciers in southern Norway show different Holocene histories, some general patterns emerge (Nesje *et al.* 2005). Glaciers in the early Holocene mainly retreated but with some significant glacier readvances at about 10 000, 9700, and 8000–8500 years ago. Most, if not all, glaciers may have melted completely at least once during the mid-Holocene. Most glaciers appear to have reformed between 4000 and 6000 years ago and advanced between 3000 and 1500 years ago ("Neoglaciation"). Most glaciers reached their maximum position relative to their position at 10 000 years ago during the "Little Ice Age", ranging in time from the early 18th century to the 1940s. Reconstructed changes in winter precipitation through the Holocene show striking millennial-scale fluctuations between periods with predominantly mild and wet winter conditions (similar to a positive North Atlantic Oscillation (NAO) index weather-mode) and periods with mainly cold and dry winters (negative NAO index weather-mode) (Nesje *et al.* 2005). The causes of these striking changes in winter precipitation are unclear but they suggest that there may have been major broad-scale variability in winter atmospheric circulation over north-west Europe during the Holocene.

Studies on Holocene glacial history using a combination of geologic approaches to reconstruct ELA and biologic approaches to reconstruct summer temperatures have provided a new paradigm in Holocene climate research, namely the detection of millennial-scale fluctuations in winter precipitation. Alpine glaciers thus have the potential to record not only changes in summer temperature but also changes in winter precipitation.

PAGES (Past Global Changes) and PEP (Pole–Equator–Pole) transects

Von Post (1946) emphasized that the reconstruction of Holocene climate history at a global scale could "scarcely be carried out without organized international collaboration". COHMAP (see above) was the first major international collaboration in Holocene climate research. The PAGES (Past Global Changes) core project of the International Geosphere–Biosphere Program has greatly stimulated further international research collaboration. The idea of Pole–Equator–Pole transects in the Americas (PEP I), Asia and Australasia (PEP II), and Europe and Africa (PEP III) was proposed and developed by Ray Bradley, Vera Markgraf, Herman Zimmerman, and others in the mid-1990s (Bradley *et al.* 1995). The idea was to study and reconstruct Holocene climate history using a variety of records along broad-scale transects that cut across major climatic gradients in insolation, hydrologic regimes, and atmospheric circulation patterns.

The results of these PEP transects are now published (Markgraf 2001; Battarbee *et al.* 2004; Dodson *et al.* 2004). These publications plus the PAGES synthesis volume (Alverson *et al.* 2003) and Frank Oldfield's (2005) vision of environmental change based on his experiences as Executive Director of PAGES between 1996 and 2001 provide major overviews of the amazing scientific achievements of the international PAGES research community over the past decade. The PAGES research summarized in Alverson *et al.* (2003) shows the enormous advances that have been made in terms of the range of proxies studied, improved temporal resolution, deep interpretative insights, and increased spatial coverage, even though the editors describe the book as "a progress report in the search for the past". They emphasize that

> "The scientific findings give cause for both exhilaration and concern. The exhilaration lies in appreciating the remarkable increase in our understanding of the complexity and elegance of the Earth System. The concern is rooted in recognising that we are now pushing the planet beyond anything experienced naturally for many thousands of years. The records of the past show that climate shifts can appear abruptly and be global in extent, while archaeological and other data emphasize that such shifts have had devastating consequences for human societies. In the past, therefore, lies a lesson. And as this book illustrates, we should heed it." (Alverson *et al.* 2003, p. iv)

Data synthesis and databases

One result of PAGES and PEP transect research and other related research on Holocene climate change has been the production of large amounts of Holocene paleoclimatic data. The systematic compilation of Holocene paleoclimatic data started in the COHMAP project (COHMAP Members 1988; Wright *et al.* 1993) and has been continued as part of the PAGES research agenda (Anderson 1995; PAGES 1998; Eakin *et al.* 2003).

It has long been recognized that meaningful and rigorous synthesis needs access to primary data (site details, chronologies, proxy records, etc.) and hence there is a major need for well-designed and well-maintained databases. Despite this need, research funding for database development and maintenance continues to be difficult to obtain, even for well-established and internationally recognized centers such as the World Data Center for Paleoclimatology, the World Data Center for Marine Environmental Sciences (PANGAEA), and MEDIAS-France.

The international scientific community has a responsibility not only to deposit their data in major databases but also to do all they can to support such databases and to encourage future international collaboration and funding for database development (Alverson and Eakin 2001; Anon 2001; Dittert *et al.* 2001).

Fine-resolution studies

The temporal resolution in the majority of Holocene paleoclimatic studies is about one sample per every 50–100 years with each sample representing 10–20 years. Such studies can detect broad-scale millennial changes only. Ice-core research has indicated that very abrupt, short-lived climatic changes have occurred in the Holocene. The detection of such changes and the reconstruction of centennial and decadal climate changes require fine-resolution studies. Such studies with one sample every 1–20 years and each sample representing 1–2 years are only possible from ice-cores, laminated lake or marine sediments, or speleothems, corals, tree-rings, and well-dated peat deposits. Baillie and Brown (2003), Zolitschka (2003), Cole (2003), Lauritzen (2003), and Fisher and Koerner (2003) review fine-resolution approaches based on tree-rings, laminated sediments, corals, speleothems, and ice-cores, respectively.

Many, but not all (e.g. Baldini *et al.* 2002; Snowball *et al.* 2002), fine-resolution studies have focused on climate variability in the past millennium (Bradley *et al.* 2003). The number of continuous high-resolution millennial-scale records is very low, with some ice-cores and laminated lake sediments and a few long tree-ring records mainly from high latitudes (Bradley *et al.* 2003). The most promising fine-resolution archive for the entire Holocene is probably speleothems (e.g. Linge *et al.* 2001; Dykoski *et al.* 2005; Tan *et al.* 2006). Henderson (2006) has recently proposed that "for paleoclimate, the past two decades has been the age of the ice core. The next two may be the age of the speleothem".

The Holocene $\delta^{18}O$ record from stalagmite calcite in the Dongge Cave in southern China (Dykoski *et al.* 2005) provides amazingly detailed information on fine-scale shifts in monsoon precipitation. Dramatic decreases in $\delta^{18}O$ occurred at the start of the Holocene and $\delta^{18}O$ remained low for 6000 years, suggesting high monsoon intensity until about 3500 years ago. Four positive $\delta^{18}O$ events in the early Holocene correlate, within their dating errors, with changes in the Greenland ice-cores and three of the events correlate with glacial meltwater pulses from the Agassiz and Ojibway glacial lakes. In addition, the Holocene is punctuated by numerous centennial- and multi-decadal-scale events with amplitudes up to half of the interstadial events seen in the record for the last glacial period. The Dongge Cave record shows that Holocene centennial- and multi-decadal-scale monsoon variability is significant but not as large as glacial millennial-scale variability. The $\delta^{18}O$ record shows significant periodicities suggesting its variation may have been influenced by solar forcing. There are several other significant peaks at El Niño frequencies, suggesting that changes in oceanic and atmospheric circulation patterns in addition to solar forcing have been important in controlling Holocene monsoon climate in this part of China. There is even evidence for a distinctive biennial oscillation of the Asian monsoon. The Dongge Cave record illustrates the level of detail about Holocene climate variability that can be obtained from detailed fine-resolution studies.

Such fine-resolution studies of Holocene climate indicate variability at all scales and different aspects of Holocene climate appear to have been changing in some way at all temporal and spatial scales that have been investigated. It is this variability at a range of scales that is the paradigm that is emerging from fine-resolution studies. Much remains to be discovered about this variability, its spatial and temporal patterns, and its underlying causes.

Climate change and human societies

The possible interactions between climate change and societal changes were discussed over 50 years ago by von Post (1946 – see above) and this topic continues to be a critical issue in Holocene climate research (Oldfield 2005, this volume). Recent impetus to understanding more fully the interactions between climate and humans (Oldfield 2005, this volume) has come from (i) the demonstration of major changes in the hydrologic balance in different areas, especially at low latitudes, leading to periods of extended drought (see above; Haberle and Chepstow-Lusty 2000; Verschuren and Charman, this volume), and (ii) Ruddiman's (2003, 2005b) Anthropocene hypothesis concerning the possible role of land-cover and land-use changes on atmospheric concentrations of carbon dioxide and methane and hence on Holocene climate (see above).

There are many examples where prolonged drought episodes detected by paleoecologic studies coincide with and may have been one of the dominant factors contributing to major declines or "collapses" in civilizations such as the Maya and Anasazi, and in areas such as the Atacama and Andean Antiplano, the Sahara, eastern Mediterranean, east Africa, South Africa, and China (Oldfield 2005, this volume). Several recent books explore the idea of the impacts of climate change on societies (e.g. Fagan 1999, 2000, 2004; Burroughs, 2005; Linden, 2006). Oldfield (2005, this volume) emphasizes that this type of research is extremely complex. Balanced and critical analyses are needed where climate changes and cultural perspectives are considered as an integrated whole (Diamond 2005). The influence of climate change on human societies is mediated by complex cultural and social processes. There is thus a need to try to understand the nature of these processes, especially during periods of rapid climate change and limited resources. There is currently a strong polarization of ideas in this area. Some interpret past changes in human societies such as the "collapse" of different civilizations to be the direct consequence of climate change, particularly extreme and/or rapid events (e.g. deMenocal 2001). Others see these societal changes as a result of changes in societal organization and of human processes that may have developed independently of climatic influences (e.g. Redman 1999). This strong polarization of ideas has resulted in the complex nature of human–climate interactions in the past being ignored and a failure to improve our understanding of the effects of these inter-actions on ecosystems and the consequences of these effects on past, present, and future systems (Oldfield 2005, this volume). Balanced paleoecologic analyses such as Haberle and Chepstow-Lusty (2000), Verschuren *et al.* (2000), Huntley *et al.* (2002), Berglund (2003), Haug *et al.* (2003), Tinner *et al.* (2003), and Haberle and David (2004) illustrate how Holocene paleoclimatic reconstructions can provide valuable hypotheses about climate–land-use and climate–society relation-ships. Oldfield (this volume) discusses how such hypotheses can be incorporated into ideas and models about societal structure and processes.

The possible roles of land-cover and land-use change on Holocene climate-change are central to Ruddiman's (2003, 2005b) Anthropocene hypothesis (see above). This challenging hypothesis raises the problems of accounting for changes in Holocene carbon budgets. There are currently major differences between estimates of carbon budgets based on paleoecologic evidence and those based on models (e.g. Pedersen *et al.* 2003; Joos *et al.* 2004). These unresolved questions highlight the importance of considering land-cover–climate interactions and of comparing model-based simulations and estimates with empirical evidence for past land-cover change. As discussed above and by Oldfield (2005, this volume), empirical approaches for estimating the extent of deforestation and carbon seques-tration and release as a result of human activity need greater rigor and quanti-fication and improved robustness.

The possible impact of climate changes on human societies in the past and the effects of past land-use and land-cover changes on climate directly via changes in albedo, moisture retention, and dust fluxes and on atmospheric gas concentrations and thus on climate are rapidly developing paradigms in Holocene climate research. As Oldfield (this volume) emphasizes, these are topics where much new, critical, and innovative research is urgently needed.

Problems

Considerable advances have been made over the past 200 years in reconstructing the temporal and spatial patterns of variation in Holocene climate, in quantifying the magnitude and rate of climate change, and in assessing the role of forcing factors such as orbital forcing and solar variability on Holocene climate. These advances have mainly occurred as a result of improved methodologies, of studying many different proxies in a wide range of geographic areas, of improved project design, of increasingly finer resolution studies, of greater concern for data quality, detail, and interpretation, and of increasing interactions between paleoecologists, earth scientists, climatologists, and climate earth-system modelers. Despite these enormous advances, several critical problems continue to pervade Holocene climate research. These problems relate to dating, data interpretation, basic gaps in our knowledge, and data availability.

The single most critical problem in establishing the temporal and spatial patterns of Holocene climate changes is chronology. An absolute chronology, mainly provided in the Holocene by radiocarbon dating, is the major means for deriving age–depth models for individual sequences and for almost all correlations between sequences. Many Holocene paleoproxy sequences (e.g. pollen), particularly those studied from lake sediments before about 1990, have poor chronologies, often based on only four to eight radiocarbon-dates from bulk sediment. Such dates may be influenced by unknown hardwater errors and other factors that can result in erroneous ages. The small number of radiocarbon-dates per sequence greatly limits the reliability of the resulting age–depth models (Telford *et al.* 2004). Such chronologic problems inevitably limit the value of many paleoclimatic reconstructions in detailed data synthesis, correlation, and synoptic mapping.

In the interpretation of Holocene paleoecologic data (e.g. pollen), there are often several possible interpretations for the observed changes (e.g. climatic shifts, human impact, biologic factors). Multi-proxy studies (Birks and Birks 2006) where several paleoenvironmental proxies are studied on the same stratigraphic sequence are particularly important as they permit the testing of alternative hypotheses to explain the observed changes. The elegant study by Shuman *et al.* (2004) on the interpretation of New England vegetational history in terms of climate changes deduced independently of the pollen stratigraphy from stable hydrogen isotope ratios and lake-level changes shows the value of multi-proxy approaches in Holocene climate research. The use of ecologic simulation models (e.g. Anderson *et al.* 2006) is another important approach to testing alternative hypotheses about underlying factors for the observed changes (e.g. Heiri *et al.* 2006).

A further problem that arises in the use of paleoecologic data as a source for paleoclimatic reconstructions concerns the climatic sensitivity of different types of organisms. Such data are from a wide range of ecologic settings that may result in different thresholds, sensitivities to climate change, and responses of individual species and biotic assemblages. It can thus be difficult to know what changes in particular proxies (e.g. diatoms, chironomids) may mean in climatic terms when local limnologic variables (e.g. lake-water pH, nutrient status) also change with time (Battarbee 2000). A related problem concerns the quantification of realistic uncertainties in paleoclimatic reconstruction that take account of the strong temporal autocorrelation in paleoclimatic time-series and the inherent spatial autocorrelation of climate variables used in organism–climate transfer functions (Telford and Birks 2005).

In addition to these paleoecologic problems, there are problems in understanding and quantifying interactions between external forcing factors such as orbital forcing, solar variability, volcanic activity, and land–atmosphere–ocean–ice feedbacks and in evaluating and quantifying the roles of land-use and vegetation cover in influencing Holocene climate (Oldfield 2005, this volume).

A challenge facing Holocene climate research is how to reconstruct "modes of variability" (Oldfield 2005). Climatologists commonly assess natural climate variability in terms of climate modes such as the El Niño–Southern Oscillation

(ENSO), the North Atlantic and Arctic Oscillations, the Pacific Decadal Oscillation, and the Atlantic Multi-decadal Oscillation (e.g. Diaz and Markgraf 2000; Hurrell *et al.* 2003). These modes are most relevant in the late Holocene when ice-extent, sea-level, orbital forcing, and atmospheric greenhouse-gas concentrations have changed relatively little (Bradley *et al.* 2003; Oldfield 2005). Late Holocene climate with almost constant external and internal boundary conditions is likely to be close to the pre-industrial patterns under those boundary conditions. It is the natural variability in this late Holocene climate system and its temporal and spatial patterns that recent climate change can be compared to identify anthropogenic forcing on climate (Oldfield 2005). Climate modes capture a large amount of the variance in climate variability and are ideal as composite indices for comparisons in time and space. The reconstruction of the temporal variation in such indices for the past 2000–3000 years requires not only fine-resolution studies but also novel ways of identifying climate modes from proxy-climate data (e.g. Thompson *et al.* 2000; Bradley *et al.* 2003; Gray *et al.* 2004; Dykoski *et al.* 2005).

Ruddiman's (2003, 2005b) Anthropocene hypothesis has highlighted basic gaps in knowledge about the extent of past land-use and vegetation-cover changes and about methane sources. Recent reports that terrestrial plants growing in an aerobic environment may emit methane (Keppler *et al.* 2006; see also Parsons *et al.* 2006) emphasize a major gap in our understanding of methane production and hence the global carbon budget. Much remains to be discovered about methane production and the causes of the observed changes in methane concentrations.

A final problem that is particularly critical in future data syntheses, mapping of paleoclimate proxies and reconstructed climate estimates, and data–model comparisons is the continuing need for international collaboration to ensure effective storage and management of the ever increasing amounts and diversity of primary paleoecologic, paleoceanographic, ice-core, geologic, and archaeologic data, modeling-experiment outputs, and paleoclimatic reconstructions (Eakin *et al.* 2003). These and many other data types must be carefully stored, managed, and made available to all researchers as these data may well be very relevant to furthering our understanding of Holocene climate history (Alverson and Eakin 2001; Anon 2001; Dittert *et al.* 2001).

Conclusions

The current paradigm for Holocene climate history is that a wide array of different paleoclimatic records indicate that there have been significant changes in summer and winter temperature, seasonality, and precipitation in the past 11 500 years. Some of these changes have been gradual over millennia, others have been very rapid and have occurred at centennial or even decadal scales. Paleoclimatic records from high latitudes suggest that in these areas the major changes have been in temperature and in seasonality. Records from low latitudes suggest that there have been major changes in precipitation and in the overall hydrologic regime. There appears to have been climate variability at a wide range of temporal scales. In addition there appears to have been considerable spatial variability in climate change, particularly in moisture. The relative importance of external forcing factors such as orbital forcing, solar variability, and volcanic activity, and the complex interactions between these factors and their impacts on the Earth's climate system and its circulation and climate modes are not fully understood (Bradley 2003; Bradley *et al.* 2003; Labeyrie *et al.* 2003; Oldfield 2005). The role of human activities in determining vegetation cover and land-use and in influencing Holocene climate remains unresolved.

Global summaries of Holocene climate change now appear to be of limited relevance (Oldfield 2005). In contrast, assessments of the magnitude of changes at different latitudes are more relevant and appear to have considerable significance environmentally, ecologically, and possibly socially. This shift from global summaries attempted in the 1940s–1960s to the increasing realization of the spatial complexity of Holocene climate change has perhaps been one of the most important paradigm shifts in Holocene climate research. Holocene climate has been highly dynamic, has varied in both time and space, and has shown considerable natural variability. These temporal and spatial patterns provide the background against which recent climate changes can be compared and assessed.

Acknowledgments

I am indebted to many people for sharing with me their insights and ideas about Holocene climate research, in particular Ray Bradley, Rick Battarbee, Hilary Birks, the late Knut Fægri, Ulrike Herzschuh, Jan Mangerud, Atle Nesje, Frank Oldfield, Heikki Seppä, Richard Telford, Tom Webb, and Herb Wright. I am, as always, extremely grateful to Cathy Jenks for her invaluable and skilful help in preparing this chapter.

References

Aaby, B. (1976) Cyclic climatic variations in climate over the past 5500 yr reflected in raised bogs. *Nature*, **263**, 281–284.

Aaby B. & Tauber H. (1975) Rates of peat formation in relation to degree of humification and local environment as shown by studies of a raised bog in Denmark. *Boreas*, **4**, 1–17.

Alley R.B. (2000) *The Two-Mile Time Machine – Ice Cores, Abrupt Climate Change, and our Future.* Princeton University Press, Princeton and Oxford.

Alley R.B. & Ágústdóttir A.M. (2005) The 8k event: cause and consequences of a major Holocene abrupt climate change. *Quaternary Science Reviews*, **24**, 1123–1149.

Alley R.B., Mayewski P.A., Sowers T., Stuiver M., Taylor K.C. & Clark P.U. (1997) Holocene climatic instability: a prominent, widespread event 8200 years ago. *Geology*, **25**, 483–486.

Alley R.B., Marotzke J., Nordhaus W.D., *et al.* (2003) Abrupt climate change. *Science*, **299**, 2005–2010.

Alverson K.D. & Eakin C.M. (2001) Making sure that world's palaeodata do not get buried. *Nature*, **412**, 269.

Alverson K.D., Bradley R.S. & Pedersen T.F. (Eds) (2003) *Paleoclimate, Global Change and the Future.* Springer-Verlag, Berlin.

Andersen C., Koç N., Jennings A. & Andrews J.T. (2004) Non-uniform response of the major surface currents in the Nordic Seas to insolation forcing: implications for the Holocene climate variability. *Paleoceanography*, **19**, PA2003, doi: 10.1029/2002PA000873,2004

Anderson D.M. (Ed.) (1995) *Global Paleoenvironmental Data. A Report from the Workshop Sponsored by Past Global Changes (PAGES), August 1993.* PAGES Workshop Report Series 95-2, 114 pp. International Geosphere–Biosphere Program, Bern.

Anderson N.J., Bugmann H., Dearing J.A. & Gaillard M.-J. (2006) Linking palaeoenvironmental data and models to understand the past and to predict the future. *Trends in Ecology and Evolution* **21**, 696–704, doi: 10.1016/j.tree.200609.005

Andersson C., Risebrobakken B., Jansen E. & Dahl S.O. (2003) Late Holocene surface ocean conditions of the Norwegian Sea (Vøring Plateau). *Paleoceanography*, **18**, 1044, doi: 10.1029/2001PA000654,2003

Andersson G. (1902) Hasseln i Sverige fordom och nu. *Sveriges Geologiska Undersökning Series C*, **3**, 1–168.

Andersson G. (1909) The climate of Sweden in the Late-Quaternary period. Facts and theories. *Sveriges Geologiska Undersökning Series C*, **Årbok 3**, 1–88.

Anon. (1994) *IGBP Global Modelling and Data Activities 1994–98.* Global Change Report 30, pp. 1–87. International Geosphere–Biosphere Program, Bern.

Anon. (2001) Make the most of palaeodata. *Nature*, **411**, 1.

Baillie M.G.L. & Brown D.M. (2003) Dendrochronology and the reconstruction of fine-resolution environmental change in the Holocene. In: *Global Change in the Holocene* (Eds A. Mackay, R.W. Battarbee, H.J.B. Birks & F. Oldfield), pp. 75–91. Arnold, London.

Bakke J., Dahl S.O. & Nesje A. (2005a) Lateglacial and early Holocene palaeoclimatic reconstruction based on glacier fluctuations and equilibrium-line altitudes at northern Folgefonna, Hardanger, western Norway. *Journal of Quaternary Science*, **20**, 179–198.

Bakke J., Dahl S.O., Paasche Ø., Løvlie R. & Nesje A. (2005b) Glacier fluctuations, equilibrium-line altitudes and palaeoclimate in Lyngen, northern Norway, during the Late-glacial and Holocene. *The Holocene*, **15**, 518–540.

Baldini J.U.L., McDermott F. & Fairchild I.J. (2002) Structure of the 8200-year cold event revealed by a speleothem trace element record. *Science*, **296**, 2203–2206.

Barber K.E. & Charman D.J. (2003) Holocene palaeoclimate records from peat-lands. In: *Global Change in the Holocene* (Eds A. Mackay, R.W. Battarbee, H.J.B. Birks & F. Oldfield), pp. 210–226. Arnold, London.

Battarbee R.W. (2000) Palaeolimnological approaches to climate change, with special regard to the biological record. *Quaternary Science Reviews*, **19**, 107–124.

Battarbee R.W., Gasse F. & Stickley C.E. (Eds) (2004) *Past Climate Variability through Europe and Africa*. Springer-Verlag, Berlin.

Berglund B.E. (2003) Human impact and climate changes – synchronous events and a causal link? *Quaternary International*, **105**, 7–12.

Berglund B.E., Birks H.J.B., Ralska-Jasiewiczowa M. & Wright H.E. Jr. (Eds) (1996) *Palaeoecological Events During the Last 15000 Years: Regional Syntheses of Palaeoecological Studies of Lakes and Mires*. Wiley, Chichester.

Bianchi G.G. & McCave I.N. (1999) Holocene periodicity in North Atlantic climate and deep-ocean flow south of Iceland. *Nature*, **397**, 515–517.

Birks C.J.A. & Koç N. (2002) A high-resolution diatom record of late-Quaternary sea-surface temperatures and oceanographic conditions from the eastern Norwegian Sea. *Boreas*, **31**, 323–344.

Birks H.H. (1975) Studies in the vegetational history of Scotland. IV. Pine stumps in Scottish blanket peats. *Philosophical Transactions of the Royal Society B*, **270**, 181–226.

Birks H.H. & Birks H.J.B. (2006) Multi-proxy studies in palaeolimnology. *Vegetation History and Archaeobotany*, **15**, 235–251.

Birks H.H., Larsen E. & Birks H.J.B. (2005) Did tree-*Betula*, *Pinus*, and *Picea* survive the last glaciation along the west coast of Norway? A review of the evidence, in light of Kullman (2002). *Journal of Biogeography*, **32**, 1461–1471.

Birks H.J.B. (1989) Holocene isochrone maps and patterns of tree-spreading in the British Isles. *Journal of Biogeography*, **16**, 503–540.

Birks H.J.B. (1995) Quantitative palaeoenvironmental reconstructions. In: *Statistical Modelling of Quaternary Science Data* (Eds D. Maddy & J.S. Brew), pp. 161–254. Technical Guide 5, Quaternary Research Association, Cambridge.

Birks H.J.B. (2005) Fifty years of Quaternary pollen analysis in Fennoscandia 1954–2004. *Grana*, **44**, 1–22.

Björck S., Muscheler R., Kromer B., *et al.* (2001) High-resolution analyses of an early Holocene climate event may imply decreased solar forcing as an important climate trigger. *Geology*, **29**, 1107–1110.

Bjune A.E., Bakke J., Nesje A. & Birks H.J.B. (2005) Holocene mean July temperature and winter precipitation in western Norway inferred from palynological and glaciological lake-sediment proxies. *The Holocene*, **15**, 177–189.

Blytt A. (1876) *Essay on the Immigration of the Norwegian Flora During the Alternating Rainy and Dry Periods*. Alb. Cammermeyer, Christiania (Oslo).

Blytt A. (1881) Die Theorie der wechselnden kontinentalen und insularen Klimate. *Englers Botanisch Jahrbüch*, **2**, 1–50.

Blytt A. (1893) Zur Geschichte der Nordeuropaischen besonders der Norwegischen flora. *Botanish Jahrbüch*, **17** (41), 1–43.

Bøe A.-G., Dahl S.O., Lie Ø. & Nesje A. (2006) Holocene river floods in the upper Glomma catchment, southern Norway: a high-resolution multiproxy record from lacustrine sediments. *The Holocene*, **16**, 445–455.

Bond G., Showers W., Cheseby M., *et al.* (1997) A pervasive millennial-scale cycle in North Atlantic Holocene and glacial climates. *Science*, **278**, 1257–1266.

Bond G., Kromer B., Beer J., *et al.* (2001) Persistent solar influence on North Atlantic climate during the Holocene. *Science*, **294**, 2130–2136.

Booth R.K. & Jackson S.T. (2003) A high-resolution record of Late Holocene moisture variability from a Michigan raised bog. *The Holocene*, **13**, 865–878.

Booth R.K., Jackson S.T., Forman S.L., *et al.* (2005) A severe centennial-scale drought in mid-continental North America 4200 years ago and apparent global linkages. *The Holocene*, **15**, 321–328.

Booth R.K., Notaro M., Jackson S.T. & Kutzbach J.E. (2006) Widespread drought episodes in the western Great Lakes regions during the past 2000 years: geographic extent and potential mechanisms. *Earth and Planetary Science Letters*, **242**, 415–427.

Bowen M. (2005) *Thin Ice – Unlocking the Secrets of Climate in the World's Highest Mountains.* Henry Holt, New York.

Bradley R.S. (1999) *Paleoclimatology – Reconstructing Climates of the Quaternary*, 2nd edn. Harcourt/Academic Press, Burlington.

Bradley R.S. (2003) Climate forcing during the Holocene. In: *Global Change in the Holocene* (Eds A. Mackay, R.W. Battarbee, H.J.B. Birks & F. Oldfield), pp. 10–19. Arnold, London.

Bradley R.S., Duplessy J.-C., Dodson J., Gasse F., Liu T.-S. & Markgraf V. (1995) PANASH science and implementation plan. In: *Paleoclimates of the Northern and Southern Hemispheres: The PANASH Project: The Pole-Equator-Pole Transects.* PAGES Series 95-1, pp. 1–19. International Geosphere–Biosphere Program, Bern.

Bradley R.S., Briffa K.R., Cole J., Hughes M.K. & Osborn T.J. (2003) The climate of the Last Millennium. In: *Paleoclimate, Global Change and the Future* (Eds K.D. Alverson, R.S. Bradley & T.F.Pedersen), pp. 105–141. Springer-Verlag, Berlin.

Brøgger W.C. (1900/01) Om de senglaciale og postglaciale nivåforandringer i Kristiania-fetet (Mollusk-faunaen). *Norges Geologiske Undersøgelse*, **31**, 1–731.

Burroughs W.J. (2005) *Climate Change in Prehistory. The End of the Reign of Chaos.* Cambridge University Press, Cambridge.

Carcaillet C., Almquist H., Asnong H., *et al.* (2002) Holocene biomass burning and global dynamics of the carbon cycle. *Chemosphere*, **49**, 845–863.

Cecil L.D., Green J.R. & Thompson L. (Eds) (2004) *Earth Paleoenvironments: Records Preserved in Mid- and Low-Latitude Glaciers.* Kluwer, Dordrecht.

Chambers F.M. & Brain S.A. (2002) Paradigm shifts in late-Holocene climatology? *The Holocene*, **12**, 239–249.

Chapman M.R. & Shackleton N.J. (2000) Evidence of 550-year and 1000-year cyclicities in North Atlantic circulation patterns during the Holocene. *The Holocene*, **10**, 287–291.

Charman D.J., Blundell A., Chiverrell R.C., Hendon D. & Langdon P.G. (2006) Compilation of non-annually resolved Holocene proxy climate records: stacked Holocene peatland palaeo-water table reconstructions from northern Britain. *Quaternary Science Reviews*, **25**, 336–350.

Clark P.U., Marshall S.J., Clarke G.K.C., Hostetler S.W., Licciardi J.M. & Teller J.T. (2001) Freshwater forcing of abrupt climate change during the last deglaciation. *Science*, **293**, 283–287.

Clarke G., Leverington D., Teller J. & Dyke A. (2003) Superlakes, megafloods, and abrupt climate change. *Science*, **301**, 922–923.

Claussen M., Brovkin V., Calov R., Ganapolski A. & Kubatzki C. (2005) Did humankind prevent a Holocene glaciation? *Climatic Change*, **69**, 409–417.

CLIMAP Project Members (1976) The surface of the ice-age earth. *Science*, **191**, 1131–1137.

COHMAP Members (1988) Climatic changes of the last 18 000 years: observations and model simulations. *Science*, **241**, 1043–1052.

Cole J.E. (2003) Holocene coral records: windows on tropical climate variability. In: *Global Change in the Holocene* (Eds A. Mackay, R.W. Battarbee, H.J.B. Birks & F. Oldfield), pp. 168–184. Arnold, London.

Crane D. (1972) *Invisible Colleges. Diffusion of Knowledge in Scientific Communities*. University of Chicago Press, Chicago.

Crucifix M., Loutre M.-F. & Berger A. (2005) Commentary on "The Anthropogenic Greenhouse Era began Thousands of Years Ago." *Climatic Change*, **69**, 419–426.

Dahl S.O., Bakke J., Lie Ø. & Nesje A. (2003) Reconstruction of former equilibrium-line altitudes based on proglacial sites: an evaluation of approaches and selection of sites. *Quaternary Science Reviews*, **22**, 275–287.

Danielsen A., Fægri K. & Henningsmoen K.E. (2000) Kvartærbotanikere vi møtte. *AmS-Varia*, **37**, 11–20.

Dansgaard W., Johnsen S.J., Moller J. & Langway Jr. C.C. (1969) One thousand centuries of climate record from Camp Century on the Greenland ice sheet. *Science*, **166**, 377–381.

Dansgaard W., Johnsen S.J., Clausen H.B. & Langway Jr. C.C. (1971) Climatic record revealed by the Camp Century ice core. In: *The Late Cenozoic Glacial Ages* (Ed. K.K. Turekian), pp. 37–56. Yale University Press, New Haven and London.

Dau J.H.C. (1829) *Allerunterhänigster Bericht an die Königliche Dänische Rentekammer über die Torfmoore Seelands*. Copenhagen and Leipzig.

Davis M.B. (1976) Pleistocene biogeography of temperate deciduous forests. *Geoscience and Man*, **13**, 13–26.

Deevey E.S. (1967) Introduction. In: *Pleistocene Extinctions* (Eds P.S. Martin & H.E. Wright Jr.), pp. 63–72. Yale University Press, New Haven and London.

DeMenocal P.B. (2001) Cultural responses to climate change during the Late Holocene. *Science*, **292**, 667–673.

Denton G.H. & Karlén W. (1973) Holocene climatic variations – their pattern and possible cause. *Quaternary Research*, **3**, 155–205.

Diamond J. (2005) *Collapse: How Societies Choose to Fail or Succeed*. Penguin, New York.

Diaz H.F. & Markgraf V. (Eds) (2000) *El Niño and the Southern Oscillation: Multiscale Variability and Global and Regional Impacts*. Cambridge University Press, Cambridge.

Digerfeldt G. (1972) The post-glacial development of Lake Trummen. Regional vegetation history, water level change and palaeolimnology. *Folia Limnologica Scandinavica*, **16**, 104 pp.

Digerfeldt G. (1988) Reconstruction and regional correlation of Holocene lake-level fluctuations in Lake Bysjön, south Sweden. *Boreas*, **17**, 165–182.

Digerfeldt G., Almendinger J.E. & Björck S. (1992) Reconstruction of past lake levels and their relation to groundwater hydrology in the Parkers Prairie sandplain, west-central Minnesota. *Palaeogeography, Palaeoclimatology, Palaeoecology*, **94**, 99–118.

Dittert N., Diepenbroek M. & Grobe H. (2001) Scientific data must be made available to all. *Nature*, **414**, 393.

Dodson J.R., Taylor D., Ono Y. & Wang P. (Eds) (2004) Climate, human and natural systems of the PEP II transect. *Quaternary International*, **118/119**, 1–203.

Dykoski C.A., Edwards R.L., Cheng H., *et al.* (2005) A high-resolution, absolute-dated Holocene and deglacial Asian monsoon record from Dongge Cave, China. *Earth and Planetary Science Letters*, **233**, 71–86.

Eakin C.M., Diepenbroek M. & Hoepffner M. (2003) The PAGES data system. In: *Palaeoclimate, Global Change and the Future* (Eds K.D. Alverson, R.S. Bradley & T.F. Pedersen), pp. 175–179. Springer-Verlag, Berlin.

Fægri K. (1940) Quartärgeologicshe Untersuchungen im westlichen Norwegen. I. Über zwei präboreale Klimaschwankungen im südwestlichsten Teil. *Bergens Museum Årbok 1933, Naturvitenskapelig rekke*, **8**, 1–40.

Fægri K. (1950) On the value of palaeoclimatological evidence. *Centenary Proceedings of the Royal Meteorological Society*, **1950**, 188–195.

Fægri K. (1981) Some pages of the history of pollen analysis. *Striae*, **14**, 42–47.

Fagan B. (1999) *Floods, Famines, and Emperors – El Niño and the Fate of Civilizations*. Basic Books, New York.

Fagan B. (2000) *The Little Ice Age – How Climate made History 1300–1850*. Basic Books, New York.

Fagan B. (2004) *The Long Summer – How Climate Changed Civilization*. Granta, London.

Fischer H., Kull C. & Kiefer T. (Eds) (2006) Ice core science. *PAGES News*, **14**(1), 44 pp.

Fisher D.S. & Koerner R.M. (2003) Holocene ice-core climate history – a multivariable approach. In: *Global Change in the Holocene* (Eds A. Mackay, R.W. Battarbee, H.J.B. Birks & F. Oldfield), pp. 281–293. Arnold, London.

Fries R.E. (1950) *A Short History of Botany in Sweden*. Almqvist & Wiksells, Uppsala.

Fritts H.C. (1976) *Tree Rings and Climate*. Academic Press, New York.

Fritts H.C., Blasing T.J., Hayden B.P. & Kutzbach J.E. (1971) Multivariate techniques for specifying tree-growth and climate relationships and for reconstructing anomalies in paleoclimate. *Journal of Applied Meteorology*, **10**, 845–864.

Fritz S.C. (2003) Lacustrine perspectives on Holocene climate. In: *Global Change in the Holocene* (Eds A. Mackay, R.W. Battarbee, H.J.B. Birks & F. Oldfield), pp. 227–241. Arnold, London.

Gams H. & Nordhagen, R. (1923) Post-glaciale Klimaänderungen und Erdkrustenbewegungen im Mitteleurope. *Landeskundliche Förschungen herausgegeben von der Geographischen Gesellschaft in München*, **25**, 336 pp.

Gasse F. (2000) Hydrological changes in the African tropics since the last glacial maximum. *Quaternary Science Reviews*, **19**, 189–211.

Gates W.L. (1976) Modeling the ice-age climate. *Science*, **191**, 1138–1144.

Giesecke T. (2005) Moving front or population expansion: how did *Picea abies* (L) Karst. become frequent in central Sweden? *Quaternary Science Reviews*, **24**, 2495–2509.

Giesecke T. & Bennett K.D. (2004) Holocene spread of *Picea abies* (L) Karst in Fennoscandia and adjacent areas. *Journal of Biogeography*, **31**, 1523–1548.

Godwin H. (1960) Radiocarbon dating and Quaternary history in Britain. *Proceedings of Royal Society B*, **153**, 287–320.

Godwin H. (1966) Introductory Address. In: *World Climate from 8000 to 0 BC* (Ed. J.S. Sawyer), pp. 3–14. Royal Meteorological Society, London.

Godwin H. (1978) *Fenland: its Ancient Past and Uncertain Future*. Cambridge University Press, Cambridge.

Godwin H. (1981) *The Archives of the Peat Bogs*. Cambridge University Press, Cambridge.

Granlund E. (1932) De Svenska högmossarnas geologi. *Sverige Geologiska Undersökning Series C*, **373**, 193 pp.

Gray S.T., Graumlich L.T., Betancourt J.L. & Pedersen G.T. (2004) A tree-ring based reconstruction of the Atlantic Multidecadal Oscillation since 1567 AD. *Geophysical Research Letters*, **31**, L12205, doi: 10.1029/2004GL019932,2004

Haberle S.G. & Chepstow-Lusty A. (2000) Can climate influence cultural development? A view through time. *Environment and History*, **6**, 349–369.

Haberle S.G. & David B. (2004) Climates of change: human dimensions of Holocene environmental change in low latitudes of the PEP II transect. *Quaternary International*, **118/119**, 165–179.

Hafsten U. (1970) A sub-division of the Late Pleistocene period on a synchronous basis, intended for global and universal usage. *Palaeogeography, Palaeoclimatology, Palaeoecology*, **7**, 279–296.

Haug G.M., Günther D., Peterson L.C., Sigman D.M., Hughen K.A. & Aeschlimann B. (2003) Climate and the collapse of the Maya civilization. *Science*, **299**, 1731–1735.

Hays J.D., Imbrie J. & Shackleton N.J. (1976) Variations in the Earth's orbit: pacemaker of the ice ages. *Science*, **194**, 1121–1132.

Heiri C., Bugmann H., Tinner W., Heiri, O. & Lischke H. (2006) A model-based reconstruction of Holocene tree line dynamics in the Central Swiss Alps. *Journal of Ecology*, **94**, 206–216.

Henderson G.M. (2006) Caving in to new chronologies. *Science*, **313**, 620–622.

Hintikka V. (1963) Über das Grossklima einiger Pflanzenareale in zwei Klima-koordinatensystemen dargestellt. *Annales Botanici Societatis Zoologicæ Botanicæ Fennicæ 'Vanamo'*, **34** (5), 1–64.

Holmboe J. (1921) Prof. dr. A.G. Nathorst. *Naturen*, **February 1921**, 33–40.

Hughes P.D.M., Blundell A., Charman D.J., *et al.* (2006) An 8500 cal. year multi-proxy climate record from a bog in eastern Newfoundland: contributions of meltwater discharge and solar forcing. *Quaternary Science Reviews*, **25**, 1208–1227.

Huntley B. & Birks H.J.B. (1983) *An Atlas of Past and Present Pollen Maps for Europe: 0–13 000 years ago*. Cambridge University Press, Cambridge.

Huntley B., Baillie M., Grove J.M., *et al.* (2002) Holocene palaeoenvironmental changes in North-West Europe: climatic implications and the human dimension. In: *Climate Development and History of the North Atlantic Realm* (Eds G. Wefer, W.H. Berger, K.-E. Behre & E. Jansen), pp. 259–298. Springer-Verlag, Berlin.

Hurrell J.W., Kushnir Y., Ottersen G. & Visbeck M. (Eds) (2003) *The North Atlantic Oscillation – Climatic Significance and Environmental Impact*. Geophysical Monograph 134, American Geophysical Union, Washington, DC.

Hyvärinen H. (1975) Absolute and relative pollen diagrams from northernmost Fennoscandia. *Fennia*, **142**, 23 pp.

Imbrie J. & Kipp N.G. (1971) A new micropaleontological method for quantitative paleoclimatology: application to a Late Pleistocene Caribbean core. In: *The Late Cenozoic Glacial Ages* (Ed. K.K. Turekian), pp. 71–181. Yale University Press, New Haven and London.

Imbrie J. & Imbrie K.P. (1979) *Ice Ages: Solving the Mystery*. Enslow Publishers, Short Hills, New Jersey.

Iversen J. (1944) *Viscum, Hedera* and *Ilex* as climate indicators. *Geologiska Föreningens i Stockholm Förhandlingar*, **66**, 463–483.

Iversen J. (1973) The development of Denmark's nature since the last glacial. *Danmarks Geologiske Undersøgelse Series V*, **number 7-C**, 1–126.

Jackson S.T. & Overpeck J.T. (2000) Responses of plant populations and communities to environmental changes of the late Quaternary. *Paleobiology*, **26** (Supplement), 194–220.

Jackson S.T. & Williams J.W. (2004) Modern analogs in Quaternary paleoecology: here today, gone yesterday, gone tomorrow? *Annual Review of Earth and Planetary Sciences*, **32**, 495–537.

Jansen E., DeMenocal P. & Grousset F. (Eds) (2004) Holocene Climate Variability – A marine perspective. *Quaternary Science Reviews*, **23**, 2061–2268.

Jansen E., Andersson C., Moros, M., Nisancioglu K.H., Nyland B.F. & Telford R.J. (this volume) The early to mid-Holocene thermal optimum in the North Atlantic. In: *Global Warming and Natural Climate Variability: a Holocene Perspective* (Eds R.W. Battarbee & H.A. Binney), pp. 123–137. Blackwell, Oxford.

Joos F., Gerber S., Prentice I.C., Otto-Bliesner B.L. & Valdes P.J. (2004) Transient simulations of Holocene atmospheric carbon dioxide and terrestrial carbon since the Last Glacial Maximum. *Global Biogeochemical Cycles*, **18**, GB2002, doi: 10.1029/2003GB002156,2004

Karlén W. (1976) Lacustrine sediments and tree-limit variations as indicators of Holocene climatic fluctuations in Lappland, northern Sweden. *Geografiska Annaler*, **58A**, 1–34.

Karlén W. & Kuylenstierna J. (1996) On solar forcing of Holocene climate: evidence from Scandinavia. *The Holocene*, **6**, 359–365.

Keppler F., Hamilton J.T.G., Brass M. & Röckmann T. (2006) Methane emissions from terrestrial plants under aerobic conditions. *Nature*, **434**, 187–191.

Kuhn T. (1970) *The Structure of Scientific Revolutions*. University of Chicago Press, Chicago.

Kullman L. (1998a) Palaeoecological, biogeographical and palaeoclimatological implications of early Holocene immigration of *Larix sibirica* Ledeb. into the Scandes Mountains, Sweden. *Global Ecology and Biogeography Letters*, **7**, 181–188.

Kullman L. (1998b) The occurrence of thermophilous trees in the Scandes Mountains during the early Holocene: evidence for a diverse tree flora from macroscopic remains. *Journal of Ecology*, **86**, 421–428.

Kullman L. (2000) The geoecological history of *Picea abies* in northern Sweden and adjacent parts of Norway. A contrarian hypothesis of postglacial tree migration patterns. *Geoöko*, **21**, 141–172.

Kullman L. (2002) Boreal tree taxa in the central Scandes during the Late-glacial: implications for Late-Quaternary forest history. *Journal of Biogeography*, **29**, 1117–1124.

Labeyrie L., Cole J., Alverson K. & Stocker T. and 12 contributors (2003) The history of climate dynamics in the Late Quaternary. In: *Paleoclimate, Global Change and the Future* (Eds K.D. Alverson, R.S. Bradley & T.F. Pedersen), pp. 33–61. Springer-Verlag, Berlin.

Laird K.R., Fritz S.C. & Cumming B.F. (1998a) A diatom-based reconstruction of drought intensity, duration, and frequency from Moon Lake, North Dakota: a sub-decadal record of the last 2300 years. *Journal of Paleolimnology*, **19**, 161–179.

Laird K.R., Fritz S.C., Cumming B.F. & Grimm E.C. (1998b) Early-Holocene limnological and climate variability in the Northern Great Plains. *The Holocene*, **8**, 275–285.

Lang G. (1994) *Quartäre Vegetationsgeschichte Europas*. Gustav Fischer Verlag, Jena.

Lauritzen S.-E. (2003) Reconstructing Holocene climate records from speleothems. In: *Global Change in the Holocene* (Eds A. Mackay, R.W. Battarbee, H.J.B. Birks & F. Oldfield), pp. 242–263. Arnold, London.

Linden E. (2006) *The Winds of Change – Climate, Weather, and the Destruction of Civilizations*. Simon & Schuster, New York.

Linge H., Lauritzen S.-E., Lundberg J. & Berstad I.M. (2001) Stable isotope stratigraphy of Holocene speleothems: examples from a cave system in Rana, northern Norway. *Palaeoceanography, Palaeoclimatology, Palaeoecology*, **167**, 209–224.

Lundqvist J. (1965) The Quaternary of Sweden. In: *The Quaternary, Volume 1* (Ed. K. Rankama), pp. 139–198. Interscience Publishers, New York.

MacDonald G.M., Beilman D.W., Kremenetski K.V., Sheng Y., Smith L.G. & Velichko A.A. (2006) Rapid early development of circumarctic peatlands and atmospheric CH_4 and CO_2 variations. *Science*, **314**, 285–288.

Mackay A., Battarbee R.W., Birks H.J.B. & Oldfield F. (Eds) (2003) *Global Change in the Holocene*. Arnold, London.

Manabe S. & Hahn D.G. (1977) Simulation of the tropical climate of an ice age. *Journal of Geophysical Research*, **82**, 3889–3911

Mangerud J. (1982) The chronostratigraphical subdivision of the Holocene in Norden: a review. *Striae*, **16**, 65–70.

Mangerud J., Andersen S.T., Berglund B.E. & Donner J.J. (1974) Quaternary stratigraphy of Norden, a proposal for terminology and classification. *Boreas*, **3**, 109–128.

Manten, A.A. (1967a) Lennart von Post and the foundation of modern palynology. *Review of Palaeobotany and Palynology*, **1**, 11–22.

Manten A.A. (1967b) Lennart von Post's pollen diagram series of 1916. *Review of Palaeobotany and Palynology*, **4**, 9–13.

Markgraf V. (Ed.) (2001) *Interhemisphere Climate Linkages*. Academic Press, San Diego.

Mason B. (2004) The hot hand of history. *Nature*, **427**, 582–583.

Maxwell H. (1915) *Trees: a Woodland Notebook*. Maclehose, Glasgow.

Mayewski P.A., Rohling E.E., Stager J.C., *et al.* (2004) Holocene climate variability. *Quaternary Research*, **62**, 243–255.

McAndrews J.H. (1966) Postglacial history of prairie, savanna, and forest in northwestern Minnesota. *Memoirs of the Torrey Botanical Club*, **22**, 1–72.

McLachlan J.S., Clark J.S. & Manos P.S. (2005) Molecular indicators of tree migration capacity under rapid climate change. *Ecology*, **86**, 2088–2098.

Mitchell G.F. (1976) *The Irish Landscape*. Collins, London.

Mitchell G.F. (1990) *The Way That I Followed*. Country House, Dublin.

Moe D. (1970) The post-glacial immigration of *Picea abies* into Fennoscandia. *Botaniska Notiser*, **23**, 61–66.

Moros M., Emeis K., Risebrobakken B., *et al.* (2004) Sea surface temperatures and ice rafting in the Holocene North Atlantic: climate influences on northern Europe and Greenland. *Quaternary Science Reviews*, **23**, 2113–2126.

Nathorst A.G. (1870) Om några arktiska växtlämningar i en sötvattenslera vid Alnarp i Skåne. *Lunds Universitets Årsskrift*, **7**, 17 pp.

Nathorst A.G. (1892) Über den gegenwärtigen Standpunkt unserer Kenntnis von dem Vorkommen fossiler Glazialpflanzen. Bihang t. K. *Svenska Vetenskaps Akademiens Handlingar 17*, **Series III**, **5**, 32 pp.

Nesje A. & Dahl S.O. (2000) *Glaciers and Environmental Change*. Arnold, London.

Nesje A. & Dahl S.O. (2003) Glaciers as indicators of Holocene climate change. In: *Global Change in the Holocene* (Eds A. Mackay, R.W. Battarbee, H.J.B. Birks & F. Oldfield), pp. 264–280. Arnold, London.

Nesje A., Dahl S.O., Matthews J.A. & Berrisford M.S. (2001) A c. 4500-yr record of river floods obtained from a sediment core in Lake Atnsjøen, eastern Norway. *Journal of Paleolimnology*, **25**, 329–342.

Nesje A., Jansen E., Birks H.J.B., *et al.* (2005) Holocene climate variability in the Northern North Atlantic region: a review of terrestrial marine evidence. In: *The Nordic Seas: an Integrated Perspective* (Eds H. Drange, T. Dokken, T. Furevik, R. Gerdes & W. Berger), pp. 289–322. Geophysical Monograph 158, American Geophysical Union Washington, DC.

O'Brian S.R., Mayewski P.A., Meeker L.D., Meese D.A., Twickler M.S. & Whitlow S.I. (1995) Complexity of Holocene climate as reconstructed from a Greenland ice core. *Science*, **270**, 1962–1964.

Oldfield F. (2005) *Environmental Change: Key Issues and Alternative Approaches.* Cambridge University Press, Cambridge.

Oldfield F. (this volume) The role of people in the Holocene. In: *Global Warming and Natural Climate Variability: a Holocene Perspective* (Eds R.W. Battarbee & H.A. Binney), pp. 58–97. Blackwell, Oxford.

PAGES (1998) Paleodata. *PAGES News*, **6**(2), 16 pp.

Parsons A.J., Newton P.C.D., Clark H. & Kelliher F.M. (2006) Scaling methane emissions from vegetation. *Trends in Ecology and Evolution*, **21**, 423–424.

Pedersen T.F., François R., François L., Alverson K. & McManus J. (2003) The late Quaternary history of biogeochemical cycling of carbon. In: *Paleoclimate, Global Change and the Future* (Eds K. Alverson, R.S. Bradley & T.F. Pedersen), pp. 63–79. Springer-Verlag, Berlin.

Prentice I.C. & Webb III T. (1998) BIOME 6000: reconstructing global mid-Holocene vegetation patterns from palaeoecological records. *Journal of Biogeography*, **25**, 997–1005.

Ralska-Jasiewiczowa M., Latałowa M., Wasylikowa K., *et al.* (2004) *Late Glacial and Holocene History of Vegetation in Poland Based on Isopollen Maps.* W. Szafer Institute of Botany, Kraków.

Raynaud D., Blunier T., Ono Y. & Delmas R.J. (2003) The late Quaternary history of atmospheric trace gases and aerosols: interactions between climate and biogeochemical cycles. In: *Paleoclimate, Global Change and the Future* (Eds K. Alverson, R.S. Bradley & T.F. Pedersen), pp. 13–31. Springer-Verlag, Berlin.

Redman C.J. (1999) *Human Impact on Ancient Environments.* University of Arizona Press, Tucson.

Rind D., deMenocal P., Russell G.J., *et al.* (2001) Effects of glacial meltwater in the GISS coupled atmosphere-ocean model: Part I: North Atlantic deep water response. *Journal of Geophysics Research*, **106**, 335–27, 354.

Risebrobakken B., Jansen E., Andersson C., Mjelde E. & Hevrøy K. (2003) A high-resolution study of paleoceanographic changes in the Nordic Sea. *Paleoceanography*, **18**, 1017, doi: 10.1029/2002PA000764, 2003

Rohling E.J. & Pälike H. (2005) Centennial-scale climate cooling with a sudden cool event around 8200 years ago. *Nature*, **434**, 975–979.

Ruddiman W.F. (2003) The anthropogenic greenhouse era began thousands of years ago. *Climatic Change*, **61**, 261–293.

Ruddiman W.F. (2005a) The early anthropogenic hypothesis a year later. *Climatic Change*, **69**, 427–434.

Ruddiman W.F. (2005b) *Plows, Plagues, and Petroleum. How Humans took Control of Climate.* Princeton University Press, Princeton and Oxford.

Ruddiman W.F. & Thomson J.S. (2001) The case for human causes of increased atmospheric CH_4 over the last 5000 years. *Quaternary Research*, **20**, 1769–1777.

Ruddiman W.F., Vavrus S.J. & Kutzbach J.E. (2005) A test of the overdue-glaciation hypothesis. *Quaternary Science Reviews*, **24**, 1–10.

Samuelsson G. (1910) Scottish peat mosses, a contribution to the knowledge of the late Quaternary vegetation and climate of north-west Europe. *Bulletin of the Geological Institute of Uppsala*, **10**, 197–260.

Samuelsson G. (1916) Über den Rückgang der Haselgrenze und anderer pflanzen-geographischer Grenzlinien in Skandinavien. *Bulletin of the Geological Institute of Uppsala*, **13**, 93–114.

Schmidt G.A. & LeGrande A.N. (2005) The Goldilocks abrupt climate change event. *Quaternary Science Reviews*, **24**, 1109–1110.

Schulz M. & Paul A. (2002) Holocene climate variability on centennial-to-millennial time scales. 1 Climate records from the North-Atlantic realm. In: *Climate Development and History of the North Atlantic Realm* (Eds G. Wefer, W.H. Berger, K.-E. Behre & E. Jansen), pp. 41–54. Springer-Verlag, New York.

Schwander J., Eicher U. & Ammann B. (2000) Oxygen isotopes of lake marl at Gerzensee and Leysin (Switzerland), covering the Younger Dryas and two minor oscillations and their correlations to the GRIP ice core. *Palaeogeography, Palaeoclimatology, Palaeoecology*, **159**, 203–214.

Seppä H. (1996) Post-glacial dynamics of vegetation and tree-lines in the far north of Fennoscandia. *Fennia*, **174**, 1–96.

Sernander R. (1889) Om växtlämningar i Skandinaciens marina bildningar. *Botaniska Notiser*, **1889**, 190–199.

Sernander R. (1890) Om förekomsten af subfossila stubbar på svenska insjöars bottem. *Botaniska Notiser*, **1890**, 10–20.

Sernander R. (1893) Om Litorina-tidens klimat och vegetation. *Geologiska Föreningens i Stockholm Förhandlingar*, **15**, 345–377.

Sernander R. (1894) *Studier öfver den Gotländiska vegetationens utvecklingshistoria.* Akademiska afhandling, Uppsala.

Sernander R. (1908) On the evidence of post-glacial changes of climate furnished by the peat mosses of northern Europe. *Geologiska Föreningens i Stockholm Förhandlingar*, **30**, 365–378.

Sernander R. (1909) De scanodaniska torfmossarnas stratigrafi. *Geologiska Föreningens i Stockholm Förhandlingar*, **31**, 423–448.

Sernander R. (1910) Die Schwedischen Torfmoore als Zeugen Postglazialer Klimaschwankungen. *Proceedings of the 11th International Geological Congress, Stockholm*, pp. 197–246.

Shuman B., Newby P., Huang Y. & Webb III T. (2004) Evidence for the close climatic control of New England vegetation history. *Ecology*, **85**, 1297–1310.

Sletten K., Blikra L.H., Ballantyne C.K., Nesje A. & Dahl S.O. (2003) Holocene debris flows recognised in a lacustrine sedimentary succession: sedimentology, chronostratigraphy and cause of triggering. *The Holocene*, **13**, 907–920.

Smith A.G. & Pilcher J.R. (1973) Radiocarbon dates and vegetational history of the British Isles. *New Phytologist*, **72**, 903–914.

Smol J.P. (2002) *Pollution of Lakes and Rivers: a Paleoenvironmental Perspective.* Arnold, London.

Snowball I., Zillén L & Gaillard M.-J. (2002) Rapid early-Holocene environmental changes in northern Sweden based on studies of two varved lake-sediment sequences. *The Holocene*, **12**, 7–16.

Steenstrup J.J.S. (1841) *Geognostisk-geologisk Undersögelse af Skovmoserne Vidnesdam- og Lillemose i det Nordlige Sjælland, Ledsaget af Sammenlignende Bemærkninger, Hentede fra Danmarks Skov- Kjær- og Lyngmoser Lalmindelighed.* Copenhagen.

Street-Perrott F.A. & Harrison S.P. (1985) Lake levels and climate reconstruction. In: *Paleoclimate Analysis and Modelling* (Ed. A.D. Hecht), pp. 291–340. Wiley, New York.

Sugita S., Gaillard M.-J. & Broström A. (1999) Landscape openness and pollen records: a simulation approach. *The Holocene*, **9**, 409–421.

Sugita S., MacDonald G.M. & Larsen C.P.S. (1997) Reconstruction of fire disturbance and forest succession from fossil pollen in lake sediments: potential and limitations. In: *Sediment Records of Biomass Burning and Global Change* (Eds J.S. Clark, H. Cachier, J.G. Goldammer & B.J. Stocks), pp. 387–412. Springer-Verlag, Berlin.

Szafer W. (1935) The significance of isopollen lines for the investigation of the geographic distribution of trees in the post-glacial period. *Bulletin de l'Academie Polonaise des Sciences B*, **3**, 235–239.

Tait C. (1794) An account of the peat-mosses of Kincardine and Flanders in Perthshire. *Transactions of the Royal Society of Edinburgh*, **3**, 266–279.

Tan M., Baker A., Genty D., Smith C., Esper J. & Cai B. (2006) Applications of stalagmite laminae to paleoclimate reconstructions: comparison with dendrochronology/climatology. *Quaternary Science Reviews*, **25**, 2103–2117.

Telford R.J. & Birks H.J.B. (2005) The secret assumption of transfer functions: problems with spatial autocorrelation in evaluating model performance. *Quaternary Science Reviews*, **24**, 2173–2179.

Telford R.J., Heegaard E. & Birks H.J.B. (2004) All age–depth models are wrong: but how badly? *Quaternary Science Reviews*, **23**, 1–5.

Thompson L.G., Mosley-Thompson E., Davis M.E., Henderson K.A. & Lin P.N. (2000) The tropical ice core record of ENSO. In: *El Niño and the Southern Oscillation: Multiscale Variability and Global and Regional Impacts* (Eds H.F. Diaz & V. Markgraf), pp. 325–356. Cambridge University Press, Cambridge.

Tinner W. & Lotter A.F. (2006) Holocene expansions of *Fagus sylvatica* and *Abies alba* in central Europe: Where are we after eight decades of debate? *Quaternary Science Reviews*, **25**, 526–549.

Tinner W., Lotter A.F., Ammann B., *et al.* (2003) Climatic change and contemporaneous land-use phases north and south of the Alps 2300 BC to 800 AD. *Quaternary Science Reviews*, **22**, 1447–1460.

Valdes P.J. (2003) An introduction to climate modelling of the Holocene. In: *Global Change in the Holocene* (Eds A. Mackay, R.W. Battarbee, H.J.B. Birks & F. Oldfield), pp. 20–35. Arnold, London.

Vaupell C. (1857) *Bögens Indvandring i de Danske Skove.* C.A. Reitzels Bo & Arvinger, Copenhagen.

Verschuren D. & Charman D. (this volume) Latitudinal linkages in late Holocene moisture-balance variation. In: *Global Warming and Natural Climate Variability: a Holocene Perspective* (Eds R.W. Battarbee & H.A. Binney), pp. 189–231. Blackwell, Oxford.

Verschuren D., Laird K.R. & Cumming B.F. (2000) Rainfall and drought in equatorial east Africa during the past 1100 years. *Nature*, **403**, 410–414.

Von Post L. (1920) Postarktiska klimattyper i södra Sverige. *Geologiska Föreningens i Stockholm Förhandlingar*, **42**, 231–241.

Von Post L. (1924) Ur de sydsvenska skogarnas regionala historia under postarktisk tid. *Geologiska Föreningens i Stockholm Förhandlingar*, **46**, 83–128.

Von Post L. (1933) Den svenska skogen efter istiden. *Studentföreningen Verdandis Småskrifter*, **357**, 1–64.

Von Post L. (1946) The prospect for pollen analysis in the study of the Earth's climatic history. *New Phytologist*, **45**, 193–217.

Von Post L. (1967) Forest tree pollen in south Swedish peat bog deposits (translated by M.B. Davis and K. Fægri with introduction by K. Fægri and J. Iversen). *Pollen et Spores*, **9**, 375–401.

Von Post L. & Sernander, R. (1910) *Pflanzen-physiognomische Studien auf Torfmooren in Närke.* International Geological Congress Stockholm, Excursion A7, 48 pp.

Walther G.-R., Berger S. & Sykes M.T. (2005) An ecological "footprint" of climate change. *Proceedings of the Royal Society B*, **272**, 1427–1432.

Webb III T. (1988) Eastern North America. In: *Vegetation History* (Eds B. Huntley & T. Webb, III), pp. 385–414. Kluwer Academic Publishers, Dordrecht.

Webb III T. & Bryson R.A. (1972) Late- and post-glacial climate change in Northern Midwest, USA: Quantitative estimates derived from fossil pollen spectra by multivariate statistical analysis. *Quaternary Research*, **2**, 70–115.

Williams J.R.G., Barry R.G. & Washington W.M. (1974) Simulations of the atmospheric circulation using the NCAR global circulation model with ice age boundary conditions. *Journal of Applied Meteorology*, **13**, 305–317.

Wright Jr. H.E. (1966) Stratigraphy of lake sediments and the precision of the paleoclimatic record. In: *World Climate from 8000 to 0 BC* (Ed. J.S. Sawyer), pp. 157–173. Royal Meteorological Society, London.

Wright Jr. H.E., Kutzbach J.E., Webb III T., Ruddiman W.F., Street-Perrott F.A. & Bartlein P.J. (Eds) (1993) *Global Climates Since the Last Glacial Maximum.* University of Minnesota Press, Minneapolis and London.

Wright W.B. (1936) *The Quaternary Ice Age*, 2nd edn. Macmillan, London.

Wu H., Guo Z. & Peng C. (2003) Land use induced changes of organic carbon storage in soils of China. *Global Change Biology*, **9**, 305–315.

Zolitschka B. (2003) Dating based on freshwater- and marine-laminated sediments. In: *Global Change in the Holocene* (Eds A. Mackay, R.W. Battarbee, H.J.B. Birks & F. Oldfield), pp. 92–106. Arnold, London.

3 The role of people in the Holocene

Frank Oldfield

Keywords

Human–environment interactions, past human impacts, biosphere–climate feedbacks, erosion histories, past sustainability

Introduction

Any current definition of Earth System Science sees human activities as integral to the way in which the Earth System functions. People are now key players in the system, not passive recipients of the consequences of natural processes. The emerging consensus is that this applies, above all, to what has become characterized as the Anthropocene (Crutzen and Stoermer 2001) – the past two to three centuries, and, most significantly, the past 60 years (Steffen *et al.* 2003). Human impact on the environment, however, has a much longer history than this. What do we know and what more do we need to know about this longer term history? It is especially appropriate to address these questions in the context of the HOLIVAR (Holocene Climate Variability) initiative, linked as it is to the PAGES (Past Global Changes) PEP (Pole–Equator–Pole) III Transect (Battarbee *et al.* 2004), for some of the clearest evidence for past human impacts has come through research in Europe. Moreover, as we turn southward from Europe to Africa, that continent includes some of the people most vulnerable to the combined effects of climate variability and human activities.

The scope and significance of past human–environment interactions

We have known since the 1940s that, in northern Europe, human activities have led to forest clearance and changes in land-cover from early Neolithic times onwards (Iversen 1941). Over the succeeding decades, several publications have shown that some of the resulting changes in ecosystem function led to early transformation of land-cover that has persisted to the present day (Godwin 1944). The role of environmental changes, including climatic change, as major factors in past human welfare has often captured the imagination of both scientists and the general public (e.g. Diamond 2005). Human–environment interactions are complex and work in several directions; how, within these complex, interactive fields can we, as paleoscientists, with the tools currently at our disposal, define and explore the key themes that are important both for understanding the present-day environment and for best addressing the likely course of future environmental change? Can we establish and exemplify a research agenda that is timely, tractable, and makes a significant contribution to Earth System Science?

Five interlinked themes may be addressed under the heading of this chapter:

1 The impact of past human activities, acting alongside and interacting with Holocene climate change, on past and present ecosystems.
2 The effects of past land-use/land-cover changes on climate at regional scale and beyond.
3 The combined effects of the above on the hydrologic cycle and on erosion regimes.
4 The effects of past land-use/land-cover changes on atmospheric greenhouse gas concentrations and the consequent climatic implications.
5 The impacts of climate change on human societies in the past and the ways in which the effects of these changes have been mediated through the responses of the societies affected.

Each of these is considered below.

The impacts of past human activities on ecosystems

Although there has been a growing emphasis on the interpretation of pollen analytic and other biologic records as proxies for past climate change (e.g. Birks 2003, this volume), the paleoecological evidence for major human impacts on past vegetation continues to grow, especially in Europe, where pollen taxonomy favors the identification of species and genera unambiguously indicative of human activities such as deforestation and the development of pastoralism and agriculture. Gaining some sense of the timing, sequence, and degree of alteration of previously forested landscapes is fundamental to addressing the subsequent themes.

In some environments, the millennia from the early Neolithic through to the early Medieval period were marked by successive alternations between episodes of deforestation and farming, and periods of forest recovery. This appears to have been the case in the Black Forest region of south-western Germany where Roesch (2000) records nine phases of forest clearance between the early Neolithic ca. 7600 years BP and the early Medieval period that saw the start of major, sustained deforestation. This record is strongly reminiscent of that from parts of north-west

England (e.g. Oldfield 1963). Such long-term records of recurrent recovery suggest that the forest ecosystems of these regions, hence the soils also, have been remarkably resilient (*sensu* Lal 1997) in the face of repeated phases of deforestation. This is despite the fact that several pollen diagrams from the north of England suggest that the area deforested during Late Iron Age and Romano-British times, between around 2500 and 1500 years BP, was as extensive as during the peak deforestation in the 17th century (Oldfield and Statham 1965; Dumayne-Peaty and Barber 1998; Wimble *et al.* 2000; Oldfield *et al.* 2003a; Figure 3.1). Strong evidence for the importance of deforestation during the period of Roman occupation also comes from many studies in France (e.g. Noel *et al.* 2001; Cyprien *et al.* 2004).

In some other parts of Europe, the evidence points to earlier and somewhat more progressive conversion of forest to less productive ecosystems as a result of human activities. In parts of Denmark, for example, there are indications of irreversible deforestation from ca. 500 BC onwards (Bradshaw *et al.* 2005a). Odgaard and Rasmussen (2000) reinforce this view by statistical comparisons between the mosaic of land-cover types documented in 1800 and those inferred from successive pollen records spanning the period 4500 BC to AD 1800 at 20 sites (Figure 3.2). In-depth, multi-disciplinary studies of ecological and cultural history at the regional scale are relatively rare, but one such is that of Berglund (1991) in the Ystad region of southern Sweden (Figure 3.3). As in the Danish study, he records the progressive deforestation of the landscape during prehistory. Intermediate between the studies showing progressive deforestation without significant recovery, and those marked by recurrent re-afforestation, are sites such as Lough Neagh in Northern Ireland where pulses of forest clearance, marked by peaks in weeds and bracken, are superimposed on evidence for a progressive increase in grassland (O'Sullivan *et al.* 1973).

Heathlands are distinctive habitats often of high conservation value in northern and western Europe. Understanding their origins and history is germane to any future management designed to preserve the special habitats and high biodiversity that they contain (Walker *et al.* 2003). They have often been seen as, in part at least, the result of forest clearance combined with subsequent soil degradation and/or sustained management practices favoring the persistence of ericaceous shrubs and acidophilous herbs. Fyfe *et al.* (2003) interpret their pollen records from Exmoor in south-west England as indicating the expansion of heathland from Neolithic times onwards, rather as did Godwin in his early study of the Breckland heaths of East Anglia (Godwin 1944), although they regard the use of fire as important for the maintenance of heathland. Savukynien *et al.* (2003) present evidence from Lithuania that points to the development of heathland from 1200 years BP onwards, as well as its maintenance largely through burning until recent times when its extent has been reduced. Following the earlier work of Kaland (1986), Prosch-Danielson and Simonsen (2000) show that the expansion of coastal heathland in south-west Norway took place from 4000 to 200 BC, but was mostly completed by the end of the Bronze Age. They suggest that the expansion can best be explained as a result of the interaction between land-use history, topography, and edaphic conditions, within a climatic regime that favored heathland development.

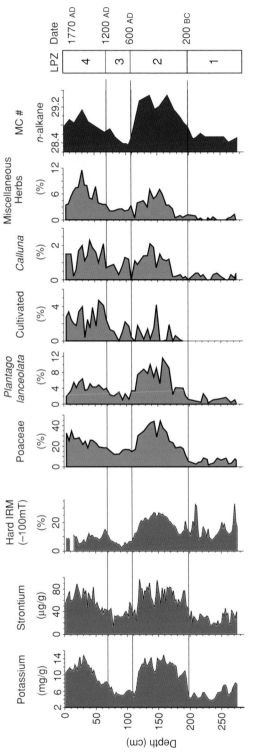

Figure 3.1 The impact of forest clearance during the late Iron Age and Romano-British period. Evidence from Gormire, North Yorkshire, UK. The chemical elements and "Hard-IRM" (a rock magnetic parameter) are indicative of the erosion of inorganic mineral soil. The pollen record links deforestation to evidence for human activities – both pastoralism and cultivation – as well as to the spread of moorland over the past 2000 years. The *n*-alkane record reflects the relative abundance in the sediments of a biomarker specific to open grassland. (Based on Oldfield *et al.* 2003a.)

Figure 3.2 Plot of the extent to which forest composition inferred from pollen diagrams from 10 sites in Denmark corresponds with forest composition at AD 1800. Note how the percentage of between-site variance explained increases steeply along with the significance of the relationship (*** = p < 0.001) from middle Neolithic to late Bronze Age times. The results show that both the differentiation between sites and the macroscale land-cover patterns were established some 3000 years ago. LEB, late Ertebølle; EN, early Neolithic; MNA, MNB, middle Neolithic A and B; LN, late Neolithic; EBA, early Bronze Age; BA, Bronze Age; PRIA, Pre-Roman Iron Age; RA, Roman; LIA, Late Iron Age; MED, Medieval. (From Odgaard B.V. & Rasmussen P. (2000). With permission from Blackwell Publishing Ltd.)

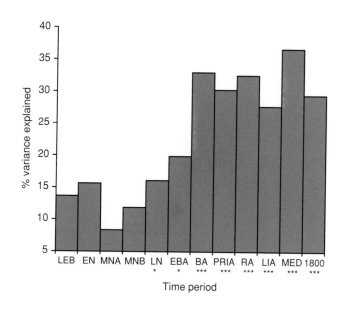

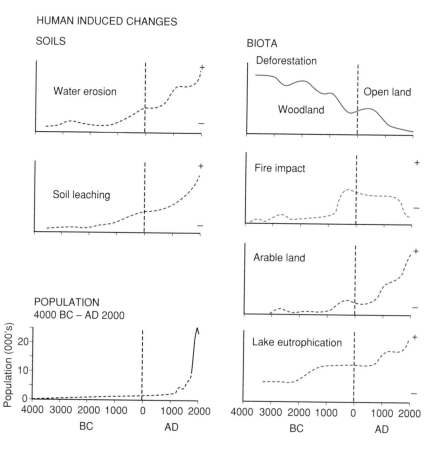

Figure 3.3 Some of the human-induced changes inferred from the multi-disciplinary Ystad Project in southern Sweden. (Based on Berglund 1991; Dearing *et al.* 2006.)

Bunting's (1996) evidence for the development of heathland in the Orkneys also points to a complex interaction between human, edaphic, and climatic factors. In less oceanic areas such as south-west Sweden, extensive heathlands were not formed until historical times (Malmer 1965; Björkman 2001), mainly through intensive grazing subsequent to deliberate deforestation.

It is entirely possible that anthropogenic transformation of upland ecosystems in Britain began before the Neolithic period. Evidence from sub-peat charcoal layers and from organic biomarkers (Chiverrell *et al.* in press) confirms that fire was prevalent in upland regions of northern England currently covered by moorland or blanket peat. Several authors (Jacobi *et al.* 1976; Simmons 1996; Innes and Simmons 2000; Innes and Blackford 2003) have suggested Mesolithic peoples used fire as part of their woodland management strategy in order to increase animal abundance. Drier climatic conditions during the Mesolithic period, however, may also have led to more widespread natural fires (Bradshaw *et al.* 1996; Brown 1997; Tipping and Milburn 2000).

Holocene pollen diagrams from east European sites that are outside the zone of Mediterranean climate record deforestation ranging in age from older than 7000 years BP in north-eastern Bulgaria (Bozilova and Tonkov 1998) to the 13th century AD in Russian Karelia (Vuorela *et al.* 2001). Kremenetski *et al.* (1999) find evidence for major human impacts from Bronze Age times onwards in southern Russia, whereas in Estonia, clear signs of human impact are delayed until around 1500 years BP (Niinemets and Saarse 2006).

Several studies in the Alps point to human activities, notably grazing and the use of fire, as the main factors responsible for shifts in the tree line and changes in forest composition from Neolithic times onwards (e.g. Gobet *et al.* 2003; Carcaillet and Muller 2005; Tinner *et al.* 2005). In the Lake Garda region of northern Italy, Valsecchi *et al.* (2005) estimate that over 60 percent of the pollen source area was deforested during the Bronze Age between 2000 and 1100 BC, whereas in the East Pyrenees, Guiter *et al.* (2005) point to significant anthropogenic reduction in forest cover from ca. 2500 years BP, accelerating during the past 2000 years.

The relative importance of the roles Holocene climate change and human activity played in the development of the present-day plant cover in those parts of Europe experiencing a Mediterranean-type climate remains a matter of debate. There is strong evidence from both marine and terrestrial archives for extensive desiccation beginning between 5000 and 4000 years BP and continuing at least until 3000 to 2000 years BP (e.g. Bar-Matthews and Ayalon 2004; Kallel *et al.* 2004; Roberts *et al.* 2004). Many pollen diagrams document the spread of xerophytic scrub and steppe biomes during this period and some authors stress climate change as the main cause (e.g. Jalut *et al.* 2000; Sadori and Narcisi 2001; Pantaléon-Cano *et al.* 2003). At the same time, many others, while acknowledging the importance of climate change, note widespread evidence for the degradation of Mediterranean ecosystems through human activities, beginning mainly during Bronze Age times towards the end of the third millennium BC, and often peaking during the Medieval period (van Andel *et al.* 1986; Jahns 1993, 2005; Brochier *et al.* 1998; Atherden and Hall 1999; Ramrath *et al.* 2000; Carrion *et al.* 2001; Oldfield

Figure 3.4 Evidence for human impact over the past 4000 years from an Adriatic core (RF 93-30) integrating evidence from a large area along the eastern flank of Italy. The pollen record shows deforestation from 3500 years BP onwards, with periods of peak human activity in the Bronze Age (ca. 3500–2000 years BP) and Medieval (post-800 years BP) periods. The magnetic properties (IRM 300 and χ_{fd}, both also influenced by tephra layers) mainly record terrigenous input during these periods. The benthic foraminifers show distinctive faunal changes at the onset of both periods of erosion. The alkenone-inferred surface water temperature highlights the likelihood that some of the changes, especially between 3500 and 3000 years BP, may also reflect climatic forcing. A–D refer to the main horizons of change as identified in the original reference. (Based on Oldfield *et al.* 2003b.)

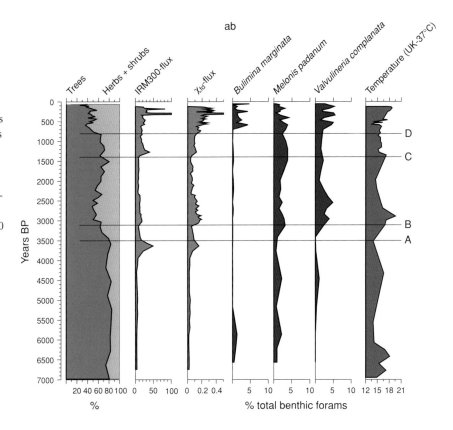

et al. 2003b, Sobrino *et al.* 2004; Butzer 2005; Figure 3.4). As noted in a later section, feedback from the creation of extensive garrigue and bare ground in place of woodland, through overgrazing and soil erosion, would have tended to reinforce any summer drought initially resulting from climate change.

Summarizing the evidence from Europe as a whole, the following generalizations are proposed.

- Paleoecological evidence distinguishes between landscapes characterized by relatively resilient woodland ecosystems within which repeated forest recovery occurred over most of the past 6000 years until the Middle Ages and open landscapes reflecting progressive deforestation from prehistoric times onwards.
- Among the latter are the heathlands of north-western Europe and Mediterranean garrigue. In both cases, human activities and climate change have acted synergistically in edaphically sensitive environments to promote or hasten the transformation of ecosystems. The extent to which persistence of these has depended on their maintenance through sustained human impact and the use of fire has varied from place to place.
- Throughout Europe, from the western edge of Ireland (Molloy and O'Connell 1995; O'Connell *et al.* 2000) to the Urals, all but the most remote or northerly sites show episodes of human impact through deforestation from Neolithic or Bronze Age times onwards (cf. Berglund 2000). Even in the central Swedish forest region as

far as 62° north, there is clear evidence of deforestation for sedentary farming from ca. AD 100 onwards (Emanuelsson 2001).

- Mesolithic cultures may also have had a significant impact on upland ecosystems in Britain at least, through the use of fire.
- At many sites in the Mediterranean region and western Europe there is evidence for extensive and, in some cases, permanent deforestation from an early stage in the past 2000 to 3000 years.
- Many studies confirm the importance of fire in the creation and maintenance of deforested ecosystems.
- In the Mediterranean region, it seems likely that climatically induced summer drought during the second half of the Holocene may have been reinforced by feedback from land-cover transformed by human impact, for example, from woodland to garrigue.

Even in Europe, where palynological evidence for agriculture and pastoralism is relatively unambiguous and there have been many studies of Holocene climate variability (Verschuren and Charman, this volume), the relations between human activities and climate change remain open to much debate (see e.g. Berglund 2003). There are cases where favorable climatic conditions appear to have encouraged the expansion of agriculture, as during Romano-British times in Britain. Equally, one may argue that challenging environmental conditions triggered critical advances in adaptive technology (e.g. Rosen 1995). These issues are explored further below, as is the challenge of developing more quantitative expressions of landscape openness from palynological data.

Looking briefly beyond Europe to other regions where a long history of human occupation and land management raises questions of early, long-term, and sustained human impact on ecosystems, the picture is less clear, partly because much less evidence is available and partly because pollen-analytic data are often less amenable to interpretation in terms of human impact. Hoelzmann *et al.* (2004), summarizing evidence from the arid and sub-arid parts of Africa, including the Sahara, point to climate change as the primary cause of mid-Holocene desiccation. All authorities now accept, however, that feedback from changing land-cover was critical in reinforcing the effects of externally driven changes in climate (Claussen *et al.* 1999; deMenocal *et al.* 2000). What remains in doubt, partly because of the absence of unambiguous pollen indicators of human activity (Waller and Salzmann 1999), is the degree to which human activities were responsible for some of the land-cover changes that helped to trigger the shift to aridification in parts of the region. Similar difficulties lie in the way of establishing the role of past human activities in the spread of savannah (Saltzmann 2000), although Sowunmi (2004) claims that human activities were responsible for a significant expansion of savannah from 3000 years BP onwards in coastal Nigeria. Indeed, Ballouche (2004) states that "most of the West African savannas are cultural landscapes which have been strongly shaped by humans". Evidence from East Africa also points to significant human impacts on land-cover in pre-colonial times (Taylor 1990; Mworia-Maitima 1997), though the evidence for strong hydrologic variability and recurrent drought is incontrovertible (Verschuren and Charman, this volume). In southern Africa (Scott and Lee-Thorp 2004), the main spread of agriculture and grazing associated with increasing population densities and deforestation appears to have taken place during the past 2000 years, peaking around 800 years BP. This is consistent with evidence from marine cores summarized by Shi *et al.* (1998).

Although the evidence from western, south-eastern, and eastern Asia for the early origins of different forms of agriculture is well documented (see summaries in Yasuda 2002), it is much more difficult to determine the timing and extent of human-induced land-cover change for most regions. Some of the strongest evidence is emerging from China, for early deforestation (Ren and Zhang 1998; Ren 2000), the long-term maintenance and expansion of open landscapes, and the likely impact of these processes on the climate (Fu 2003). Huang *et al.* (2002) find evidence for deforestation and cultivation on the southern Loess Plateau from 7500 years BP onwards, and Zhou *et al.* (2002) ascribe the south-eastern displacement of the desert–loess boundary to human activity over the past 3000 years, although evidence presented by Li *et al.* (2003) suggests that the vegetation of the east-central part of the Loess Plateau to the north of Xian was dominated by steppe and grassland throughout the Holocene. In the far north-east of China, Makohonienko *et al.* (2004) document deforestation from around 900 BC onwards and show that the spread of grassland over the Manchurian Plains was the result of human activities over the past 1000 years. Li *et al.* (2006) also show that in Inner Mongolia human impact on the landscape, already significant before 5400 years BP, intensified from 4700 years BP onwards. In the region of the Yellow River delta, successive episodes of deforestation and cultivation took place from 4000 years BP onwards (Yi *et al.* 2003). Ji *et al.* (2006) show that deforestation by human populations began as early as 7000 years ago around Erhai Lake in Yunnan Province, south-west China.

In the case of the Indian sub-continent, it is relatively easy to reconstruct in outline the main sequences and patterns of cultural changes over the past 6000 years, from early Neolithic times onwards (see e.g. Misra 2001). Many studies confirm that the establishment of settled agriculture was widespread from the fourth millennium BC onwards, not only in the Indus Valley where the Harappan Civilization flourished but also in southern India (Fuller *et al.* 2004). What is much more difficult to glean is any clear indication of the impact pre-colonial societies had on land-cover. Maloney (1980, 1981) finds evidence for early prehistoric deforestation in Indonesia.

Evidence from Oceania points to rapid deforestation on many islands spanning the past 3000 to 800 years, depending on location, with devastating human consequences in some cases (Bahn and Flenley 1992; Diamond 2005). Even in the New World, clear indications of deforestation and changed land-cover go back as far as 4000 years in central Mexico (Watts and Bradbury 1982; Bradbury 2000; Fisher 2005) and 2000 years in the Cuzco area of Peru (Chepstow-Lusty *et al.* 1998). Australian landscape evolution has been dominated by the traditional use of fire in land management for many thousands of years (see e.g. Bowman 1998), although there are a few indications of changing Holocene patterns and intensities of aboriginal fire use, which may well stretch back for over 40 000 years.

The above account, albeit partial and regionally selective, confirms the reality of extensive land-cover transformation through human activities long before AD 1700, the starting point for some recent evaluations of anthropogenic land-cover change (Ramankutty and Foley 1999; Goldewijk 2003). This is a key observation when we come to consider the likely impact of land-use and land-cover change on climate directly and on the carbon cycle. A further important conclusion to draw from this account is the extent to which many habitats of great value for amenity and biodiversity are themselves the result of human intervention, rather than a simple and direct response to "natural" climatic and edaphic conditions. This can be illustrated by studies both at regional scale (Segerström *et al.* 1996; Berglund *et al.* in press) and for individual sites (Oldfield 1969). Such studies, together with others (e.g. He *et al.* 2002; Latty *et al.* 2004) showing that legacies of disturbance events may have impacts on ecosystem processes on century time-scales and beyond, suggest that paleoresearch can contribute to developing plans for future conservation. Paleoresearch, embracing long-term perspectives, shows that it is impossible to generalize about the rates, trajectories or degree of recovery from disturbance using only a small number of isolated case studies. The degree of asymmetry or hysteresis between the sudden impact of fire or land clearance, for example, and the rates of recovery varies with a wide range of factors (see Figure 3.5). There is a need for more comparative studies taking the long-term view and oriented towards a stronger theoretical basis for understanding the ecosystem processes involved (Dearing *et al.* 2006). Such studies are needed in order to improve our understanding of vulnerability, provide realistic guidance for conservation and restoration, and add empirically derived, temporal dimensions to the evidence against which the performance of ecosystems models can be tested.

All the emphasis above has been on terrestrial ecosystems, but from the 1960s onwards it has been increasingly apparent that human activities have had a major impact on water quality, in both inland (e.g. Digerfeldt 1972; Battarbee 1978,

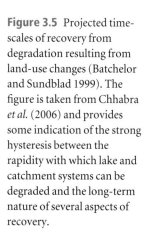

Figure 3.5 Projected time-scales of recovery from degradation resulting from land-use changes (Batchelor and Sundblad 1999). The figure is taken from Chhabra *et al.* (2006) and provides some indication of the strong hysteresis between the rapidity with which lake and catchment systems can be degraded and the long-term nature of several aspects of recovery.

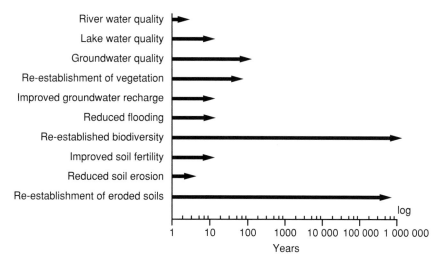

1990) and near-shore marine (Andren *et al.* 1999, 2004) environments. Although the main impacts have taken place over the past 200 years, there is growing evidence for discernable impacts in Europe at least from Bronze Age times onwards (e.g. Gaillard *et al.* 1991; Renberg *et al.* 1993; Anderson *et al.* 1995; Oldfield *et al.* 2003b; Bradshaw *et al.* 2005b).

The impacts of past land-cover changes on climate

Changes in land-cover affect climate through, for example, changes in albedo, in moisture retention and water vapor flux, and in gas and fine particulate flux into the atmosphere, with implications for temperature, cloud formation, and precipitation regimes. All these and other interlinked processes are being observed, measured, and modeled at the present day (summarized in Chhabra *et al.* 2006; Lawrence and Chase 2006). Observational (Lim *et al.* 2005) and experimental studies (Sturm *et al.* 2005) of present-day vegetation changes confirm the sensitivity of surface climate to changes in land-cover. Model sensitivity studies suggest that biomass burning can increase cloud condensation nuclei, which can seriously reduce the efficiency of rain production in convective clouds. This in turn changes the energy budget, convective heating, and vertical temperature gradients, with likely effects on atmospheric circulation (Nober *et al.* 2003). Pielke (2005) claims that at the scale at which thunderstorms are generated, for example, the effects of spatially heterogeneous land-cover may be at least as important in altering weather as changes in climate patterns associated with greenhouse gases. A growing number of modeling studies designed to estimate the likely impact of hypothesized land-cover change on future climate (Bounoua *et al.* 2002; Snyder *et al.* 2004; Fedemma *et al.* 2005; Crucifix, this volume; Claussen, this volume) highlight their importance. This carries with it the implication that these processes should be taken into account in evaluating the climatic significance of past land-cover changes.

Several studies point to the importance of past changes in land-use and land-cover as modulators of climate at least at the regional scale. Moreover, land-cover change is being increasingly incorporated in climate/Earth system models, especially those of intermediate (reduced) complexity (Claussen, this volume). Recent studies by Werth and Avissar (2005) and Schneider and Eugster (2005) attempt to assess the likely effects of past land-cover change on climate in south-east Asia and the Swiss Plateau, respectively, and identify significant local effects on temperature and precipitation regimes. Fu (2003) estimates that over the past 3000 years over 60 percent of East Asia has been affected by various forms of land-cover conversion, e.g. forest to farmland, grassland to semi-desert, and various types of land degradation. By comparing the effects of inferred "natural" and the existing, transformed land-cover in climate simulations, he suggests that major changes in the hydrologic balance have resulted, with important implications for precipitation, runoff, and soil water content. He cites these changes as likely contributors to a decline in atmospheric and soil humidity, and the consequent trend to aridification over the same period, especially in northern China. On a global scale, Myhre *et al.* (2005) use comparison of MODIS-derived land-surface cover and inferred "natural" vegetation to estimate radiative forcing due to surface albedo changes resulting from anthropogenic vegetation changes. They infer weaker cooling on a global scale than do earlier modeling studies (Bauer *et al.* 2003; Matthews *et al.* 2004), but identify regionally important effects. They claim that 25–33 percent of the temperature change forced by land-cover change pre-dated industrialization (see also Goosse *et al.* this volume). As hinted above there are conflicting views on the extent to which aridification trends in the Sahel region of Africa have been reinforced by anthropogenic changes in vegetation. Taylor *et al.* (2002) suggest that these changes are not large enough to have been the main cause of Sahel drought in recent decades, although they may have a larger impact in the future. Govaerts and Lattanzio (2007), however, use satellite-derived surface albedo data from the past two decades to show that the feedback from reduced precipitation and effects on land-cover sustains drought conditions arising initially from changes in sea-surface temperatures in the tropical Atlantic. Another view of Sahel desertification, on a longer time-scale, is presented by Foley *et al.* (2003). They envisage the possibility that nonlinear shifts from wet to dry states might have been triggered by stochastic events (e.g. a period of drought) superimposed on trends that included land degradation. One of the very few attempts to apply a modeling approach to the issue of land-cover–climate feedbacks in Europe during prehistoric times is that of Reale and Dirmeyer (2000) in their study of the effects of vegetation on Mediterranean climate during the Roman classical period.

From the research published so far, it is clear that further work on feedbacks from changing land-cover to the climate system (see e.g. Koster *et al.* 2006) have important implications for interpreting past ecosystem changes and deepening our understanding of past human–environment interactions. Given sufficiently robust reconstructions of past land-cover change (see below) there should also be scope for using them to test the credibility of models used to assess the extent to which land-cover feedback will affect future climate change, especially at the regional scale.

Impacts on hydrologic and erosion regimes

Holocene reconstructions of hydrologic and erosion regimes stretching beyond the time-span of documentary records mainly depend on lake sediments or datable, alluvial sequences, supplemented in moist, temperate environments by peat stratigraphic evidence for hydrologic variability (Verschuren and Charman, this volume). In the absence of significant human impacts on catchments, Holocene variability in hydrologic regimes and erosion processes can usually be ascribed to changes in climate (e.g. Knox 2000) or tectonic activity, though the relationships between forcing mechanisms and the resulting responses are seldom simple and direct. Once human activities are superimposed on these relationships, forcing-response linkages often become even more difficult to discern and characterize as more processes interact to introduce additional thresholds and delays in response (e.g. Trimble and Lund 1982; Wasson and Sidorchuk 2000).

Most reconstructions spanning the whole period of human impact on land-cover are based on calculations of sediment yield from lake sediment studies. In interpreting sediment yield calculations from lake sediment studies, the effects of catchment size on sensitivity to perturbations, as well as on yield as a percentage of total erosion, must be borne in mind (Dearing and Jones 2003). One implication of the summary in Dearing and Jones is that the most sensitive systems are likely to be small catchments with minimal sediment storage within the drainage net. On the other hand, these may be least representative of processes occurring at a regional scale.

Deforestation and the creation of more open, especially tilled landscapes, has the effect of reducing rainfall interception and evapotranspiration, exposing more of the land surface to rain-splash impact, and increasing hydrologic efficiency within river catchments. These effects tend to increase sediment delivery as well as to increase the volume of material derived from surface soil and slope wash, rather than from within-channel sources, although within-channel instability may also be triggered by land-cover-driven changes in hydrology. These effects are additional to those arising from changed feedback to atmospheric processes considered above. Figure 3.6 shows some European examples of mainly lake-sediment-budget studies illustrating links between sediment yield and human occupation of the catchment. In all cases, there are clear links between the history of human occupation and sediment yield, but they only emerge when the time frame exceeds the shorter time-scales on which individual flood events have been documented. Foster *et al.* (2003), in stating that "as the time scale of observation becomes shorter, changes in climate and hydro-meteorological conditions become progressively more important" echo earlier observations in a range of environmental contexts (cf. Messerli *et al.* 2000). Equally, as the time perspective lengthens, the role of human activities in transforming the landscape upon which climate variability acts often becomes more important (Lang *et al.* 2000; Sidorchuk 2000).

Moving beyond Europe to other long-settled parts of the world, a growing number of Chinese studies are shedding a similar light on the impact of human activities on erosion and sedimentation. Xu (1999) in his study of the Yellow River detects a sharp increase in mean sediment accumulation rate as a result of increased human impacts over the past two to three millennia, as do Saito *et al.* (2001) in their calculations based on deltaic sedimentation at the mouths of both the Yangtze and Yellow rivers. These inferences are strongly reinforced by He *et al.* (2006) who detect a sharp increase in erosion rates over the same period on the Loess Plateau of China, culminating in rates some four times those prevailing before significant anthropogenic impact. The impact of these changes on sediment delivery during the extreme flood event of 1998 is evaluated in a simulation study by Xu *et al.* (2005). Alongside these studies based on major regions and river systems, detailed multi-disciplinary research on individual lake catchments is

Figure 3.6 Long-term changes in sediment yields from lake sediments and alluvial sequences. (a) Frequency of OSL-dated soil-erosion-derived sediments from the loess hills of southern Germany (Lang 2003). (b) Sediment accumulation rates at Holzmaar, western Germany (Zolitschka 1998). (c) Frequency of dated alluvial units in British rivers (Macklin 1999). (d) Changes in the deposition of inorganic mineral matter at Lago di Mezzano, central Italy (Ramrath *et al.* 2000). (e) Depositional flux of magnetic minerals derived from soil erosion into the mid-Adriatic core RF 93-30 (see Figure 3.4). (f) Alluvial accumulation rates in the Yellow River, China (Xu 1999). Note the close parallels (allowing for some imprecision in chronologies) between (a) and (b) and between (d) and (e). (Reproduced from Oldfield F. (2005) *Environmental Change: Key Issues and Alternative Approaches.* Cambridge University Press, Cambridge.)

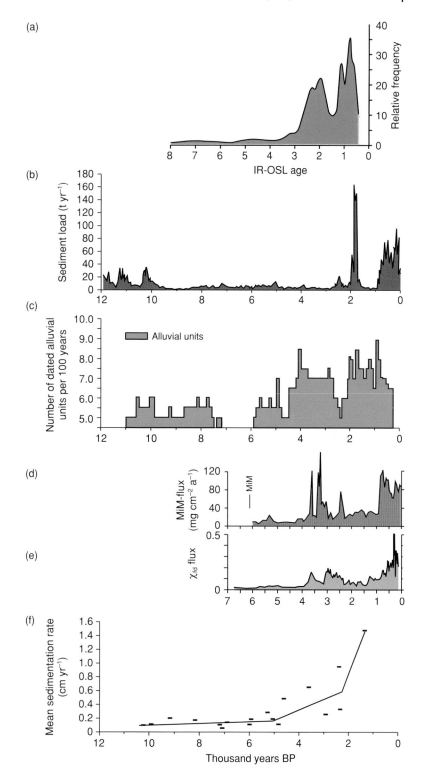

beginning to shed important light on the links between China's long history of human occupation and the effects on erosion and sediment accumulation rates. In a wide-ranging study of Erhai Lake in south-west China (see summary in Dearing, 2006) detectable human impact on the landscape is shown to stretch back at least 7000 years, with increased topsoil erosion from 2200 years BP in response to the expansion of agriculture by Han peoples. Peak erosion occurred during the 17th to 19th centuries AD. One inference from the study relates to the likely role of irrigation systems from 4300 years BP as moderators of the hydrologic regime within the catchment: evidence for minimum flood peaks declines on millennial time-scales from that date to the present day, despite evidence for a strengthening monsoon system over the past 1500 years.

Evidence for the dramatic impact on erosion by the first European settlers in North America is widespread. The classic study of Davis (1976) at Frains Lake, Michigan is paralleled by others from as far afield as coastal California, where Plater *et al.* (2006) detect an order of magnitude increase in sediment accumulation rates on the arrival of the first European immigrants. Where it has been possible to discriminate between sediment derived from topsoil and that derived from deeper in the regolith, accelerated loss of topsoil is seen to coincide with the first evidence for deforestation and tillage by European settlers (cf. Oldfield *et al.* 1985; Yu and Oldfield 1989).

In so far as the results of the many and varied studies of Holocene erosion and sedimentation encourage generalizations, they support the view that the landscape, as transformed by forest clearance and tillage for example, creates a new canvas upon which climate variability and especially hydrologic extremes act. As the land-cover is transformed, the impact of continued variability will change even without a shift in the magnitude/frequency of extreme events. The interplay through time between climate variability and changing land-cover introduces the type of nonstationarity in hydrologic response that negates the assumptions upon which magnitude/frequency projections are based. This is a significant concern for the future (Clarke 2003). Only the empirical evidence from a long-term view of the past can provide well-founded guidance. The need for long-time perspectives in order to understand the nature of human impacts on ecosystems and the hydrologic implications of these impacts is strongly reinforced by Batchelor and Sundblad's (1999) schematic representation of impacts and recovery times (Figure 3.5).

Holocene vegetation changes and atmospheric methane and carbon dioxide concentrations

To what extent have changes in land-use and land-cover modified the atmospheric concentrations of both methane and carbon dioxide during the pre-industrial parts of the Holocene? Forest clearance, followed by cultivation or pastoralism, reduces carbon sequestration in the terrestrial biosphere by generating an initial release to the atmosphere usually through burning, by diminishing the store of carbon in standing crops and soils, and by increasing the rate of carbon turnover. Moreover, Lal (2002) shows that much of the carbon transported from sites as a result of land-use-driven soil erosion is also emitted to the atmosphere. The initial changes occur rapidly and recovery from them in the event of reforestation is a much slower process. Rice paddy cultivation, especially under conditions where drainage is poorly controlled, leads to methane release.

Several types of conversion are important in changing the carbon budgets of terrestrial ecosystems, for example, from most types of "natural" plant cover to permanent cropland, slash and burn agriculture or pasture; or from undisturbed forest to commercial plantations or managed forest (see Lawrence and Chase 2006). Wu *et al.* (2003) have attempted to quantify the loss of carbon from soils as a result of anthropogenic land-cover changes. Despite indications that paddy and irrigated soils have experienced an increase in soil organic carbon, they estimate a loss of some 7.1 pg of soil organic carbon over the country as a whole as a result of human activities, land-cover conversion, and associated soil degradation in many regions, especially in north-east China. Other activities, such as drainage of swamps and peatlands, can also lead to reduced carbon sequestration in the terrestrial biosphere as well as to increased vulnerability to wildfires.

Ruddiman's proposal (2003) that human activities have had significant effects on atmospheric carbon dioxide and methane from 8000 and 5000 years BP, respectively, have attracted considerable interest and controversy (EPICA Community members 2004; Claussen *et al.* 2005; Crucifix *et al.* 2005; Ruddiman 2005a,b; Ruddimann *et al.* 2005; Schmidt and Shindell 2005; Broecker 2006; Birks, this volume). The significance of this debate goes beyond improving reconstructions of past changes in the carbon cycle. Ruddiman's hypothesis has wider implications in relation to past climate change and future climate sensitivity to greenhouse gas forcing.

Part of Ruddiman's claim was based on the fact that increases in atmospheric CO_2 and methane concentrations from 8000 and 5000 years BP onwards, respectively, appeared to be unique to the Holocene (Figure 3.7). At the time of proposing his hypothesis, data for atmospheric greenhouse concentrations were only available from Marine Isotope Stage (MIS) 9 and subsequent interglacials . The subsequent availability of data for MIS 11, the interglacial associated with orbital forcing most comparable to that prevailing during the Holocene, has provided a more appropriate basis for comparing the behavior of each greenhouse gas with and without the possibility of human intervention. For some (EPICA Community Members 2004; Broecker 2006), the comparison disposes of Ruddiman's claim that the Holocene CO_2 increase is unique. For Ruddiman (2005a) their argument is not conclusive, for he adopts a different part of the MIS 11 CO_2 curve for comparison and advances credible reasons for his choice (Figure 3.8). At the same time, and in parallel with Ruddiman, Carcaillet *et al.* (2002) provide a wealth of evidence in support of the view that fire alone may have made a significant contribution to the Holocene increase in atmospheric CO_2 (Figure 3.9).

Another dimension to the controversy over Ruddiman's hypothesis arises from the mismatch between inferences based on empirical evidence for past land-use/cover change and model-based reconstructions incorporating stable

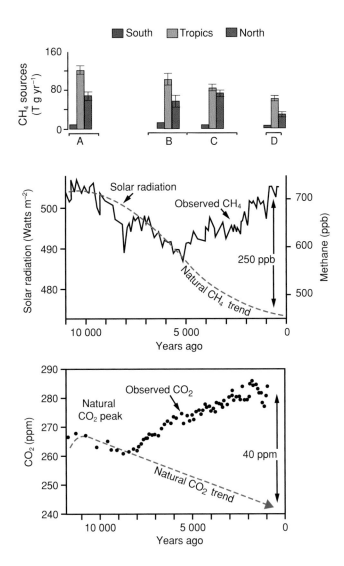

Figure 3.7 Actual and inferred "natural" trends of atmospheric concentrations of CO_2 and CH_4 during the Holocene (Ruddiman 2003). The histograms ascribing methane to different source regions during the Holocene (Flückiger *et al.* 2004) are shown at the top of the graph.

isotope evidence (Joos *et al.* 2004). Figure 3.10, which shows an estimate of the change in total biospheric carbon during the Holocene, indicates the extent to which any attempt to calculate past carbon budgets depends on the approach used. Broecker (2006) casts doubt on Ruddiman's hypothesis by claiming that for it to work "the forest biomass of 8000 years ago must have been more than double that in the year 1800 A.D.". This requirement he finds quite unlikely to have been met, but it may seem less incredible to many Old World paleoecologists. More difficult to counter are the arguments based on the latest measurements of the ^{13}C to ^{12}C ratio in Antarctic ice, quoted by Broecker (2006). These are interpreted as favoring the view that the main changes in atmospheric CO_2 were the result of changes in the world's oceans. In response to the stable isotope data, Ruddiman has revised his latest estimate of anthropogenic carbon release downwards to 14 ppm CO_2

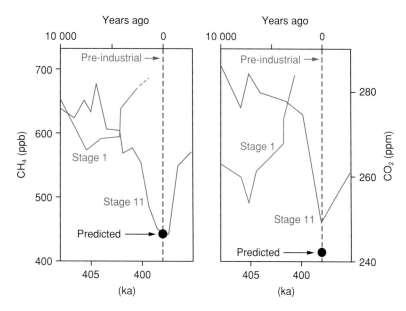

Figure 3.8 Ruddiman's interpretation of the comparison between greenhouse gas concentration trends during the Holocene (red line – from the Vostok core measurements) and those during Marine Isotope Stage 11 (blue line – from Dome C). (From Ruddiman *et al.* (2005a) with kind permission from Springer Science and Business Media.)

rather than the original 40. This would imply a much lower biomass reduction of 17–20 percent.

Ruddiman's ascription of the post-5000 year BP rise in atmospheric methane concentrations (see also Ruddiman and Thompson 2001) has provoked rather less controversy and the ascription of the main part of the increase to tropical sources (Flückiger *et al.* 2001; Figure 3.7) lends credibility to Ruddiman's hypothesis. Schmidt and Shindell (2005), however, consider that "in the absence of further studies ruling out boreal wetlands, tropical river deltas and peatlands as sources of the late Holocene increase in methane emissions, a definitive attribution of the trend to anthropogenic sources is premature".

Up to now, much of the debate over Ruddiman's hypothesis has been in terms of acceptance or rejection as it stands. It seems quite likely that any resolution will require more empirical evidence to constrain the range of possibilities and to address the issue in less extreme terms. The extent to which human activities such as deforestation and paddy cultivation contributed to the observed increases remains an open question. Even if the greater percentage of the Holocene increases proves to be unrelated to anthropogenic activities, there still remains the intriguing possibility that on shorter, more recent time-scales the fortunes of human populations may have led to sufficiently large oscillations in forest biomass to have produced significant fluctuations in atmospheric CO_2.

Human societies, climate change, and cultural collapse

How have human societies reacted to climate change in the past and what lessons may we learn for the future from the interactions between climate variability

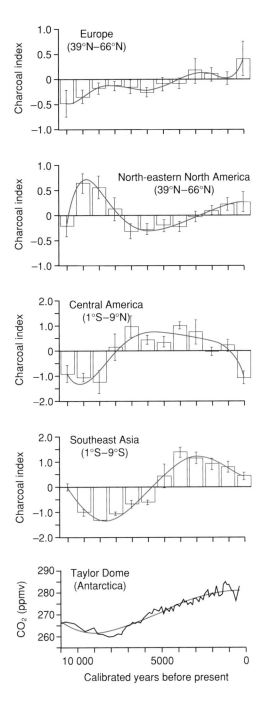

Figure 3.9 Biomass burning reconstructions for different parts of the world expressed as a charcoal index, derived from stratigraphic data, compared with the atmospheric CO_2 concentration curve from the Taylor Dome, Antarctica (Carcaillet *et al.* 2002).

(notably the incidence of damaging extremes such as severe drought) and the changing patterns of social organization developed as a response to the resulting environmental stresses? This theme too is controversial. At one extreme, those who favor a degree of environmental determinism can point to many instances

Figure 3.10 Estimates of the increase in biospheric carbon storage between the Last Glacial Maximum and the present day. Note the contrast between the range of sedimentologically derived estimates and those based on modeling and isotope studies. PMIP, Paleo-model Intercomparison Project. (Reproduced from Oldfield F. (2005) *Environmental Change: Key Issues and Alternative Approaches.* Cambridge University Press, Cambridge.)

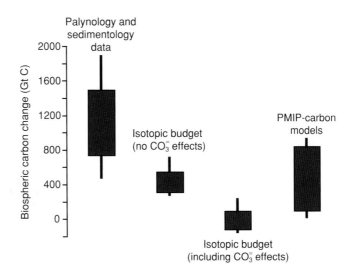

where the transformation, decline or collapse of societies, even whole civilizations, coincides with major climatic changes or periods of intense and persistent drought (e.g. Hodell *et al.* 1995; Weiss 1997; Cullen *et al.* 2000; deMenocal 2001; Weiss and Bradley 2001; Haug *et al.* 2003; Huang *et al.* 2003; Bradley, this volume; Beer and van Geel, this volume). By contrast, most authorities with a background in social sciences stress instead the degree to which poorly adapted patterns of social organization or inappropriate response strategies appear to be more direct and significant factors in social decline or collapse (e.g. Redman 1999). To a large degree, these differences reflect different academic backgrounds and training: there is a tendency to favor interpretations that lie within the scope of the disciplines with which we are familiar, although the historical rejection of determinism in the wake of its eventual discrediting during the middle decades of the 20th century may have reinforced the second type of attitude. If we take a wider and more balanced view and detach the issue from the emotive concept of "collapse", it is clearly unrealistic to suppose that climate variability has had no effect on human welfare and social viability – there are too many examples from the more recent past, for which well-documented historical evidence is available. Equally, explanations based entirely on climate determinism have been shown to be inadequate wherever interdisciplinary studies have been carried out that span biophysical and cultural/socio-economic research. The key lies in trying to unravel and learn from the interactions between the biophysical and the social systems. This poses formidable challenges.

One of the most recent attempts to present an overview is that of Diamond (2005). He examines diverse case studies of societal collapse in the light of a wide range of possible causative factors, of which climatically induced stress is only one. The others include eight broad categories of unsustainable damage to the environment, the impact of hostile neighbors, interactions with trade partners, and the nature of the societal response to environmental problems. In almost all the cases

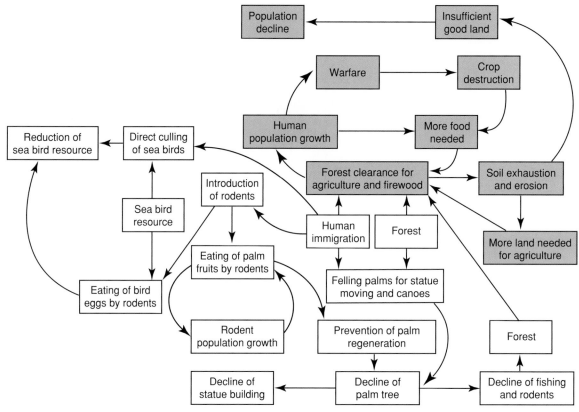

Figure 3.11 Bahn and Flenley's (1992) conceptual model of the interactions between human activities and their environment that could have led to the dramatic population decline on Easter Island. The shaded boxes identify processes and interactions that are less specific to Easter Island. (Based on Oldfield 2005.)

he considers, environmental damage and societal reactions form a constant thread. The role of climate varies from case to case, with sustained drought a key factor in many, including those considered by the more "determinist" authors referred to in the previous paragraph. In each case, however, the notion of a single, simple response leading to total demise never provides a fully satisfactory explanation. Moreover, in the most extreme case, that of Easter Island (Bahn and Flenley 1992, 2003; Figure 3.11), it seems unlikely that climate played any significant role. Although much of Bahn and Flenley's scheme is particular to Easter Island, the processes and interactions portrayed in, and linking, the shaded boxes are not without parallels in the present day, for example, in southern Sudan where climate variability imposes additional stresses.

Although it is tempting to explain the demise of Norse settlements in Greenland by climatic deterioration during the early stages of the Little Ice Age, it is clear that many other factors played an important role, including land degradation, adoption of a life-style too dependent on European contact and supplies (both of which dwindled as sea-ice in the North Atlantic made navigation more difficult), the replacement of walrus ivory by Africa elephant ivory, and failure to broaden and adapt diets in parallel with the Inuit who survived through the period of Norse abandonment (Barlow *et al.* 1997; Pringle 1997; Diamond 2005; Orlove 2005). Similar evidence for a complex of interactive processes, both biophysical and cultural, lies at the heart of many other analyses of cultural transformation or demise. Rosen's (1995) analysis of the collapse of early Bronze Age societies in southern Levant stresses an inability to adapt appropriate technological responses to a diminished water supply. Hassan's (1986, 1997) recent studies show that the failure of the Nile floods around 2150 BC appears to have been the main reason for the demise of the Old Kingdom, but out of the crisis that drought created came a period of radical social change of long-term importance for emerging concepts of social justice (http://www.bbc.co.uk/history/).

Severe drought still seems to be the dominant reason for the collapse of prehistoric communities in the Saharan region (Hassan 1986, 1997; Cremaschi and Di Lernia 1999), although there is no simple link between climate and cultural dynamics and one of the most significant responses to desertification appears to have been the movement of people into the Nile Valley, leading to the appearance of Neolithic sites in the Delta and central Sudan. Response via migration has also characterized some sub-Saharan societies. Tyson *et al.* (2002) point to southward migration by Iron Age agriculturalists following changed rainfall patterns in southern Africa, notably the southern migration of the Sotho-Tswana speaking people during the first few centuries of the past millennium and earlier. These migrations appear to be linked to the anti-phase incidence of rainfall north and south of the Equator documented by records from Lake Naivasha in Kenya and Cold Air cave in the Makapansgat Valley, South Africa. Verschuren *et al.* (2000) link their reconstruction of hydrologic change in Lake Naivasha to Webster's (1979) reconstruction of periods of prosperity, famine, and migration, but Robertshaw *et al.*'s (2004) evaluation of the records from the region once more stress multi-causation and the importance of the administrative structures in place during different periods of famine and disease.

One example of inferred collapse is that of the Mapungubwe agro-pastoralist society in the Limpopo Valley, which lasted for some 300 years before its demise around AD 1280–90. Huffman (1996) ascribes this mainly to drought and O'Connor and Kiker (2004), using a modeling approach, suggest that both crop failure and destabilization of pastoralism may have been involved. Scott and Lee-Thorp (2004), however, point out that the evidence for societal collapse is not unambiguous and the suggestion that drought may have been responsible is not borne out by the latest paleoclimatic evidence.

Two well-documented examples of societal collapse come from the New World, that of the Anasazi in the south-west of the USA and the Maya in Yucatan during the 12th and 10th centuries AD, respectively. Drought has been invoked in both cases (see e.g. Hoddell *et al.* 1995), but a fuller review of the evidence (Diamond 2005) reveals, in both cases, discrepancies in a simple climate-collapse hypothesis. In the case of the Maya, using numerical values expressing estimates of different types and degrees of vulnerability, both climatically and anthropogenically related,

Me-Bar and Valdez (2005) infer an 80 percent increase in vulnerability during the 9th century AD (the period of cultural demise) compared with the level of vulnerability during the late pre-Classical period some 600 years earlier.

For the Anasazi culture, severe drought may have been the "last straw" but earlier there were severe droughts that the Anasazi survived. There is also evidence of evacuation of sites before drought set in. The demise of the Anasazi is now believed to have also involved other factors such as overexploitation of local resources, overextended trade routes, internecine strife, religious/ideological conflict, and the possible pull of Kachina-based religion to the south in the areas now occupied by the related Hopi and Zuni peoples (see e.g. Kohler 1988, 1992; van West 1991). There have been several attempts to deepen understanding of the processes involved through model-based simulations that are of wider interest for their methodological implications. In Dean *et al.*'s (1999) study, households act as the agent units. The authors conclude that this approach allows some testing of hypotheses (or at least their reinforcement or negation in light of the degree of compatibility with the available empirical evidence), sheds light on the importance of interaction between social and environmental factors, identifies examples of equifinality, permits experimental manipulation of behavioral modes and their effects on responses to environmental variability, and makes it possible to explore previously ignored, unspecified or discounted factors. They claim only partial success in relation to their main goal, to generate target outcomes consistent with the archaeological evidence and identify rules of agent behavior that generate these. Axtell *et al.* (2002) claim to have come closer to achieving this goal.

There is drama in the collapse of past civilizations, for the remains – the cliff dwellings of Mesa Verde, or the great temples of the Mayan or Khmer cultures – are evocative and their desertion is inexplicable without much research. In so far as generalizations are possible, the key to their understanding lies in unraveling the complex relationships between environmental change and human cultures. Some poorly adaptive behaviors recur in many of the case studies now available – the evolution of exploitative hierarchies out of touch with the cultural and environmental consequences of their roles in society and demands of the rest of society, overexploitation of a limited resource base in response to population growth, environmental damage resulting from this, short-term responses to shortage without regard to long-term consequences, and failure to incorporate memory of previous periods of environmental stress into the repertoire of survival strategies employed. Couple a combination of these behavior types to the stresses imposed by severe drought especially, and collapse has often been the result. Climatic factors have not been implicated in every case, but there are relatively few in which they can be ruled out.

Fascinating though the consideration of past collapse is, there is surely an urgent need to balance these studies with ones devoting at least as much attention to documenting successful survival strategies during periods of climatically imposed stress, as well as to learning more from situations where irreversible damage to the resource base of a society has been avoided and long-term productivity maintained (see e.g. Crumley 2000).

A role for Holocene paleoresearch

We turn now to an examination of the role paleoresearch may play in addressing the issues raised in the foregoing sections. The first four themes above hinge, in large part, on the reconstruction of changes in the structure, function, and extent of past ecosystems, of the disturbance regimes to which they have been subjected, and, in the case of the third theme, of the hydrologic and erosional responses to such changes. It is my belief that, amongst other things, addressing these issues calls for a shift in emphasis from using paleobiological techniques, such as pollen or macro-fossil analysis, primarily for climate reconstruction towards using them primarily as the basis for ecological reconstruction. This would allow other evidence for past climate change (chironomids, tree rings, and stable isotopes, for example) to generate trajectories of variability independent of the evidence for ecosystem changes. This is a prerequisite of any attempt to disentangle evidence for climate change from that portraying ecosystem responses (see also Birks, this volume). It is also essential if we are to deepen our understanding of past human impacts on ecosystems. This requires further efforts to characterize and detect the imprint of human activities in the results of paleobiological research, and to link such evidence to the archaeological record.

These are also the important components of any contribution paleoscientists may make to the unraveling of problems posed by the fifth theme, in view of the frequency with which ecological degradation has been invoked as a contributory factor in societal collapse. In addition though, there is a need for a much higher degree of interdisciplinarity, spanning the biophysical and social sciences.

Achieving better empirically based quantification of feedbacks from land-cover change both to climate directly and via the impact on atmospheric greenhouse gas concentrations depends, in the first instance, on developing a more robust framework for inferring past land-cover and, in so far as possible, ecosystem structure and function from palynological data. One step in the right direction is the attempt to improve the basis for estimating landscape openness from pollen data (Gaillard et al. 1998; Sugita et al. 1999; Broström et al. 2004). Nielsen and Odgaard (2005) summarize a parallel approach comparing the degree of success achieved by different approaches to quantitative, pollen-based vegetation reconstruction by comparing the results for AD 1800 with cartographic evidence from the period. They claim that their Extended R-value (ERV) model captures the main trends in vegetation and land-cover, and also allows extrapolation, incorporates background pollen explicitly, and may be more applicable to large lakes than an alternative partial least-squares regression approach. There is an urgent need to extend the application of pollen-based land-cover reconstructions on a much wider spatial scale than achieved hitherto and this is currently part of the agenda of the POLLANDCAL initiative using the Europe Pollen Data Base (Gaillard, personal commmunication). A further step must be the application of similar approaches to other regions, especially in areas of long-standing human impact in Eurasia. Given the pivotal role of fire in the regeneration and disturbance histories of many ecosystems, there is also a need to integrate more reconstructions of land-cover changes with reconstructions of fire incidence. There is also a continued need for improved chronologies to establish the phasing and precise links between changes in biophysical and social systems. More studies need to go beyond the demonstration of compelling coincidences between extreme climatic events and cultural changes.

One of the crucial issues concerns the role of modeling in these areas of paleo-research. In some cases, the possible roles for modeling are relatively easy to define. Given sufficiently robust reconstructions of changes in past ecosystem structure and function, the estimation of consequences for climate and for atmospheric greenhouse gas concentrations using models derived from present-day observations of the relevant processes becomes a logical next step. Indeed, there are already examples of this type of linkage, but the empirical basis for inferring changes through time in past vegetation and land-cover is still relatively weak. The potential value of studies at the interface between pollen-based reconstructions of land-cover and models of vegetation–atmosphere interactions works in two directions. Good quality empirical reconstructions of both vegetation and regional climate could be used to test the likely reliability of models designed to explore the future impact of land-cover change on projected future climate. Equally, well-constrained models of the role of past land-cover change in driving regional climate change could be of great value in establishing the role of anthropogenic land-cover transformation in past desertification.

Given the susceptibility of human societies to changes governed by indeterminacy, what, realistically, is the role of modeling in the area of human–environment interactions described above? Several authors use agent-based modeling techniques, which are, to some degree, common to the methodologies of modelers studying past societies and those using integrated impact assessment to develop scenarios of future human–environment relationships (e.g Holman *et al.* 2005; Matthews 2005). Holman *et al.*'s approach (Figure 3.12) rests in part on the

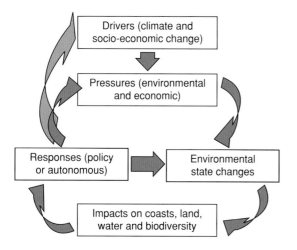

Figure 3.12 The "Drivers–Pressure–State–Impact–Response" (DPSIR) conceptual framework used by Holman *et al.* (2005) in developing future regional integrated impact assessments for the UK, but equally applicable as a framework within which to study past interactions between societies and changing environmental pressures. (From Holman *et al.* (2005) with kind permission from Springer Science and Business Media.)

identification of "sensitivity of impacts and interactions using historical analogs, e.g. previous droughts, changes in crop subsidies, increases in housing stock". Matthews' article is a future-oriented study using multi-agent modeling applied to simulated subsistence systems. One important criterion for successful models that seek to link biogeophysical and cultural processes is the capacity to incorporate adaptive learning in the model as simulations evolve through time. One possible approach uses cellular automaton-type models (Dearing *et al.* 2006). Wirtz and Lemmen (2003) apply, with impressive success, a rule-based multi-agent model with adaptive capacity to simulate the spread of Neolithic agriculture. If we take the view that human–environment systems are special cases of self-organizing and emergent systems, there may be potentially applicable models developed within ecology (e.g. Breckling *et al.* 2005).

Conclusions

In many parts of the world, especially those long settled by societies with agricultural or urban components, an evaluation of present-day ecosystems and future sustainability should recognize the importance of past human activities and the roles they have played in creating and modifying the environmental systems upon which sustainability depends.

Many of the environmental impacts of past human activities are mediated through changes in land-use/cover. These impacts include feedbacks into the climate system, both directly and via changes in atmospheric greenhouse gas concentrations. Land-cover changes arising from human exploitation are also often implicated in changes in hydrologic and erosional regimes: they alter the canvas upon which the interplay between climatic and hydrologic variability is expressed. They are also central to the creation, maintenance or destruction of habitats upon which biodiversity depends.

Many past human societies have been strongly affected by climate change and it is unrealistic to explore the long-term fate of past cultures and civilizations without acknowledging this. That said, acknowledging the role played by climate change should not lead to interpretations of cultural expansion or demise in terms of a simple causative link between climate and the fate of societies. The nature of social organization, its flexibility, and adaptive capacity are critical in determining the success or otherwise of human responses to any external stresses.

From the above, it would seem important to use the record of past human–environment interactions, in all its complexity, as part of the basis for planning a sustainable future in the face of projected climate change, with all its daunting implications. For as long as a methodological divide exists between researchers dealing with contemporary and future change and paleoscientists concerned with longer time perspectives, the desired synergy between the two types of study will never be fully realized. The divide springs in part from training and the consequent differences in the methodological repertoire acquired, and in part from the extent to which much of paleoscience is seen as "history" rather than part of a temporal continuum. The climate community has succeeded in bridging the gap between paleoresearch and future projections by viewing reconstructions of past climatic variability as insights into the nature of processes, rather than simply as retrospective snapshots of "history". From this springs recognition of the value of reconstructions of past variability and change on a wide range of time-scales. Let this serve as a model for the wider research community that embraces human as well as biophysical aspects of environmental change.

Acknowledgments

I should like to thank Sandra Mather and Suzanne Yee for all their help in preparing the figures, also John Dearing, Richard Bradshaw, and Gina Hannon for fruitful discussions. Yiyin Li and Liping Zhou provided indispensable help with Chinese evidence for Holocene land-cover change. Björn Berglund and Arlene Rosen provided helpful guidance in their reviews of the chapter.

References

Anderson N.J., Renberg I. & Segerström U. (1995) Diatom production responses to the development of early agriculture in a boreal lake-catchment (Kassjon, northern Sweden). *Journal of Ecology*, **83**, 809–822.

Andren E., Shimmield G. & Brand T. (1999) Environmental changes of the last three centuries indicated by siliceous microfossil records from the Baltic Sea. *The Holocene*, **9**, 25–38.

Andren E., Andren T. & Kunzendorf H. (2000) Holocene history of the Baltic Sea as a background for assessing records of human impact in the sediments of the Gotland basin. *The Holocene*, **10**, 687–702.

Atherden M.A. & Hall J.A. (1999) Human impact on vegetation in the White Mountains of Crete since AD 500. *The Holocene*, **9**, 183–193.

Axtell R.L., Epstein J.M., Dean J.S., *et al.* (2002) Population growth and collapse in a multi-agent model of the Kayenta Anasazi in Long House valley. *Proceedings of the National Academy of Sciences*, **99**, 7275–7279.

Bahn P. & Flenley J.R. (1992) *Easter Island, Earth Island: a Message from our Past for the Future of our Planet*. Thames and Hudson, London.

Bahn P. & Flenley J.R. (2003) *Enigmas of Easter Island*. Oxford University Press, Oxford.

Ballouche A. (2004) Fire and burning in West Africa Holocene savanna palaeoenvironment. Anthropogenic and natural processes in environmental changes. In: Abstracts for *Rapid and Catastrophic Environmental Changes in the Holocene and Human Response*, IGCP 490 and ICSU field conference. International Geological Correlation Programme and International Council of Scientific Unions.

Barlow L.K., Sadler J.P., Ogilvie A.E.J., *et al.* (1997) Interdisciplinary investigations of the end of the Norse Western Settlement in Greenland. *The Holocene*, **7**, 489–499.

Bar-Matthews M. & Ayalon A. (2004) Speleothems as palaeoclimatic indicators, a case study from Soreq Cave located in the eastern Mediterranean region, Israel. In: *Past Climate Variability through Europe and Africa* (Eds R.W. Battarbee, F. Gasse & C.A. Stickley), pp. 363–392. Springer-Verlag, Berlin.

Batchelor C.H. & Sundblad K. (1999) Reversibility and the time scales of recovery. In: *A Reflection of Land Use. Options for Counteracting Land and Water Mismanagement* (Eds M. Falkenmark, L. Andersson, R. Castensson & K. Sundblad), pp. 79–91. Swedish Natural Science Research Council, Stockholm.

Battarbee R.W. (1978) Observations on the recent history of Lough Neagh and its drainage basin. *Philosophical Transactions of the Royal Society B*, **281**, 303–345.

Battarbee R.W. (1990) The causes of lake acidification, with special reference to the role of acid deposition. *Philosophical Transactions of the Royal Society of London B*, **327**, 339–347.

Battarbee R.W., Gasse F. & Stickley C.A. (Eds) (2004) *Past Climate Variability through Europe and Africa*. Springer-Verlag, Berlin.

Bauer E., Claussen M. & Brovkin V. (2003) Assessing climate forcings of the Earth system or the past millennium. *Geophysical Research Letters*, **20** (6), 9.

Beer J. & van Geel B. (this volume) Holocene climate change and the evidence for solar and other forcings. In: *Global Warming and Natural Climate Variability: a Holocene Perspective* (Eds R.W. Battarbee & H.A. Binney), pp. 135–162. Blackwell, Oxford.

Berglund B.E. (Ed.) (1991) *The Cultural Landscape during 6000 Years in Southern Sweden – the Ystad Project*. Munksgaard, Copenhagen.

Berglund B.E. (2000) The Ystad Project: a case study for multidisciplinary research on long-term human impact. *PAGES Newsletter*, **8** (3), 6–7.

Berglund B.E. (2003) Human Impact and Climate Changes – Synchronous Events and a Causal Link? *Quaternary International*, **105**, 7–12.

Berglund B.E., Gaillard M-J., Björkman L. & Persson T. (in press). Long-term changes in floristic diversity in southern Sweden – palynological richness, vegetation dynamics, and land-use. *Vegetation History and Archaeobotany*.

Birks H.J.B. (2003) Quantitative palaeoenvironmental reconstructions from Holocene biological data. In: *Global Change in the Holocene* (Eds A.W. Mackay, R.W. Battarbee, H.J.B. Birks & F. Oldfield), pp. 107–123. Arnold, London.

Birks H.J.B. (this volume) Holocene climate research – progress, paradigms, and problems. In: *Global Warming and Natural Climate Variability: a Holocene Perspective* (Eds R.W. Battarbee & H.A. Binney), pp. 7–57. Blackwell, Oxford.

Björkman L. (2001) The role of human disturbance in late Holocene vegetation changes on Kullaberg, southern Sweden. *Vegetation History and Archaeobotany*, **10**, 201–210.

Bounoua L., DeFries R., Collatz G.J., Sellers P. & Khan H. (2002) Effects of land Conversion on Surface Climate. *Climatic Change*, **52**, 29–64.

Bowman D.M.J.S. (1998) The impact of Aboriginal landscape burning on the Australian Biota. *New Phytologist*, **140**, 385–410.

Bozilova E. & Tonkov S. (1998) Towards the vegetation and settlement history of the southern Dobrudza coastal region, north-eastern Bulgaria: a pollen diagram from Lake Durankulak. *Vegetation History and Archaeobotany*, **7**, 141–148.

Bradbury J.P. (2000) Limnological history of Lago de Patzcuaro, Michoacan, Mexico for the past 48 000 years: impacts of climate and man. *Palaeogeography, Palaeoclimatology, Palaeoecology*, **163**, 69–95.

Bradley R.S. (this volume) Holocene perspectives on future climate change. In: *Global Warming and Natural Climate Variability: a Holocene Perspective* (Eds R.W. Battarbee & H.A. Binney), pp. 254–268. Blackwell, Oxford.

Bradshaw R.H.W., Tolonen K. & Tolonen M. (1996) Holocene records of fire from the boreal and temperate zones of Europe. In: *Sediment Records of Biomass Burning and Global Change* (Eds J.S. Clark, H. Cachier, J.G. Goldammer & B. Stocks), pp. 347–365. Springer-Verlag, Berlin.

Bradshaw E.G., Rasmussen P. & Odgaard B.V. (2005a) Mid- to late-Holocene land-use change and lake development at Dallund So, Denmark: synthesis of multiproxy data linking land and lake. *The Holocene*, **15**, 1152–1162.

Bradshaw E.G., Rasmussen P., Nielsen H. & Anderson N.J. (2005b) Mid- to late-Holocene land-use change and lake development at Dallund So, Denmark: trends in lake primary productivity as reflected by algal and macrophyte remains. *The Holocene*, **15**, 1130–1142.

Breckling B., Mueller F., Reuter H., Hoelker F. & Fraenzle O. (2005) Emergent properties in individual-based ecological models – introducing case studies in an ecosystem research context. *Ecological Modelling*, **186**, 376–88.

Brochier J.E., Claustre F. & Heinz C. (1998) Environmental impact of Neolithic and Bronze age farming in the eastern Pyrenees forelands, based on multidisciplinary investigations at La Caune de Belasta (Belasta Cave), near Perpignan, France. *Vegetation History and Archaeobotany*, **7**, 1–9.

Broecker W.S. (2006) The Holocene CO_2 rise: anthropogenic or natural? *Eos (Transactions of the American Geophysical Union)*, **87**, 27.

Broström A., Sugita S. & Gaillard M.-J. (2004) Pollen productivity estimates for the reconstruction of past vegetation cover in the cultural landscape of southern Sweden. *The Holocene*, **14**, 368–381.

Brown A. (1997) Clearances and clearings: deforestation in mesolithic/neolithic Britain. *Oxford Journal of Archaeology*, **16**, 133–146.

Bunting J. (1996) The development of heathland in Orkney, Scotland: pollen records from Loch of Knitchen (Rousay) and Loch of Torness (Hoy). *The Holocene*, **6**, 193–212.

Butzer K.W. (2005) Environmental history in the Mediterranean world: cross-disciplinary investigation of cause-and-effect for degradation and soil erosion. *Journal of Archaeological Science*, **32**, 1773–1800.

Carcaillet C. & Muller S.D. (2005) Holocene tree-limit and distribution of *Abies alba* in the inner French Alps: anthropogenic or climatic change. *Boreas*, **34**, 468–476.

Carcaillet C., Almquist H. & Asnong H., *et al.* (2002) Holocene biomass burning and global dynamics of the carbon cycle. *Chemosphere*, **49**, 845–863.

Carrion J.S., Andrade A., Bennett K.D., Navarro C. & Munuera M. (2001) Crossing forest thresholds: inertia and collapse in a Holocene sequence from south-central Spain. *The Holocene*, **11**, 635–653.

Chepstow-Lusty A.J., Bennett K.D., Fjelds A.J., Kendall, A., Galiano, W. & Herrera, A.T. (1998) Tracing 4000 years of environmental history in the Cuzco area, Peru, from the pollen record. *Mountain Research and Development*, **18**, 159–172.

Chhabra A., Geist H., Houghton R., *et al.* (2006) Multiple impacts of land use/ cover change. In: *Land-Use and Land-Cover Change: Local Processes and Global Impacts* (Eds E.F. Lambin & H.J. Geist), pp. 70–116. Springer-Verlag, Berlin.

Chiverrell R.C., Oldfield F., Appleby P.G., *et al.* (in press) Evidence for changes in Holocene sediment flux in Semer Water and Raydale, North Yorkshire, UK. *Geomorphology*.

Clarke R.T. (2003) Frequencies of extreme events under conditions of changing hydrological regime. *Geophysical Research Letters*, **30** (3), 24.

Claussen M. (this volume). Holocene rapid land-cover changes – evidence and theory. In: *Global Warming and Natural Climate Variability: a Holocene Perspective* (Eds R.W. Battarbee & H.A. Binney), pp. 232–253. Blackwell, Oxford.

Claussen M., Kubatzki C., Brovkin V., Ganapolski, A., Hoelzmann, P. & Pachur, H-J. (1999) Simulation of an abrupt change in Saharan vegetation at the end of the mid-Holocene. *Geophysical Research Letters*, **24**, 2037–2040.

Claussen M., Brovkin V., Calov R., Ganapolski A. & Kubatzki C. (2005) Did humankind prevent a Holocene glaciation. *Climatic Change*, **69**, 417–419.

Cremaschi M. & Di Lernia S. (1999) Holocene climatic changes and cultural dynamics in the Libyan Sahara. *African Archaeological Review*, **16**, 211–238.

Crucifix M. (this volume). Modelling the climate of the Holocene. In: *Global Warming and Natural Climate Variability: a Holocene Perspective* (Eds R.W. Battarbee & H.A. Binney), pp. 98–122. Blackwell, Oxford.

Crucifix M., Loutre M.-F. & Berger A. (2005) Commentary on "The Anthropogenic greenhouse era began thousands of years ago". *Climatic Change*, **69**, 419–426.

Crumley C.L. (2000) From garden to globe: linking time and space with meaning and memory. In: *The Way the Wind Blows: Climate, History, and Human Action* (Eds R.J. McIntosh, J.A. Tainter & S.K. McIntosh), pp. 193–208. New York: Columbia University Press.

Crutzen, P.J. & Stoermer, E.F. (2000) The "Anthropocene". *Global Change Newsletter*, **41**, 12–13.

Cullen H.M., deMenocal P.D., Hemming S., *et al.* (2000) Climate change and the collapse of the Akkadian empire: evidence from the deep sea. *Geology*, **28**, 379–382.

Cyprien A-L., Visset L. & Carcaud N. (2004) Evolution of vegetation landscapes during the Holocene in the central and downstream Loire basin (Western France). *Vegetation History and Archaeobotany*, **13**, 181–196.

Dean J.S., Gumerman G.J., Epstein J.M., *et al.* (1999) Understanding Anasazi culture change through agent-based modeling. In: *Dynamics of Human and Primate Societies* (Eds T. Kohler & G. Gumerman), pp. 179–204. Oxford University Press, Oxford.

Dearing J.A. (2006) Climate–human–environment interactions: resolving our past. *Climate of the Past*, **2**, 187–203.

Dearing J.A. & Jones R.T. (2003) Coupling temporal and spatial dimensions of global sediment flux through lake and marine sediment records. *Global and Planetary Change*, **39**, 147–168.

Dearing J.A., Battarbee R.W., Dikau R., Larocque I & Oldfield F. (2006) Human–environment interactions: towards synthesis and simulation. *Regional Environmental Change*, **6**, 115–123.

DeMenocal P.B. (2001) Cultural responses to climate change during the Late Holocene. *Science*, **292**, 667–673.

DeMenocal P.B., Ortiz J., Guilderson T., *et al.* (2000) Abrupt onset and termination of the African Humid Period: rapid climate responses to gradual insolation forcing. *Quaternary Science Reviews*, **19**, 347–61.

Diamond J. (2005) *Collapse: How Societies Choose to Fail or Succeed*. Viking Penguin, New York.

Digerfeldt G. (1972) The Post-Glacial development of Lake Trummen. Regional vegetation history, water level changes and palaeolimnology. *Folia Limnologica Scandinavica*, **16**, 1–104.

Dumayne-Peaty L. & Barber K.E. (1998) Late Holocene vegetational history, human impact and pollen representivity variations in northern Cumbria, England. *Journal of Quaternary Science*, **13**, 147–164.

Emanuelsson M. (2001) Settlement and land-use history in the central Swedish forest region. *Acta Universitatis Agriculturae Sueciae*, **223**, Umea, Sweden.

EPICA Community Members (2004) Eight glacial cycles from an Antarctic ice core. *Nature*, **429**, 623–628.

Fisher C.T. (2005) Demographic and landscape change in the Lake Pátzcuaro Basin, Mexico: abandoning the garden. *American Anthropologist*, **107**, 87–95.

Flückiger, J., Blunier, T., Stauffer, B., *et al.* (2004) N_2O and CH_4 variations during the last glacial epoch: Insight into global processes, *Global Biogeochemical Cycles*, **18**(1) GB1020, doi: 10.1029/2003GB002122.

Foley J.A., Coe M.T., Scheffer M. & Wang G. (2003) Regime shifts in the Sahara and Sahel: Interactions between ecological and climatic systems in Northern Africa. *Ecosystems*, **6**, 524–532.

Foster G.C., Dearing J.A., Jones R.T., *et al.* (2003) Meteorological and land use controls on past and present hydro-geomorphic processes in the pre-alpine environment: an integrated lake-catchment study at the Petit Lac d'Annecy, France. *Hydrological Processes*, **17**, 3287–3305.

Fu C. (2003) Potential impacts of human-induced land cover change on East Asia monsoon. *Global and Planetary Change*, **37**, 219–229.

Fuller D., Korisettar P.C., Vankatrasubbaiah P.C. & Jones M.K. (2004) Early domestications in southern India: some preliminary archaeobotanical results. *Vegetation History and Archaeobotany*, **13**, 115–129.

Fyfe R.M., Brown A.G. & Rippon S.J. (2003) Mid- to late-Holocene vegetation history of Greater Exmoor, UK: estimating the spatial extent of human-induced vegetation change *Vegetation History and Archaeobotany*, **12**, 215–232.

Gaillard M.-J., Dearing J.A., El-Daoushy F., Enell M. & Håkansson H. (1991) A late Holocene record of land-use history, soil erosion, lake trophy and lake level fluctuations at Bjäresjösjön (South Sweden). *Journal of Paleolimnology*, **6**, 51–58.

Gaillard M.-J., Birks H.J.B., Ihse M. & Runborg S. (1998) Pollen/landscape calibration based on modern pollen assemblages from surface sediments samples and landscape mapping – a pilot study in south Sweden. In: *Palaeoklimaforschung/ Palaeoclimate Research, Quantification of land Surface Cleared Forest During the Holocene-Modern/Vegetation/Landscape Relationships as an Aid to Interpretation of Fossil Pollen* data (Eds M.-J. Gaillard, B. Berglund, B. Frenzel & U. Huckriede), pp. 31–55. Fisher, Stuttgart.

Gobet E., Tinner W., Hochuli P.A., van Leeuwen J.F.N. & Ammann B. (2003) Middle to Late Holocene vegetation history of the Upper Engadine (Swiss Alps): the role of man and fire. *Vegetation History and Archaeobotany*, **12**, 143–63.

Godwin H (1944) Age and origin of the Breckland Heaths. *Nature*, **154**, 6–8.

Goldewijk K.K. (2003) Estimating global land use change over the past 300 years: the HYDE database. *Global Biogeochemical Cycles*, **15**, 417–34.

Goosse H., Mann M.E. & Rensen H. (this volume). Climate of the past millennium: combining proxy data and model simulations. In: *Global Warming and Natural Climate Variability: a Holocene Perspective* (Eds R.W. Battarbee & H.A. Binney), pp. 163–188. Blackwell, Oxford.

Govaerts Y. & Lattanzio A. (2007) Surface albedo responses to Sahel precipitation changes. *Eos (Transactions of the American Geophysical Union)*, **88**(3), 25–26.

Guiter F., Andrieu-Ponel V., Digerfeldt G., Reille, M., de Beaulieu, J-L. & Ponel, P. (2005) Vegetation history and lake-level changes from the Younger Dryas to the present in the Easter Pyrenees (France): pollen, plant macrofossils and lithostratigraphy from lake Racou (2000 m a.s.l.). *Vegetation History and Archaeobotany*, **14**, 99–118.

Hassan F. (1986) Desert environment and origins of agriculture in Egypt. *Norwegian Archaeological Review*, **19**, 63–76.

Hassan F. (1997) Holocene palaeoclimates of Africa. *African Archaeological Review*, **14**, 213–30.

Haug G.H., Gunther D., Peterson L.C., Sigman D.M., Hughen K.A. & Aeschlimann B. (2003) Climate and the Maya. *PAGES News*, **11**, 28–30.

He H.S., Hao Z., Larsen D.R., Dai L., Hu Y. & Chang Y. (2002) A simulation study of landscape scale forest succession in northeastern China. *Ecological Modelling*, **156**, 153–166.

He X., Zhou J., Zhang X. & Tang K. (2006) Soil erosion response to climatic changes and human activity during the Quaternary on the Loess Plateau, China. *Regional Environmental Change*, **6**, 62–70.

Hodell D.A., Curtis J.H. & Brenner M. (1995) Possible role of climate in the collapse of Classic Maya civilization. *Nature*, **375**, 391–394.

Hoelzmann P., Gasse F., Dupont L.M., *et al.* (2004) Palaeoenvironmental changes in the arid and sub-arid belt (Sahara–Sahel–Arabian peninsula) from 150 kyr to present. In: *Past Climate Variability through Europe and Africa* (Eds R.W. Battarbee, F. Gasse & C.A. Stickley), pp. 219–256. Springer-Verlag, Berlin.

Holman I.P., Rounsevell M.D.A., Shackley S., *et al.* (2005) A regional, multi-sectoral and integrated assessment of the impact of climate and socio-economic change in the UK. Part 1. Methodology. *Climatic Change*, **71**, 9–41.

Huang C.C., Zhao S., Pang J., *et al.* (2003) Climatic aridity and the relocations of the Zhou Culture in the Southern Loess Plateau of China. *Climatic Change*, **61**, 361–378.

Huang C.C., Pang J., Huang P., Hou C. & Han Y. (2002) High-resolution studies of the oldest cultivated soils in the southern Loess Plateau of China. *Catena*, **47**, 29–42.

Huffman T.N. (1996) Archaeological evidence for climatic change during the last 2000 years in southern Africa. *Quaternary International*, **33**, 55–60.

Innes J.B. & Blackford J.J. (2003) The ecology of Late Mesolithic woodland disturbances: model testing with fungal spore assemblage data. *Journal of Archaeological Science*, **30**, 185–194.

Innes J.B. & Simmons I.G. (2000) Mid-Holocene charcoal stratigraphy, fire history and palaeoecology at North Gill, North York Moors, UK. *Palaeogeography, Palaeoclimatology, Palaeoecology*, **164**, 151–165.

Iversen J. (1941) Landnam i Danmark's stenalder (Land occupation in Denmark's Stone Age). *Danmarks Geologiske Undersögelse*, **Series II**, 1–68.

Jacobi R.M., Tallis J.H. & Mellars P.A. (1976) The southern Pennine Mesolithic and the ecological record. *Journal of Archaeological Science*, **3**, 307–320.

Jahns S. (1993) On the Holocene vegetation history of the Argive Plain (Peloponnese, southern Greece). *Vegetation History and Archaeobotany*, **2**, 187–203.

Jahns S. (2005) The Holocene history of vegetation and settlement at the coastal site of lake Voulkaria in Acarnania, western Greece. *Vegetation History and Archaeobotany*, **14**, 55–66.

Jalut G., Amat A.E., Bonnet L., Gauquelin T. & Fontugne M. (2000) Holocene climatic changes in the Western Mediterranean, from south-east France to south-east Spain, *Palaeogeography, Palaeoclimatology, Palaecology*, **160**, 255–290.

Ji S., Jones R.T., Yang X., Dearing J.A. & Wang S. (2006) The Holocene vegetation history of Erhai Lake, Yunnan Province southwestern China: the role of climate and human forcings. *The Holocene.*

Joos F., Gerber S., Prentice I.C., Otto-Bliesner B.L. & Valdes P. (2004) Transient simulations of Holocene atmospheric carbon dioxide and terrestrial carbon since the Last Glacial Maximum. *Global Biogeochemical Cycles*, **18**, GB2002, 1–18.

Kaland P.E. (1986) The origin and management of Norwegian coastal heaths as reflected by pollen analysis. In: *Anthropogenic Indicators in Pollen Diagrams* (Ed. K-E. Behre), pp. 19–36. Balkema, Rotterdam.

Kallel N., Duplessy J., Labeyrie L., Fontugne M. & Paterne M. (2004) Mediterranean Sea palaeohydrology and pluvial periods during the Late Quaternary. In: *Past Climate Variability through Europe and Africa* (Eds R.W. Battarbee, F. Gasse & C.A. Stickley), pp. 307–324. Springer-Verlag, Berlin.

Knox J. (2000) Sensitivity of modern and Holocene floods to climate change. *Quaternary Science Reviews*, **19**, 439–457.

Kohler T.A. (1988) Long-term Anasazi land use and forest reduction: a case study from southwest Colorado. *American Antiquity*, **53**, 537–564.

Kohler T.A. (1992) Prehistoric human impact on the environment in upland North American Southwest. *Population and Environment*, **13**, 255–268.

Koster R.D., Guo Z., Dirmeyer P.A., *et al.* (2006) GLACE: The Global Land–Atmosphere Coupling Experiment. 1. Overview and results. *Journal of Hydrometeorology*, **7**, 590–610.

Kremenetski C.V., Chichagova O.A. & Shishlina N.I. (1999) Palaeoecological evidence for Holocene vegetation, climate and land-use change in the low Don basin and Kalmuk area, southern Russia. *Vegetation History and Archaeobotany*, **8**, 233–246.

Lal R. (1997) Degradation and resilience of soils. *Philosophical Transactions of the Royal Society of London B*, **352**, 997–1010.

Lal R. (2002) Soil carbon dynamics in cropland and rangeland. *Environmental Pollution*, **116**, 353–362.

Lang, A. (2003) Phases of soil erosion-derived colluviation in the loess hills of South Germany. *Catena*, **51**, 209–221.

Lang A., Preston N., Dickau R., Bork H-R. & Maeckel R. (2000). Examples from the Rhine catchment. *PAGES Newsletter*, **8**(3), 11–13.

Latty E.F., Canham C.D. & Marks P.L. (2004) The effects of land-use history on soil properties and nutrient dynamics in northern hardwood forests of the Adirondack Mountains. *Ecosystems*, **7**, 193–207.

Lawrence P.J. & Chase T.N. (2006) Climate impacts. In: *Our Earth's Changing Land*. Vol. 1. (Ed. H. Geist), pp. 115–24. Greenwood, Westport.

Li X., Zhou J. & Dodson J. (2003) The vegetation characteristics of the "Yuan" area at Yaoxian on the Loess Plateau in China over the last 12 000 years. *Review of Palaeobotany and Palynology*, **124**, 1–7.

Li Y.Y., Willis K.J., Zhou L.P. & Cui H.T. (2006) The impact of ancient civilization on the northeastern Chinese landscape: palaeoecological evidence from the Western Liaohe River Basin, Inner Mongolia. *The Holocene*, **16**, 1109–1121.

Lim Y-K., Cai M., Kalnay E. & Zhou L. (2005) Observational evidence of sensitivity of surface climate changes to land types and urbanization. *Geophysical Research Letters*, **32**, L22712, 1–4.

Macklin M.G. (1999) Holocene river environments in prehistoric Britain: human interaction and impact. *Quaternary Proceedings*, **7**, 521–530.

Makohonienko M., Kitigawa H., Naruse T., *et al.* (2004) Late-Holocene natural and anthropogenic vegetation changes in the Dongbei Pingyuan (Manchurian Plain), northeastern China. *Quaternary International*, **123–125**, 71–88.

Malmer N. (1965) The southwestern dwarf shrub heaths. *Acta Phytogeographica Suecica*, **50**, 123–130.

Maloney B.K. (1980) Pollen analytical evidence for early forest clearance in North Sumatra. *Nature*, **287**, 324–326.

Maloney B.K. (1981) A pollen diagram from Tao Sipinggan, a lake site in the Batak Highlands of North Sumatra, Indonesia. *Modern Quaternary Research in South East Asia*, **6**, 57–76.

Matthews H.D., Weaver A.J., Meissner K.J., Gillett N.P. & Eby M. (2004) Natural and anthropogenic climate change: incorporating historical land cover change, vegetation dynamics and the global carbon cycle. *Climate Dynamics*, **22**, 461–479.

Matthews R. (2005) The People and Landscape Model (PALM): towards full integration of human decision-making and biophysical simulation models. *Ecological Modelling*, **194**, 329–343.

Misra V.N. (2001) Prehistoric human colonization of India. *Journal of Biosciences*, **26**(4), 491–531.

Me-Bar Y. & Valdez F. (2005) On the vulnerability of the ancient Maya society to natural threats. *Journal of Archaeological Science*, **32**, 813–825.

Molloy K. & O'Connell M. (1995) Palaeoecological investigations towards the reconstruction of environment and land-use changes during prehistory at Céide Fields, western Ireland. *Probleme der Küstenforschung im südlichen Nordseegebiet*, **23**, 187–225.

Mworia-Maitima J. (1997) Prehistoric fires and land-cover change in western Kenya: evidence from pollen, charcoal, grass cuticles and grass phytoliths. *The Holocene*, **7**, 409–417.

Myhre G., Kvalevag M.M. & Schaaf C.B. (2005) Radiative forcing due to anthropogenic vegetation change based on MODIS surface albedo data. *Geophysical Research Letters*, **32**, L21410, 1–4.

Nielsen A.B. & Odgaard B.V. (2005) Reconstructing land cover from pollen assemblages from small lakes in Denmark. *Review of Palaeobotany and Palynology*, **133**, 1–21.

Niinemets E. & Saarse L. (2006) Holocene forest dynamics and human impact in southeastern Estonia. *Vegetation History and Archaeobotany*, **16**(1), 1–13.

Nober F.J., Graf H-F. & Rosenfeld D. (2003) Sensitivity of the global circulation to the suppression of precipitation by anthropogenic aerosol. *Global and Planetary Change*, **37**, 57–80.

Noel H., Garbolino E., Brauer A., Lallier-Verges E., de Beaulieu J-L. & Disnar R. (2001) Human impact and soil erosion during the last 5000 yrs as recorded in lacustrine sedimentary organic matter at Lac d'Annecy, the French Alps. *Journal of Paleolimnology*, **25**, 229–244.

O'Connell M., Molloy K., Saarinen T., *et al.* (2000) Human impact and climate change at the western fringe of Europe: multidisciplinary studies of calcareous sediments from An Loch Mór, Aran Islands, W. Ireland. *Terra Nostra*, **2000**, 77–81.

O'Connor T.G. & Kiker G.A. (2004) Collapse of the Mapungubwe society: vulnerability of pastoralism to increasing aridity. *Climatic Change*, **66**, 49–66.

Odgaard B.V. & Rasmussen P. (2000) Origin and temporal development of macroscale vegetation patterns in the cultural landscape of Denmark. *Journal of Ecology*, **88**, 733–748.

Oldfield F. (1963) Pollen-analysis and man's role in the ecological history of the south-east Lake District. *Geografiska Annaler*, **45**, 23–49.

Oldfield F. (1969) The ecological history of Blelham Bog National Nature Reserve. In: *Studies in Vegetation History of the British Isles* (Eds R.G. West & D. Walker), pp. 141–157. Cambridge University Press, Cambridge.

Oldfield F. (2005) *Environmental Change: Key Issues and Alternative Approaches.* Cambridge University Press, Cambridge.

Oldfield F. & Statham D.C. (1965) Stratigraphy and pollen-analysis on Cockerham and Pilling Mosses, North Lancashire. *Memoirs and Proceedings of the Manchester Literary and Philosophical Society*, **107**, 1–16.

Oldfield F., Maher B.A., Donoghue J. & Pierce J. (1985) Particle-size related mineral magnetic source–sediment linkages in the Rhode River Catchment, Maryland, U.S.A. *Journal of Geology*, **142**, 1035–1046.

Oldfield F., Wake R., Boyle J., *et al.* (2003a) The late-Holocene history of Gormire Lake (NE England) and its catchment: a multiproxy reconstruction of past human impact. *The Holocene*, **13**, 677–690.

Oldfield, F., Asioli, A., Accorsi, C. A., *et al.* (2003b) A high resolution Late-Holocene palaeo-environmental record from the Central Adriatic Sea. *Quaternary Science Reviews*, **22**, 319–342.

Orlove B. (2005) Human adaptation to climate change: a review of three historical cases and some general perspectives. *Environmental Science and Policy*, **8**, 589–600.

O'Sullivan P.E., Oldfield F. & Battarbee R.W. (1973) Preliminary studies of Lough Neagh sediments I. Stratigraphy, chronology and pollen analysis. In: *Quaternary Plant Ecology* (Eds H.J.B. Birks & R.G. West), pp. 267–278. Blackwell, Oxford.

Pantaléon-Cano J., Errikarta-Imanol Y., Perez-Obiol R. & Roure J.M. (2003) Palynological evidence for vegetational history in semi-arid areas of the western Mediterranean (Almeria, Spain). *The Holocene*, **13**, 109–119.

Pielke R.A. (2005) Land use and climate change. *Science*, **310**, 1625–1626.

Plater A.J., Boyle J.F., Mayers C., Turner S.D. & Stroud R.W. (2006) Climate and human impact on lowland lake sedimentation in Central Coastal California: the record from c.650 AD to the present. *Regional Environmental Change*, **6**, 77–85.

Pringle H. (1997) Death in Norse Greenland. *Science*, **275**, 924–926.

Prosch-Danielson L. & Simonsen A. (2000) Palaeoecological investigations towards the reconstruction of the history of forest clearances and coastal heathlands in south-western Norway. *Vegetation History and Archaeobotany*, **9**, 189–204.

Ramankutty N. & Foley J.A. (1999) Estimating historical changes in global land cover: croplands from 1700 to 1992. *Global Biogeochemical Cycles*, **13**, 997–1027.

Ramrath A., Sadori L. & Negendank J.F.W. (2000) Sediments from Lago di Mezzano, central Italy: a record of Late glacial/Holocene climatic variations and anthropogenic impact. *The Holocene*, **10**, 87–95.

Reale O. & Dirmeyer P. (2000) Modeling the effects of vegetation on Mediterranean climate during the Roman classical period. Part 1: History and model sensitivity. *Global and Planetary Change*, **25**, 163–184.

Redman C.L. (1999) *Human Impact on Ancient Environments.* The University of Arizona Press.

Ren G. (2000) Decline of the mid- to late Holocene forests in China: climatic change or human impact. *Journal of Quaternary Science*, **15**, 273–281.

Ren G. & Zhang L. (1998) A preliminary mapped summary of Holocene pollen data for Northeast China. *Quaternary Science Reviews*, **17**, 669–688.

Renberg I., Korsman T. & Birks H.J.B. (1993) Prehistoric increases in the pH of acid-sensitive lakes caused by land-use changes. *Nature*, **362**, 824–826.

Roberts N., Stevenson A., Davis B., Cheddadi R., Brewer S. & Rosen A. (2004) Holocene climate, environment and cultural change in the Mediterranean region. In: *Past Climate Variability through Europe and Africa* (Eds R.W. Battarbee, F. Gasse & C.A. Stickley), pp. 343–362. Springer-Verlag, Berlin.

Robertshaw P., Taylor D., Doyle S. & Marchant R. (2004) Famine, climate and crisis in western Uganda. In: *Past Climate Variability through Europe and Africa* (Eds R.W. Battarbee, F. Gasse & C.A. Stickley), pp. 535–550. Springer-Verlag, Berlin.

Roesch M. (2000) Long-term human impacts as registered in an upland pollen profile from the southern Black Forest, south-western Germany. *Vegetation History and Archaeobotany*, **9**, 205–218.

Rosen A.M. (1995) The social response to environmental change in early Bronze Age Canaan. *Journal of Anthropological Archaeology*, **14**, 26–44.

Ruddiman W.F. (2003) The anthropogenic greenhouse era began thousands of years ago. *Climatic Change*, **61**, 261–293.

Ruddiman W.F. (2005a) The early anthropogenic hypothesis a year later. *Climatic Change*, **69**, 427–434.

Ruddiman W.F. (2005b) Cold climate during the closest Stage 11 analog to recent Millennia. *Quaternary Science Reviews*, **24**, 1111–1121.

Ruddiman W.F. & Thomson J.S. (2001) The case for human causes of increased atmospheric CH_4 over the last 5000 years. *Quaternary Science Reviews*, **20**, 1769–1777.

Ruddiman W.F., Vavrus S.J. & Kutzbach J.E. (2005) A test of the overdue-glaciation hypothesis. *Quaternary Science Reviews*, **24**, 1–10.

Sadori L. & Narcisi B. (2001) The Postglacial record of environmental history from Lago di Pergusa, Sicily. *The Holocene*, **11**, 655–670.

Saito Y., Yang Z. & Hori K. (2001) The Haunghe (Yellow River) and Changjiang (Yangtze River) deltas: a review on their characteristics, evolution and sediment discharge during the Holocene. *Geomorphology*, **41**, 219–231.

Saltzmann U. (2000) Are modern savannas degraded forests? A Holocene pollen record from the Sudanian vegetation zone of NE Nigeria. *Vegetation History and Archaeobotany*, **9**, 1–15.

Savukynien N., Moe D. & Usaityte D. (2003) The occurrence of former heathland vegetation in the coastal areas of the south-east Baltic sea, in particular Lithuania: a review. *Vegetation History and Archaeobotany*, **12**, 165–175.

Schmidt G.A. & Shindell D.T. (2005) Reply to comment by W.F. Ruddiman on "A note on the relationship between ice core methane concentrations and insolation". *Geophysical Research Letters*, **32**, L15704.

Schneider N. & Eugster W. (2005) Historical land use changes and mesoscale summer climate on the Swiss Plateau. *Journal of Geophysical Research*, **140**, MO, D19102.

Scott L. & Lee-Thorp J.A. (2004) Holocene climatic trends and rhythms in Southern Africa. In: *Past Climate Variability through Europe and Africa* (Eds R.W. Battarbee, F. Gasse & C.E. Stickley), pp. 69–91. Springer-Verlag, Berlin.

Segerström U., Hörnberg G. & Bradshaw R. (1996) The 9000-year history of vegetation development and disturbance patterns of a swamp-forest in Dalarna, northern Sweden. *The Holocene*, **6**, 37–48.

Shi N., Dupont L.M., Beug H-J. & Schneider R. (1998) Vegetation and climate changes during the last 21 000 years in S.W. Africa based on a marine pollen record. *Vegetation History and Archaeobotany*, **7**, 127–140.

Sidorchuk A. (2000) Past erosion and sedimentation within drainage basins on the Russian plain. *PAGES Newsletter*, **8**(3), 19.

Simmons I.G. (1996) *The Environmental Impact of Later Mesolithic Cultures.* Edinburgh University Press, Edinburgh.

Snyder P.K., Delire C. & Foley J.A. (2004) Evaluating the influence of different vegetation biomes on the global climate. *Climate Dynamics*, **23**, 279–302.

Sobrino C.M., Ramil-Rego P. & Gomez-Orellana L. (2004) Vegetation of the Lago de Sanabria area (NW Iberia) since the end of the Pleistocene: a palaeoecological reconstruction on the basis of two new pollen sequences. *Vegetation History and Archaeobotany*, **13**, 1–22.

Sowunmi D.A. (2004) Aspects of Nigerian coastal vegetation in the Holocene: some recent insights. In: *Past Climate Variability through Europe and Africa* (Eds R.W. Battarbee, F. Gasse & C.A. Stickley), pp. 199–218. Springer-Verlag, Berlin.

Steffen W., Sanderson A., Tyson P., *et al.* (2004) *Global Change and the Earth System: a Planet Under Pressure.* Springer-Verlag, Berlin.

Sturm M., Douglas T., Racine C. & Liston G.E. (2005) Changing snow and shrub conditions affect albedo with global implications. *Journal of Geophysical Research*, **110**, G01004.

Sugita S., Gaillard M.-J. & Broström A. (1999) Landscape openness and pollen records: a simulation approach. *The Holocene*, **9**, 409–421.

Taylor C.M., Lambin E.F., Stephenne N., Harding R.J. & Essery R.L.H. (2002) The influence of land use change on climate in the Sahel. *Journal of Climate*, **15**, 3615–3629.

Taylor D. (1990) Late Quaternary pollen diagrams from two Ugandan mires: evidence for environmental change in the Rukiga Highlands of southwest Uganda. *Palaeogeography, Palaeoclimatology, Palaeoecology*, **80**, 283–300.

Tinner W., Conedra M., Ammann B. & Lotter A. (2005) Fire ecology north and south of the Alps since the last ice age. *The Holocene*, **15**, 1214–1226.

Tipping R. & Milburn P. (2000) The mid-Holocene charcoal fall in southern Scotland: spatial and temporal variability. *Palaeogeography, Palaeoclimatology, Palaeoecology*, **164**, 193–209.

Trimble S.W. & Lund S.W. (1982) Soil conservation and the reduction of erosion and sedimentation in the Coon Creek basin, Wisconsin, 1975–1992. *Science*, **285**, 1244–1246.

Tyson P., Lee-Thorp J., Holmgren K. & Thackery J.F. (2002) Changing gradients of climatic change in southern Africa during the past millennium: implications for population movements. *Climatic Change*, **52**, 129–135.

Valsecchi V., Tinner W., Finsinger W. & Ammann B. (2005) Human impact during the bronze Age and the vegetation at Lago Lucone (northern Italy). *Vegetation History and Archaeobotany*, **15**, 99–113.

Van Andel T.H., Runnels C.N. & Pope K.O. (1986) Five thousands years of land use and abuse in the Southern Argolid, Greece. *Hesperia*, **55**, 103–128.

Van West C.R. (1991) Reconstructing prehistoric climatic variability and agricultural production in southwestern Colorado, A.D. 901–1300: a G.I.S. approach. In: *Proceedings of the Anasazi Symposium 1991* (Eds A. Hutchinson, J.E. Smith & J. Usher), pp. 25–34. Mesa Verde Museum Association, Inc., Mesa Verde, Co.

Verschuren D. & Charman D. (this volume) Latitudinal linkages in late Holocene moisture-balance variation. In: *Global Warming and Natural Climate Variability: a Holocene Perspective* (Eds R.W. Battarbee & H.A. Binney), pp. 189–231. Blackwell, Oxford.

Verschuren D, Laird K.R. & Cumming B.F. (2000) Rainfall and drought in equatorial east Africa during the past 1100 years. *Nature*, **403**, 410–414.

Vuorela I., Saarnisto M., Lempiainen T. & Taavitsainen J-P. (2001) Stone Age to recent land-use history at Pegrema, northern Lake Onega, Russian Karelia. *Vegetation History and Archaeobotany*, **10**, 121–138.

Walker K.J., Pywell R.F., Warman E.A., Fowbert, J.A., Bhogal, A. & Chambers, B.J. (2003) The importance of former land use in defining successful re-creation of lowland heath in southern England. *Biological Conservation*, **116**, 289–303.

Waller M. & Salzmann U. (1999) Holocene vegetational changes in the Sahelian zone of NE-Nigeria: the detection of anthropogenic activity. *Palaeoecology of Africa*, **20**, 37–54.

Wasson R.J. & Sidorchuk A. (2000) History for soil conservation and catchment management. In: *Australian Environmental History: Still Settling Australia* (Ed. S.R. Dovers), pp. 97–117. Oxford University Press, Melbourne.

Watts W.A. & Bradbury J.P. (1982) Paleoecological studies at Lake Patzcuaro on the west-central Mexican Plateau and at Chalco in the Basin of Mexico. *Quaternary Research*, **17**, 56–70.

Webster J.B. (Ed.) (1979) *Chronology, Migration and Drought in Interlacustrine Africa*. Longman and Dalhousie University Press, London.

Weiss H. (1997) Late third millennium abrupt climate change and social collapse in West Asia and Egypt. In: *Third Millennium BC Climate Change and Old World Collapse* (Eds H.N. Dalfes, G. Kukla & H. Weiss), pp. 711–22. NATO ASI Series. Springer-Verlag, Berlin.

Weiss H. & Bradley R.S. (2001) What drives societal collapse? *Science*, **291**, 609–610.

Werth D. & Avissar R. (2005) The local and global effects of Southeast Asia deforestation. *Geophysical Research Letters*, **32**, L20702, 1–4.

Wimble G., Wells C.E. & Hodgkinson D. (2000) Human impact on mid- and late-Holocene vegetation in south Cumbria, UK. *Vegetation History and Archaeobotany*, **9**, 17–30.

Wirtz K.W. & Lemmen C. (2003) A global dynamic model for the Neolithic Transition. *Climatic Change*, **59**, 333–367.

Wu H., Guo Z. & Peng C. (2003) Land use induced changes of organic carbon storage in soils of China. *Global Change Biology*, **9**, 305–315.

Xu K, Chen Z., Zhao Y., *et al.* (2005) Simulated sediment flux during 1998 big-flood of the Yangtze (Changjiang) River, China. *Journal of Hydrology*, **313**, 212–233.

Xu L. (1999) From GCMs to river flow: a review of downscaling methods and hydrological modelling approaches. *Progress in Physical Geography*, **23**, 229–339.

Yasuda Y. (Ed.) (2002) *The Origins of Pottery and Agriculture*. Roli Books, New Delhi.

Yi S., Saito Y., Oshima H., Zhou Y. & Wei H. (2003) Holocene environmental history inferred from pollen assemblages in the Huanghe (Yellow River) delta, China: climatic change and human impact. *Quaternary Science Reviews*, **22**, 609–628.

Yu L. & Oldfield F. (1989) A multivariate mixing model for identifying sediment source from magnetic measurements. *Quaternary Research*, **32**, 168–181.

Zhou W.J., Dodson J., Head M.J., Li., Hou Y.J. and Lu X.F. (2002) Environmental variability within the Chinese desert–loess transition zone over the last 20 000 years. *The Holocene*, **12**, 107–112.

Zolitschka B. (1998) A 14 000 year sediment yield record from western Germany based on annually laminated sediments. *Geomorphology*, **22**, 1–17.

4 Modeling the climate of the Holocene

Michel Crucifix

Keywords

Climate variability, conceptual models, models of intermediate complexity, general circulation models, orbital forcing

Introduction

The aim of climate modeling is to understand past changes in climate that are currently unexplained and to be able to predict successfully the future evolution of climate. This chapter begins with some general considerations on climate dynamics followed by an introduction to climate models, how they are constructed, the role of different models, what needs to be specified, and the need for data–model comparison and assimilation. The second half of the chapter is dedicated to the results of two modeling applications. The first one concerns the stability of global climate during the Holocene. The astronomical theory of paleoclimates is now sufficiently well understood to enable a prediction of the next glacial inception. We will see that it is not expected for 50 000 years but, paradoxically, this conclusion does not contradict a recent postulate that glacial inception would be occurring by now in the absence of an early anthropogenic perturbation. The second result focuses on the stability of the ocean–atmosphere system. Advective and convective processes in the deep-ocean may introduce stochastic effects. As a consequence, a specific event, such as an abrupt cooling, cannot necessarily be traced back to a well-identified cause.

Climate modeling: the problem of chaos and instability

The ocean–atmosphere–cryosphere–biosphere system is a complex system in the sense that it is made of different components that may interact with each other on a very wide range of time-scales (Mitchell 1976). These interactions are generally nonlinear, that is the response is not proportionate to the amplitude of the excitation. A physical system with at least three components interacting nonlinearly with each other may be chaotic (e.g. Hilborn 2001, p. 134) (Figure 4.1). In other words,

Figure 4.1 Lorenz's model is a classic example of a chaotic system with a strange attractor (Lorenz 1963). It is made of three nonlinear equations describing the trajectory of a particle along three coordinates {x,y,z} given two parameters (*s* and *r*). The figure represents the projection of a trajectory projected on the {x,y} plane. The particle leaves its initial condition and is attracted to a certain region of the space called the attractor. Depending on the chosen parameters, the attractor may be a point, a well-defined geometric form such as a circle or an ellipse, or a "strange attractor". "The strange attractor" featured here characterizes a chaotic behavior. This means that the particle circulates around a mean path (the butterfly shape) but it never passes twice through the same point in the three-dimensional space. The shape of the attractor is a function of the parameters and can be statistically defined after a long integration time. By contrast, individual trajectories cannot be predicted because any small error on initial conditions grows exponentially with time. This is the reason why climate (the attractor) may be predicted, but weather (the location of a particle at a moment in time) cannot beyond a certain time horizon.

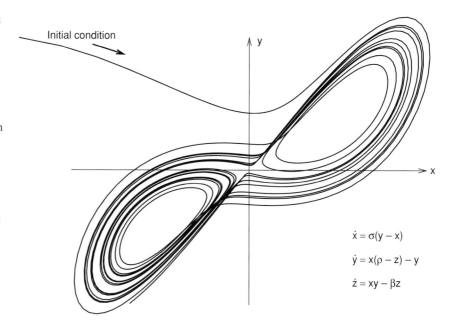

Initial condition

$$\dot{x} = \sigma(y - x)$$
$$\dot{y} = x(\rho - z) - y$$
$$\dot{z} = xy - \beta z$$

its evolution cannot be predicted accurately beyond a certain time horizon because any error on the initial conditions grows exponentially with time. The atmosphere is chaotic. This is the reason why we cannot forecast weather much beyond about 6 days. Yet, we can predict global warming. Indeed, conservation of energy, heat, and momentum makes it possible to predict the general evolution of a chaotic system in statistical terms. This statistical description of weather is nothing but the definition of climate.

A special problem about climate is that it features exchanges of energy between components characterized by radically different temporal and spatial scales. For example, a cloud has a typical spatial scale of the order of 1 km and a life cycle of a few hours to one or two days long. Cloud formation and the resulting precipitation may influence and be influenced by the topography of continental ice sheets that evolve over millennia. This is an example of an interaction between a component with a very short time-scale and one with a very long one. These exchanges may be extraordinarily complex and this is the reason why climate models are uncertain.

Nonlinear systems may also exhibit instability. A steady state is said to be unstable when a small perturbation is irremediably amplified by feedbacks internal to the system until it reaches a new mode of behavior. There may be several stable steady states and the one chosen by the system depends on its history. A typical example is glaciated versus nonglaciated Earth. Both states may be stable (Budyko 1969; Calov *et al.* 2005) and the one chosen depends on whether the preceding history is favorable or not to a glacial inception.

Finally the actual climate system has no strict steady state because oceanic and atmospheric currents vary constantly. This is why theoreticians prefer to use the notion of "attractor". The attractor is, loosely speaking, a closed trajectory that the climate system follows more or less closely (Figure 4.1).

Figure 4.2 Time- and space-scales covered by numerical models used for weather and climate prediction. Conceptual models have only a few degrees of freedom and are designed to formulate and test hypotheses in a very well-defined framework. Three examples are given here: (1) turbulence in the boundary layer; (2) stability of the ocean circulation; and (3) ice-sheet response to astronomical forcing. Comprehensive climate models (also called "general circulation models") include the largest number of degrees of freedom and are suitable to study climate dynamics on time-scales of a few decades to a few centuries. Earth models of intermediate complexity (EMICs) usually cover longer time-scales. Mesoscale and regional models simulate weather and climate over a limited domain of the globe. A mesoscale model typically covers a domain the size of the UK, and a regional model may cover Europe or the USA.

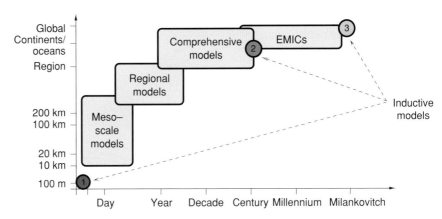

Inductive and deductive climate models

It has become customary to define three categories of global climate models (Claussen *et al.* 2002; Renssen *et al.* 2004) (Figure 4.2).

- **Conceptual models** are made of a small number of differential equations designed to represent interactions between the major climate components. They are called *inductive* because the number of adjustable parameters is of the same order of magnitude as the number of differential equations (number of degrees of freedom). Their primary purpose is to formulate a phenomenological theory of climate dynamics. This may cover problems as various as the stability of the ocean circulation (Stommel 1961) or the astronomical theory of paleoclimates (Imbrie and Imbrie 1980; Saltzman and Maasch 1990; Paillard 2001). Conceptual models can produce very complex solutions that may even be chaotic. The conceptual models that can successfully be tuned on the climate record provide a structure to observations which, according to information theory (Leung and North 1990), may confer on them a prediction skill.*
- **Comprehensive climate models** are built from first principles of physics (equations of movement, radiative transfer, etc.) numerically implemented on three-dimensional grids representing the atmosphere, the oceans, and sea-ice.† The characteristic horizontal spatial scale of the grid is of the order of 100 km and the

---

* Saltzman (2002) considered as an "act of faith" that long-term climate dynamics may be described by some low-order model, similar to thinking in physics that the cosmos is governed by a "unified theory". There is no easy demonstration of this, but Hargreaves and Annan (2002) showed that the Saltzman and Maasch (1990) model does have significant skill in predicting climate over about 100 kyr.
† Technically, the discretized equations of motion may be solved directly on the grid (grid-based models). Another possibility (spectral models) is to compute first the spherical harmonics of the physical quantities and then to resolve the equations of motion in this "conjugate" space. Differential operators, such as the Laplacian, are indeed more easily expressed in the conjugate space. The spatial resolution of a spectral model depends on the number of spherical transforms retained to perform the calculations. For example, T32 means a triangular (T) truncation to the first 32 spherical harmonics. This approximately corresponds to a resolution of 400 km × 400 km.

integration time step is a few hours (see Johns *et al.* (2006) for a recent example). Synoptic atmospheric variability is thus explicitly calculated. Comprehensive climate models are "deductive" because the number of constitutive equations is several orders of magnitude larger than the number of adjustable parameters. Phenomena occurring at spatial scales smaller than the model grid, such as convective cloud formation, are parameterized by means of phenomenological equations. Climate modelers establish these equations on the basis of local observations (soundings, aircraft measurements, surface data) and specialized models. A parameterization has to be "physically reasonable" and respect conservation principles (conservation of energy, entropy, momentum, etc.). In spite of these constraints, different mathematical formulations of a parameterization may seem to provide equally good results and there is no easy way to know which one is best. This is what is called **structural uncertainty**.

In practice, the parameters need to be tuned again so that the climate generated by the global model agrees with observed global-scale features of the climate system (e.g. the existence of a meridional overturning cell in the North Atlantic). This tuning process is too often viewed as a necessary evil about which little detail is given in the model documentation.

Fortunately, tuning tends to be better recognized as a natural step of model development. Model parameters are attributed uncertainty ranges (constrained by laboratory measurements and local observations), and the likelihood of a given parameter combination is estimated from the agreement between the model and a well-defined set of global observations (surface temperature, precipitation, satellite estimates of the radiative balance, etc.) (more detail is given below).

Nowadays, the development of comprehensive models mobilizes large multidisciplinary teams driven by the aim of providing reliable future climate predictions. These models must therefore include all processes relevant at the decadal to century time-scales with as much detail as computational power permits. This covers various aspects such as soil dynamics, vegetation dynamics, ocean biogeochemistry, ice-sheet dynamics, and river hydrology. At the time of writing, a 100-year long simulation of climate with a comprehensive climate model (e.g. $1.5 \times 1.5°$ resolution) requires more than a month of a supercomputer with powerful data storage facilities.

- **Earth models of intermediate complexity (EMICs)** fill the gap between conceptual and comprehensive climate models (Claussen *et al.* 2002). They resemble comprehensive climate models but calculations are made on longer time-scales and larger spatial scales. The degree of parameterization is higher, which may imply a larger structural uncertainty. Yet, like comprehensive models, the number of degrees of freedom in EMICs exceeds the number of adjustable parameters by several orders of magnitude. The EMIC category covers a range of models that may be used to study interdecadal to astronomical time-scales depending on the model. Examples of EMICS used to study the Holocene are given in Table 4.1.

Earth models of intermediate complexity and conceptual models are useful because they cover spatial and temporal scales for which comprehensive models

Table 4.1 Examples of Earth models of intermediate complexity (EMICs) used to study the Holocene

Model	Example of published applications over the Holocene	Reference
LLN-2D	Prediction of the next glacial inception	Loutre *et al.* 2007
MoBidiC	Vegetation–climate interactions at high latitudes	Crucifix *et al.* 2002
CLIMBER-2	Impact of freshwater input in the North Atlantic around 8000 years ago	Bauer *et al.* 2004
	Factor decomposition (see text)	Ganopolski *et al.* 1998
	Desertification of the Sahara in response to orbital forcing	Claussen *et al.* 1999; Claussen *et al.* this volume
	Changes in ocean and terrestrial carbon storages during the Holocene	Brovkin *et al.* 2002
Green McGill	Analysis of the carbon budget	Wang *et al.* 2005a
Paleoclimate Model	Existence of a climate optimum after the disappearance of the Laurentide Ice Sheet	Wang *et al.* 2005b
	Impact of freshwater input in the North Atlantic around 8000 years ago	Wang and Mysak 2005
ECBILT (coupled to a low-resolution ocean model)	The influence of changes in precipitation and temperature on the evolution of three representative glaciers	Weber and Oerlemans 2003
ECBILT–CLIO (coupled to a higher resolution ocean–sea-ice model)	See text	Renssen *et al.* 2005
	Impact of freshwater input in the North Atlantic around 8000 years ago	Renssen *et al.* 2002

are not suitable. Consider the glacial–interglacial cycles: the growth and decay of the total continental ice mass at the glacial–interglacial time-scale is of the order of a few centimeters of sea-level equivalent per year. These one or two centimeters result from a difference between total evaporation, precipitation, melting, and freezing that is so small compared with the quantities themselves that it cannot be confidently estimated by a general circulation model (Saltzman 1988, 2002). In fact, the present net accumulation rates of snow over Antarctica and Greenland are even not accurately known (Rignot and Thomas 2002). Earth models of intermediate complexity provide a solution. They may be tuned to reproduce reasonable results over a given section of a glacial–interglacial cycle (e.g. the last glacial inception, as in Gallée *et al.* 2002) and then used to study other periods (Loutre *et al.* 2007).

We have so far considered global climate models. Comparison with (paleo) data may make it necessary to resolve smaller spatial scales than those of a comprehensive climate model. The method demanding least computing time is statistical downscaling (Murphy 1998), where statistical relationships are applied between climate variations at the synoptic scale (200–300 km) and the local climate. Two more elaborate strategies are documented in the literature: nesting and zooming (Giorgi and Mearns 1999). Under "nesting", output of a global model is used to drive a regional dynamical model of the atmosphere that resolves horizontal length scales of the order of 30 to 50 km. "Regional" indicates that this high-resolution model only covers a defined region of the globe, such as Europe. Zooming is based on a comprehensive climate model featuring an irregular mesh refined over the region of interest in order to capture the smaller spatial scales.

Initial conditions, boundary conditions, and model parameters

Mathematically speaking, a climate model is a system of equations with diagnostic and prognostic equations. Diagnostic equations instantaneously link different variables to each other, for example, the hydrostatic assumption linking pressure and density. Prognostic equations need to be integrated forward in time, for example, the Navier–Stokes equations describing hydrodynamics. In theory, it should be possible to integrate prognostic equations backwards in time, but this task is almost insurmountable because many diagnostic relationships are nonbijective (i.e. cannot easily be inverted). The model is thus integrated forward in time and one must define initial conditions, boundary conditions, and parameters.

- **Initial conditions** represent the climate state from which the model equations are integrated. They are generally supplied from a previous experiment or from observations.
- **Boundary conditions** define values or fluxes at the boundary of the domain, such as the surface topography and incoming shortwave radiation at the top of the atmosphere. The latter is calculated from the geometry of the Earth's orbit.
- **Parameters** are constants used in the equations. Some are directly specified from laboratory experiments (e.g. heat capacity of water) or direct observations (albedo of sea-ice, concentrations of greenhouse gases). Others are more phenomenological and they must be calibrated by comparing model results with observations (see below).

Once initial conditions, boundary conditions, and parameters are specified, the model is run to perform either an **steady-state experiment** or a **transient experiment**.

The climate modeler's "steady-state experiment" is an integration for which boundary conditions and parameters are constant, except for the insolation seasonal cycle. Such experiments are suited to study the statistics of a quasi-steady climate, where the time-scale of the external forcing change is larger than the longest dissipative time-scale of the system. This is about a few thousand years if deep-ocean dynamics are taken into account (this number is reached by dividing the depth of the ocean to the square by the turbulent vertical diffusion coefficient). Initial conditions seem unimportant in steady-state experiments because they are eventually dissipated ("forgotten"). There are two caveats to this statement. First, it is more economical to guess initial conditions that are not too far from the solution in order to reduce as much as possible the time spent by the system to reach it (spin-up time). Second, there may be two quasi-state solutions to the model equations, with only a small probability of transition from the one to the next: For example, a green and a white Sahara (Claussen, this volume). An experiment is said to be **transient** if the boundary conditions and some other parameters (such as greenhouse gas concentrations) change during the integration. These changes constitute an **external forcing** that influences the climate trajectory. Two categories of transient experiments may be distinguished. In the first one, the external forcing varies slowly or at a rate comparable to the dissipative time-scales of the system. A good example is the change in the spatial and seasonal incoming solar radiation due to the quasi-cyclic variations in the Earth's orbit: eccentricity, precession, and tilt of the rotation axis on the ecliptic (i.e. orbital forcing). Such experiments allow us to identify and understand nonlinear processes in the response to the forcing, such as deglaciation or desertification of the Sahara (Claussen *et al.* 1999). The response is said to be abrupt if its characteristic time-scale is shorter than that of the forcing (Alley *et al.* 2003). The second category of transient experiments gathers those in which the forcing varies with a characteristic time shorter than the dissipative time-scales of the system. In this case the abrupt character of the response is due to the forcing itself. Experiments of this kind are usually performed to test hypothetical scenarios such as a discharge of freshwater in the ocean (Renssen *et al.* 2001; Bauer *et al.* 2004; LeGrande *et al.* 2006).

Model–data comparison versus data assimilation

The classic methodology assumes a clear distinction between model input (initial conditions, boundary conditions, parameters) and output (the predicted climate). Boundary conditions and parameters are considered as "known" and there is no formal uncertainty attached to them. The model output is then compared with observations. There is usually a pair of experiments. One is designed to produce a simulation of the pre-industrial climate and, in the other, boundary conditions and certain parameters (such as greenhouse gas concentrations) are modified to predict past climate.

The Paleoclimate Modeling Intercomparison Project (PMIP) framed past climate simulations (mid-Holocene and Last Glacial Maximum) with different comprehensive climate models and organized a systematic comparison between the model output and paleoclimatic data (Joussaume and Taylor 2000; Braconnot *et al.* 2007) (Figure 4.3). Appropriate **proxy** models may facilitate the model–data comparison. Proxy models are calibrated to map climate model outputs on observable quantities such as the dominant biome (Haxeltine and Prentice 1996), lake level, or glacier length (Weber and Oerlemans 2003). Climate models may also directly include the necessary equations to simulate observable features such as dust flux (Joussaume and Jouzel 1993a), oxygen (Joussaume and Jouzel 1993b), carbon (Marchal *et al.* 1999), and boron isotopic ratios (LeGrande *et al.* 2006).

It was shown that climate models correctly reproduce a number of observed features of the mid-Holocene climate: increased precipitation in the Sahel, decreased precipitation in South America, reduced sea-ice cover in the Norwegian Sea, northward advance of boreal forest in Russia, and reduced frequency of El

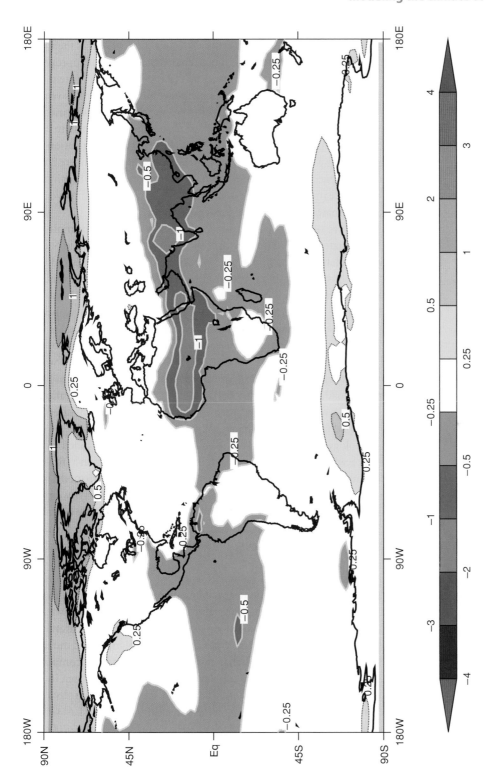

Figure 4.3 Change in annual mean surface temperature induced by switching from today's orbital forcing to that of 6000 years ago (see also Figure 4.4) as simulated by Paleoclimate Modeling Intercomparison Project (PMIP) models (the mean model response is displayed). The plot evidences well the annual polar warming and equatorial cooling. Sensitivity experiments suggest that the annual tropical cooling is mainly caused by the increase in obliquity and the resulting decrease in annual mean insolation below 43° of latitude. The polar warming results from a combination of obliquity (larger annual mean insolation, concentrated in summer) and precession (larger summer and autumn insolation). Seasonal changes in insolation are translated into an annual temperature signal by the sea-ice feedback. Cooling of northern sub-tropical deserts is a signature of enhanced summer monsoon precipitation and the resulting increase in surface evaporation. Note that the vegetation response was not taken into account in these experiments. (Data were supplied by Jean-Yves Peterschmitt and extracted from the PMIP database in Saclay, France.)

Niño events (see the recent reviews by Braconnot *et al.* 2004; Renssen *et al.* 2004; Cane *et al.* 2006; pioneering work is covered in Wright *et al.* 1993). It is then possible to decrypt the mechanisms of these climate changes by means of appropriate sensitivity experiments. One method, known as "factor separation" (Stein and Alpert 1993), consists in sequentially freezing certain components such as vegetation or sea-ice distribution normally calculated by the model. It was used by Ganopolski *et al.* (1998) to show that hemispheric warming in response to mid-Holocene orbital parameters results from feedbacks between boreal vegetation and sea-ice (in the CLIMBER model) (see also Harvey 1988; Crucifix and Loutre 2002). More generally, feedback analysis of mid-Holocene experiments has highlighted the importance of ocean dynamics and vegetation and justified including vegetation dynamics in climate models used for future climate prediction.

Climate models are never in "perfect agreement" with data. For example, the "IPSL" (i.e. the climate model of the "Institut Pierre et Simon Laplace" in Saclay, France) model simulation of the mid-Holocene climate indicates, compared with the present-day, increased aridity in central Eurasia and a northward advance of the boreal forest's northern limit in Canada, contrary to observations (Wohlfahrt *et al.* 2004). Climate modelers tend to be very defensive when it comes to model–data comparisons because discrepancies call the model performance into question. This attitude is unfortunate because model–data differences contain particularly useful information. There are at least two things the modeler would like to know about them. First, what is their cause? Are they due to a process badly accounted for, an inadequate boundary condition (e.g. incorrect specification of ice sheets) or a data misinterpretation? Second, does this disagreement affect the model's prediction of future climate change?

A new methodology is being formalized to address these questions, called "probabilistic inference with climate models" or "climate data assimilation" (Rougier 2006). The fundamental idea is to attribute explicitly uncertainty ranges to model parameters, which are explored by performing large ensembles of sensitivity experiments. A likelihood is attributed to a parameter value depending on (i) the difference between the model prediction obtained with this parameter value and the data-estimate, (ii) the data uncertainty, and (iii) prior knowledge on the parameter. This is a data assimilation because observations (temperature, precipitation, etc.) provide explicit constraints on model parameters and/or boundary conditions.

Climate data assimilation may help us to refine estimates on phenomenological parameters used in parameterizations. It therefore provides a formal framework to "tuning" by clarifying which information is being used to constrain parameters and estimate probability distribution functions on both model input and output.

The trouble with this process is that it puts a high demand on computing resources because it requires numerous experiments. It has therefore been implemented in only a few cases using modern climatic observations (Murphy *et al.* 2004), plus a small number of studies using large-scale estimates of the Last Glacial Maximum temperature (Annan *et al.* 2005; Schneider von Deimling *et al.* 2006).

Examples above are based on steady-state experiments. Data assimilation also applies to transient experiments. In short experiments (i.e., time length shorter than the dissipative time-scales of the system), data assimilation is used to provide estimates of climate variables for which there is no direct observation (e.g. Goosse *et al.* this volume) by constraining initial conditions (cf. Jones and Widdmann 2004; van der Schrier and Barkmeijer 2005). In long transient simulations, data assimilation may be used to constrain model parameters (Hargreaves and Annan 2002) (see below).

How long will the Holocene last?

The Holocene is a particularly long episode of stable climatic conditions compared with the three previous interglacial periods (Sirocko *et al.* 2007). Its stability certainly favored the establishment of modern civilizations. What explains the stability of the Holocene, and how long will it be?

Modeling the long-term evolution of climate requires taking into account changes in the spatial and seasonal distribution of incoming shortwave radiation (insolation) at the top of the atmosphere (Figure 4.4). The latter is fully determined by three parameters: eccentricity, obliquity, and climatic precession (Berger 1978). **Eccentricity** is a geometric measure of the stretching of the Earth's orbit. It varies between about 0.002 and 0.040 on periods of 400 000 and 100 000 years. It is presently small (0.016) and a minimum will be reached in 27 000 years.

Figure 4.4 Month–latitude distribution of incoming shortwave radiation (insolation) received from the Sun at the top of the atmosphere between 9000 years BP and 6000 years AP (after present). A mean distribution assuming no eccentricity and a mean obliquity of 23°20′ was subtracted in order to highlight the effects of precession and obliquity changes. Precession redistributes heat across the seasons (positive anomaly around July 9000 years BP and positive anomaly around January at present). The decrease in obliquity during the Holocene contributes to reduce summer insolation in both hemispheres.

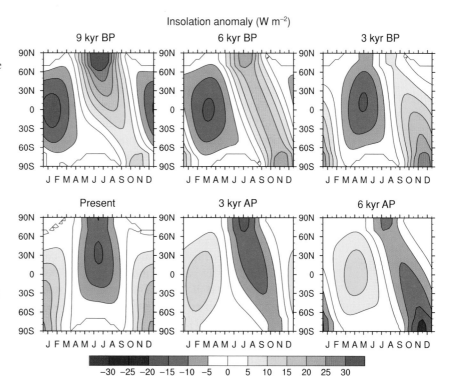

Obliquity is the angle between the Equator and the orbital plane (ecliptic). It varies between 22 and 25° with a period of 40 000 years. Obliquity has decreased by 0.8° during the past 9000 years, and it will continue to decrease over the next 10 000 years. As a consequence, the distribution of insolation is being slightly modified with (i) less insolation available to the summer hemisphere, with highest differences at high latitudes, and (ii) a decrease in annual mean insolation polewards of 43°N and 43°S, symmetric about the Equator, which compensates for a corresponding increase equatorwards of 43°. Changes in annual mean insolation are small (1 to 2 W m^{-2}) but provisional calculations confirmed by sensitivity experiments with a comprehensive climate model reveal that the corresponding thermal forcing easily explains sea-surface temperature changes by 0.5 to 1°C (Liu *et al.* 2002; Loutre *et al.* 2004). **Precession** is the cycling of the angle formed by the position of the Earth on 21 March, the Sun, and the point of the orbit closest to the Sun (perihelion). The cycle takes about 21 000 years. Perihelion was reached in July 11 000 years ago. It then drifted from July to later in the year. It occurs in January today. Given that insolation decreases with the square of the Earth–Sun distance, precession during the Holocene caused a decrease in June insolation and a corresponding increase in January. Precession does not alter annual mean insolation at any latitude. The seasonal redistributing action of precession makes it an efficient modulator of seasonal weather systems, such as tropical monsoons (Kutzbach and Otto-Bliesner 1982; Harrison *et al.* 2003; Braconnot *et al.* 2004). These effects naturally become less important when eccentricity decreases as at present.

All comprehensive model experiments so far have demonstrated that orbital forcing induces significant and measurable changes in temperature, precipitation, and atmospheric circulation. These changes constitute the "fast" climate response to orbital forcing, which drives – and may be altered by – the slow components of the climate system (ice sheets, deep-ocean dynamics, ocean biogeochemistry, vegetation) over several millennia.

Conceptual models provide a convenient theoretical framework to study free and forced interactions between the slow components of the climate system. A particularly significant development is Saltzman's model of the Late Cenozoic Ice Ages (Saltzman and Maasch 1990). This three-equation model with nine parameters represents the interactions between the carbon cycle, ice sheets, and ocean circulation. Probabilistic inference with this model (Hargreaves and Annan 2002) indicates an immediate end to the Holocene, with ice volume reaching a maximum in around 60 kyr (assuming no anthropogenic perturbation). But is this prediction correct? For example, it is noted that the Saltzman–Maasch model fails to reproduce the steadily increasing trend in CO_2 concentration during marine isotope stage (MIS) 11 (after termination V, 400 000 years ago) (Raynaud *et al.* 2005; Siegenthaler *et al.* 2005). Therefore, some stabilizing mechanisms may have been ignored in this model. We therefore turn to another conceptual model presented by Paillard (2001).

Paillard's conceptual model features three possible climate regimes (glacial, mild glacial, interglacial) to which the system is successively attracted depending on insolation and ice volume. Contrary to Saltzman's, Paillard's model succeeds in predicting the correct length for MIS 11 (two precession cycles). Paillard's model can then be used to predict the length of the Holocene. The result is ambiguous. The prediction can either be a short Holocene (glacial inception already begun) or a very long one (glacial inception in 50 kyr) depending on the model parameters. Yet the tested parameter values seem equally reasonable: the model provides a satisfactory fit to data of the past 800 kyr (Imbrie *et al.* 1984) in both cases. Only the observation that we are presently not undergoing a glaciation allows us to reject the first solution. In other words, the length of the Holocene would not have been predictable 9000 years ago with Paillard's model.

There is presently no comprehensive model or even an EMIC (see above) capable of representing the interactions between the slow components of the climate system satisfactorily enough to predict the evolution of ice volume and greenhouse gas concentrations over several glacial–interglacial cycles. There are a few EMICs, however, that are able to simulate the evolution of the atmosphere–ocean–ice-sheet system on those time-scales: the LLN-2D model (Gallée *et al.* 1991, 1992), CLIMBER-2 (Petoukhov *et al.* 2000; Calov *et al.* 2005), and the Toronto climate–ice-sheet model (Tarasov and Peltier 1999). The LLN-2D model is particularly suitable to study the evolution of ice volume in response to hypothetical CO_2 scenarios and orbital forcing because it successfully reproduces several past glacial–interglacial cycles assuming that the evolution of greenhouse gas concentrations is correctly prescribed (Loutre and Berger 2003).

Both CLIMBER and LLN-2D consistently show no glacial inception during the Holocene when observed CO_2 concentrations are prescribed (Claussen *et al.* 2005; Loutre *et al.* 2007). Other CO_2 scenarios were then tested. Neither LLN-2D nor CLIMBER predicted a modern glacial inception as long as CO_2 remained above 240 ppmv during the Holocene. The LLN-2D further shows that even if CO_2 concentration decreased in the future down to glacial levels, glaciation will not occur before 50 000 years (Loutre *et al.* 2007).

The quasi-absence of precessional forcing due to the weak eccentricity may therefore explain the exceptional stability of the Holocene. This statement, however, calls for a few clarifications. First, sensitivity experiments with the LLN-2D model suggest that there is actually a window, roughly between −5000 and +5000 years from now, during which a low enough CO_2 concentration (below 240 ppmv) induces a glacial inception (Figure 4.5, red line). This is consistent with sensitivity experiments with a more comprehensive model calibrated on the present-day climate showing accumulation of perennial snow for 240 ppmv CO_2 and 450 ppbv CH_4 (Ruddiman *et al.* 2005) (the methane feedback is implicitly taken into account in the LLN-2D via the model calibration on previous glacial–interglacial cycles). These results confirm Paillard's prediction that the present orbital configuration may be compatible with a glacial inception (Paillard 2001), but they also show that the inception scenario is no longer possible given the present CO_2 concentration.

Second, the LLN-2D predicts a three-precession cycle duration for MIS 11 when forced by the Vostok CO_2 concentrations (Petit *et al.* 1999), but sensitivity experiments with slightly lower CO_2 concentrations result in a short MIS 11 (Loutre and Berger 2003). Paillard (2001) also found that small parameter changes may make

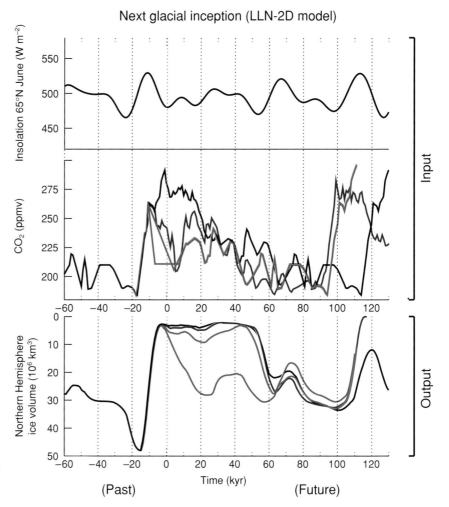

Figure 4.5 Prediction of the next glacial inception with the LLN-2D climate–ice-sheet model. The LLN-2D model is a model of intermediate complexity representing the interactions between the atmosphere, the ocean mixed-layer, and ice sheets. The CO_2 concentration needs to be prescribed. This model correctly reproduces several glacial–interglacial cycles. Here, it is shown that the next glacial inception is not expected to occur before 50 000 years from present under all possible scenarios of CO_2 concentration (green, blue, black) except in one scenario (red) where CO_2 decreases to 210 ppmv as early as 7000 years BP. (Data are from Loutre *et al.* 2007.)

MIS 11 either short or long in his model. We note that MIS 11 eccentricity was almost as small as today (0.019 versus 0.017) with the implication that in the late Quaternary background climate small values of eccentricity are times when small climate perturbations may decide whether the climate system is sent to a long or a short interglacial. In an extreme formulation of this working hypothesis, the decisive perturbation may be so small that it is unpredictable. Another possible consequence is that glacial inception may crucially depend on the phase of obliquity when eccentricity is small (Masson-Delmotte *et al.* 2007).

These latter two remarks are relevant to a consideration of Ruddiman's recent proposal that the absence of an ongoing glaciation is due to anthropogenic factors (Ruddiman 2003). Millennia of moderate greenhouse gas emissions due to forest clearance and rice plantations before industrialization would have been large enough to cause a significant deviation in the natural evolution of the slow components of climate (see also Oldfield, this volume). According to Ruddiman, the natural evolution was a glacial inception, with CO_2 and CH_4 concentrations decreasing to 240 ppmv and 450 ppbv, respectively (they were 280 ppmv and 710 ppbv respectively before the industrialization). Ruddiman argues in later articles (Ruddiman 2005, 2006, 2007) that the major part of these anomalies is actually a feedback to a small perturbation: the ocean did not absorb as much CO_2 as it would otherwise have because it warmed up. This explains the lack of a marked isotopic signature in the Holocene ice record of CO_2 (noted by Joos *et al.* (2004) and Broecker and Stocker (2006)). We therefore arrive at an important conclusion (expressed by Crucifix and Berger 2006): the early anthropogenic theory implies – if it is correct – that there was a bifurcation point during the past 6000 years during which the climate system hesitated between opting for a glacial inception or staying interglacial. The anthropogenic perturbation gave it the necessary kick to opt for a long interglacial.

In light of the modeling works discussed above, this hypothesis cannot easily be proved or disproved. Only refined models properly calibrated by means of past data assimilation will allow us to determine if anthropogenic emissions over the past six millennia significantly increased the probability of a long interglacial. This statement should not be confused with another one: it is now granted that we are in this long interglacial and that glacial inception is no longer expected for about 50 000 years. Actually, Archer and Ganopolski (2005) estimate that 25 percent of present anthropogenic emissions will remain in the atmosphere for thousands of years and about 7 percent will remain beyond 100 000 years. They conclude that the CO_2 concentration threshold below which a glacial inception occurs may not be reached before 500 000 years.

Ocean stability during the Holocene

So far we have examined the stability of the global climate system, which includes the ice sheets and carbon cycle. We now focus on the ocean–atmosphere system and anything external to this component of the climate system will be considered as a boundary condition or a forcing.

The time-scale of ocean advective processes is of the order of a few thousand years. This provides a rough upper limit to the period of quasi-periodic oscillations that the ocean–atmosphere system may exhibit in the absence of external forcing. Most known oscillatory modes resulting from the coupling between ocean and atmosphere dynamics have periods ranging from a few weeks (e.g. the Maiden Julian oscillation) to a few years (e.g. ENSO). The nonlinear nature of the ocean–atmosphere system and its number of degrees of freedom can bring about aperiodic changes in regime manifested by abrupt transitions.

These elementary considerations demonstrate that the ocean–atmosphere system may exhibit a vast range of nontrivial variations without even the presence of external forcing. An external forcing, however, may modulate the amplitude of the frequency of the quasi-periodic oscillations, influence the statistical distribution of the time spent in different regimes (e.g. positive versus negative North Atlantic

Oscillation index), and increase or decrease the transition probability between the different regimes. For example, at least two ocean–atmosphere models show a reduction in the amplitude of eastern tropical Pacific temperature inter-annual variability when mid-Holocene insolation is prescribed (Otto-Bliesner 1999; Liu *et al.* 2000). Clement *et al.* (2000) show that reduced El Niño during the mid-Holocene is consistent with our understanding of the Bjerknes feedback and this also appears to be in agreement with the interpretation of an alluvial record in Ecuador (Rodbell *et al.* 1999). By contrast, analysis of the influence of orbital forcing on the North Atlantic Oscillation is, so far, inconclusive (Gladstone *et al.* 2005).

Accurate modeling of sub-decadal modes of variability requires relatively high-resolution models run over several thousands of years, which is unachievable with current technology. There is therefore a need to separate the modeling of the longer and shorter modes of variability despite the possible interactions linking them. The shortest modes may be parameterized by means of diagnostic equations or stochastic noise in lower-resolution climate models.

Concentrating on long-period modes, two nonlinear terms in the constitutive equations of ocean motion may cause instability or sustained oscillations. These are advection and convection.

Advection designates the transport of heat and salt by the main ocean currents. Given that ocean currents are determined by the distribution of density, it is conceivable that advective processes cause oscillations or instabilities. An important contribution to this theory is Stommel's (1961) "box-model" of the thermohaline circulation made up of four differential equations. This model presents an instability featuring a sharp decrease in advective transport, balanced by an increase in diffusive transport. This transition is more commonly known as the shut-down of the thermohaline circulation. We know that such a shut-down is very unlikely to have occurred during the Holocene.

A refined version of Stommel's model incorporating realistic propagation times suggests that advective processes may also cause sustained oscillations (Welander 1986), which are absent in Stommel's original model. This proposition was further investigated by analyzing the eigenvectors of a three-dimensional ocean model (Weijer and Dijkstra 2003). Three modes of free-damped oscillations were identified that are associated with the propagation of temperature and salinity anomalies through the conveyor belt and their periods range between two and four millennia.

Weijer and Dijkstra suggested that advective oscillations may explain the millenia cycle observed in the North Atlantic record of lithic fraction (Bond *et al.* 1997). Advective oscillations do tend to occur in EMICs (which account for the interactions with the atmosphere and sea-ice), but only for certain parameter ranges. A well-documented example is the 200-year oscillation featured by the ECBILT atmosphere model coupled to a low-resolution three-dimensional ocean with a flat bottom (Weber *et al.* 2004). The 200-year oscillation mode may be efficiently excited and phase-locked by solar forcing, giving rise to a resonant response of the ocean–atmosphere system at the corresponding frequency. The response is manifested by variations in the North Atlantic meridional overturning cell as well as in temperature in the North Atlantic region.

Convection is the vigorous vertical mixing of water that occurs when denser water lies on lighter water. Its role is to restore gravitational stability. It is a self-maintained process: convection acts as a heat pump and prevents formation of sea-ice. The resulting heat exchange between the ocean surface and the atmosphere contributes to the increase in density of the surface water, and this promotes further convection. Deep ocean convection occurs today in the Norwegian Sea, the Labrador Sea, and in the Ross and the Weddell Seas. There is geochemical evidence that convection did not take place in the Labrador Sea before 6000 years ago (Solignac *et al.* 2004).

What may be the effect of deep-ocean convection on climate variability? A conceptual model with two degrees of freedom (Welander 1982) shows that convective instability may induce spontaneous and repeated shut-off and restarts of convection, termed flip-flops. Convective feedback may therefore constitute a plausible explanation for abrupt cooling and warming observed in the Norwegian Sea during the Holocene (Andersen *et al.* 2004).

This hypothesis seems verified in the view of a few long experiments with low-resolution ocean–atmosphere models. A long steady-state experiment with the climate model of the Geophysical Fluid Dynamics Laboratory (GFDL; ocean resolution of about 400 km) exhibits a spontaneous abrupt cooling in the Denmark Strait after about 2000 years of integration. The anomaly, which has no apparent cause, persists for around 30 years (Hall and Stouffer 2001). Similar events were then observed with other models, in particular the ECBILT-CLIO model (ocean resolution of about 300 km but the atmosphere is coarser; Goosse *et al.* 2002). A way to describe these events is to say that climate is trapped for some time in an infrequently visited region of the climatic attractor. Let us call it the "cold state". The trapping process is, in ECBILT-CLIO, the persistence of a cyclonic pressure anomaly around Scandinavia causing southward advection of sea-ice in response to the convection shut-down. This is a trap because the southward advection maintains the convection shut-down and the cyclonic anomaly.

The probability of visiting the cold state may be increased in the presence of external forcing, such as a decrease in solar irradiance (enhances sea-ice formation), a volcanic eruption (Goosse and Renssen 2004), or a freshwater discharge (Renssen *et al.* 2002). This probability also increases during the Holocene because of the gradual cooling of the Norwegian Sea induced by orbital forcing (Figure 4.6).

It is important to realize that the time spent by climate in the "cold state" is stochastic. In other words, the system's response may be analyzed statistically but may not be predicted precisely. If this is correct, the cold events found in diatom-based temperature reconstructions in the sub-Polar Atlantic (Rekjanes Ridge, Andersen *et al.* 2004) cannot be ascribed to a precise forcing event, but they are expected to occur more frequently at the end of the Holocene, after a volcanic eruption or during periods of reduced solar irradiance.

Convective instability is also a potential player in the outstanding negative temperature recorded in Greenland, North Atlantic, and Europe around 8200 years

Norwegian Sea

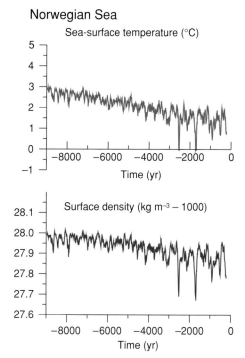

Labrador Sea

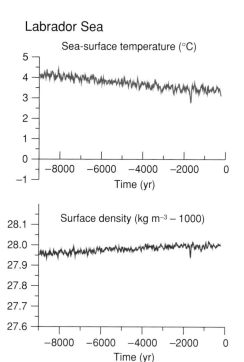

Figure 4.6 Sea-surface temperature and surface water density predicted by the atmosphere–ocean–vegetation climate model of intermediate complexity ECBILT–CLIO–VECODE. Only the orbital forcing is taken into account. Other boundary conditions (including ice sheets and land-sea mask) are as today. The model features a cooling in these two regions but convective regimes exhibit contrasting trends. Convection shut-downs occur increasingly frequently in the Norwegian Sea with, in particular, two sharp cold events around 2400 years BP and 1800 years BP. These events are triggered by stochastic sea-ice advances maintained by cyclonic anomalies in the atmospheric circulation caused by the cooling itself. By contrast, convection is very stable in the Labrador Sea. Surface density increases during the Holocene in response to enhanced heat exchanges between the surface and increasingly cold winds blowing from Canada. (Data are from Renssen *et al.* 2005.)

ago (von Grafenstein *et al.* 1998). It is intriguing that the cooling persisted more than a century (Thomas *et al.* 2007) but its hypothetical cause, an outburst of a proglacial lake complex shortly termed Lake Agassiz, occurred in a few years at most (Teller *et al.* 2002). In fact, such "cold-state trapping" explains that a given amount of freshwater discharged in the Labrador Sea generally induces a more persistent cooling (up to 500 years) in ECBILT–CLIO when it is released in 10 years (this triggers convective instability) than in 50 years (Renssen *et al.* 2002).

Experiments with the CLIMBER-2 EMIC (Bauer *et al.* 2004) also show that a freshwater pulse of a size compatible with geologic evidence may lead the ocean circulation to a "cold state", with convection persisting south of 60°N only. As in Renssen *et al.* (2002), the cold state is metastable, which means that it is resilient to small perturbations but unstable to larger ones. As a consequence, the ocean may remain in this cold state for one or two centuries depending on the natural atmospheric variability. Note that this variability is parameterized under the form of white noise in CLIMBER-2.3 because it is not resolved explicitly .

LeGrande *et al.* (2006) provides a further argument to the Lake Agassiz hypothesis. "Model E" of the Goddard Institute for Space Studies (GISS) explicitly simulates a range of variables (dust, calcite, ocean and atmospheric water oxygen isotopic ratio, [10]Be ratios, methane emissions) and experiments that simulate the outburst are in broad agreement with the relevant climate records.

At last, a cautionary tale. So far, spontaneous transitions and persistent cooling due to the convective feedback were simulated in models that do not resolve the details of the convective process. A model resolving spatial scales of a few tens of kilometers would be more suitable to study the properties of convective instability in the presence of long-term atmospheric variability, but this presently would be very technologically demanding.

Conclusions

Understanding the Holocene climate record is an exceptional challenge because of the countless possible interactions between components characterized by hugely different spatial and temporal scales. A spectrum of climate models is therefore required. Conceptual models made of a few equations are useful to formulate general hypotheses. Earth models of intermediate complexity and comprehensive models are needed to study mechanisms of climate change in more detail.

The Holocene is a time during which no major global change in climate regime occurred. In fact, the astronomical theory of paleoclimates predicts no glacial inception before the next 50 000 years. We do not exclude the possibility, however, that climate chose the trajectory of a long interglacial by chance, or perhaps was helped by a small kick from anthropogenic activities as early as 6000 years ago. A short Holocene with a currently ongoing glacial inception was another option that is now definitively excluded given the present atmospheric carbon dioxide concentrations.

Although the Holocene is globally stable, regional changes in regime occurred. Detection of local climate instability is naturally highly relevant in the context of global warming. The Holocene records provide the opportunity to do so. Claussen (this volume) describes abrupt transitions associated with land-surface feedbacks. Here, we discussed ocean circulation instabilities. Sudden coolings in northern Europe and Greenland may be ascribed to stops or drastic reduction of deep-ocean convection in the North Atlantic. Analysis of the dynamics of the convective feedback and interactions with sea-ice and atmosphere dynamics suggest that climate may be "trapped" in this cold state for several decades to several centuries, the exact duration being ruled by the stochastic character of the climate system. This relatively new theory certainly deserves further consideration.

Acknowledgments

The author thanks the HOLIVAR Steering Committee for the invitation to the open science meeting held in London, June 2006. Thanks are also due to Martin Claussen (Max Planck Institute of Meteorology, Germany), Rick Battarbee and Heather Binney (editors) and two reviewers for constructive comments and suggestions. Jean-Yves Petershmitt (CEA, Saclay, France) produced Figure 4.3 using data supplied by different modeling teams (NCAR, Hadley Centre, Japan Meteorological Research Institute, IPSL, Hadley Centre, Bristol University, Université Catholique de Louvain and Chinese Academy of Sciences). Marie-France Loutre (Université Catholique de Louvain, Belgium) supplied Figure 4.4 while Hans Renssen (University of Amsterdam, Netherlands) provided the data for Figure 4.6. Numerous colleagues helped with the literature review and some of them provided figures that unfortunately did not all make it in the final version: Victor Brovkin (PIK, Potsdam), Nanne Weber (KNMI, Netherlands), Johannes Oerlemans (Utrecht University, Netherlands) and Allegra Legrande (GISS, New York, USA). Part of this manuscript was written when the author was affiliated to the Meteorological Office Hadley Centre, UK. He is now funded by the Belgian National Fund of Scientific Research.

References

Alley R.B., Marotzke J., Nordhaus W.D., *et al.* (2003) Abrupt climate change. *Science*, **299**, 2005–2010, doi:10.1126/science.1081056.

Andersen C., Koç N. & Moros M. (2004) A highly unstable Holocene climate in the subpolar North Atlantic: evidence from diatoms. *Quaternary Science Reviews*, **23**, 2155–2166.

Annan J.D., Hargreaves J.C., Ohgaito R., Abe-Ouchi A. & Emori S. (2005) Efficiently constraining climate sensitivity with ensembles of paleoclimate experiments. *SOLA*, **1**, 181–184, doi:10.2151/sola.2005-047.

Archer D. & Ganopolski, A. (2005) A movable trigger: fossil fuel CO_2 and the onset of the next glaciation. *Geochemistry, Geophysics, Geosystems*, **6**, Q05003.

Bauer E., Ganopolski A. & Montoya M. (2004) Simulation of the cold climate event 8200 years ago by meltwater outburst from Lake Agassiz. *Paleoceanography*, **19**, PA3014, doi:10.1029/2004PA001030.

Berger A.L. (1978) Long-term variations of daily insolation and quaternary climatic. *Journal of Atmospheric Science*, **35**, 2362–2367.

Bond G., Shower W., Cheseby N., *et al.* (1997) A pervasive millennial-scale cycle in North Atlantic Holocene and glacial climate. *Science*, **278**, 1257–1264.

Braconnot P., Harrison S.P., Joussaume S., *et al.* (2004) Evaluation of PMIP coupled ocean–atmosphere simulations of the Mid-Holocene. In: *Past Climate Variability through Europe and Africa* (Eds R.W. Battarbee, F. Gasse & C.E. Stickley), pp. 515–533. Springer-Verlag, Berlin.

Braconnot P., Otto-Bliesner B.L., Harrison S., *et al.* (2007) Results of PMIP2 coupled simulations of the Mid Holocene and Last Glacial Maximum – part 1: experiments and large-scale features. *Climate of the Past*, **3**, 261–277.

Broecker W. & Stocker T.F. (2006) The Holocene CO_2 rise: anthropogenic or natural? *Eos (Transactions of the American Geophysical Union)*, **87**, 27.

Brovkin V., Bendtsen J., Claussen M., Kubatzki C., Petoukhov V. & Andreev A. (2002) Carbon cycle, vegetation, and climate dynamics in the Holocene: Experiments with the CLIMBER-2 model. *Global Biogeochemical Cycles*, **16**, 1139, doi:10.1029/2001GB001662.

Budyko M.I. (1969) The effect of solar radiation variations on the climate of the Earth. *Tellus*, **21**, 611–619.

Calov R., Ganopolski A., Claussen M., Petoukhov V. & Greve R. (2005) Transient simulation of the last glacial inception. Part I: glacial inception as a bifurcation in the climate system. *Climate Dynamics*, **24**, 545–561, doi:10.1007/s0382-005-0007-6.

Cane M.A., Braconnot P., Clement A., *et al.* (2006) Progress in paleoclimate modeling. *Journal of Climate*, **19**, 5031–5057.

Claussen M. (this volume) Holocene rapid land-cover changes – evidence and theory. In: *Global Warming and Natural Climate Variability: a Holocene Perspective* (Eds R.W. Battarbee & H.A. Binney), pp. 232–253, Blackwell, Oxford.

Claussen M., Kubatzki C., Brovkin V., Ganopolski A., Hoelzmann P. & Pachur H.-J. (1999) Simulation of an abrupt change in Saharan vegetation in the mid-Holocene. *Geophysical Research Letters*, **26**, 2037–2040.

Claussen M., Mysak L., Weaver A., *et al.* (2002) Earth system models of intermediate complexity: closing the gap in the spectrum of climate system models. *Climate Dynamics*, **18**, 579–586.

Claussen M., Brovkin V., Calov R., Ganopolski A. & Kubatzki C. (2005) Did humankind prevent a Holocene glaciation? *Climate Change*, **69**, 409–417.

Clement A.C., Seager R. & Cane M.A. (2000) Suppression of El-Niño during the mid-Holocene by changes in the earth's orbit. *Paleoceanography*, **15**, 731–737.

Crucifix M. & Berger A. (2006) How long will our interglacial be? *Eos (Transactions of the American Geophysical Union)*, **87**, 352.

Crucifix M. & Loutre M.F. (2002) Transient simulations over the last interglacial period (126–115 kyr BP): feedback and forcing analysis. *Climate Dynamics*, **19**, 419–433.

Crucifix M., Loutre M.F., Tulkens P., Fichefet T. & Berger A. (2002) Climate evolution during the Holocene: a study with an Earth system model of intermediate complexity. *Climate Dynamics*, **19**, 43–60.

Gallée H., van Ypersele J.P., Fichefet T., Tricot C. & Berger A. (1991) Simulation of the last glacial cycle by a coupled, sectorially averaged climate–ice sheet model. Part I: The climate model. *Journal of Geophysical Research*, **96**, 13139–13161.

Gallée H., van Ypersele J.P., Fichefet T., Marsiat I., Tricot C. & Berger A. (1992) Simulation of the last glacial cycle by a coupled, sectorially averaged climate–ice sheet model. Part II: response to insolation and CO_2 variation. *Journal of Geophysical Research*, **97**, 15713–15740.

Ganopolski A., Kubatzki C., Claussen M., Brovkin V. & Petoukhov V. (1998) The influence of vegetation-atmosphere-ocean interaction on climate during the mid-Holocene. *Science*, **280**, 1916–1919.

Giorgi F. & Mearns, L.O. (1999) Introduction to special section: Regional climate modeling revisited. *Journal of Geophysical Research*, **104**, 6335–6352.

Gladstone R.M., Ross I., Valdes P.J., *et al.* (2005) Mid-Holocene NAO: a PMIP2 model intercomparison. *Geophysical Research Letters*, **32**, L16707, doi:10.1029/2005GL023596.

Goosse H. & Renssen H. (2004) Exciting natural modes of variability by solar and volcanic forcing: idealized and realistic experiments. *Climate Dynamics*, **23**, 153–163, doi:10.1007/s00382-004-0424-y.

Goosse H., Renssen H., Selten F.M., Haarsma R.J. & Opsteegh J.D. (2002) Potential causes of abrupt climate events: a numerical study with a three-dimensional climate model. *Geophysical Research Letters*, **29**, 1860, doi:10.1029/2002GL014993.

Goosse, H., Mann, M.E. & Renssen, H. (this volume) Climate of the past millennium: combining proxy data and model simulations. In: *Global Warming and Natural Climate Variability: a Holocene Perspective* (Eds R.W. Battarbee & H.A. Binney), pp. 163–188. Blackwell, Oxford.

Hall A. & Stouffer R.J. (2001) An abrupt climate event in a coupled ocean–atmosphere simulation without external forcing. *Nature*, **409**, 171–174.

Hargreaves J.C. & Annan J.D. (2002) Assimilation of paleo-data in a simple Earth system model. *Climate Dynamics*, **19**, 371–381.

Harrison S.P., Kutzbach J.E., Liu Z., *et al.* (2003) Mid-Holocene climates of the Americas: a dynamical response to changes in seasonality. *Climate Dynamics*, **20**, 663–688.

Harvey L.D.D. (1988) On the role of high latitude ice, snow and vegetations feedbacks in the climatic response to external forcing changes. *Climate Change*, **13**, 191–224.

Haxeltine A. & Prentice I.C. (1996) BIOME3: An equilibrium terrestrial biosphere model based on ecophysiological constraints, resource availability, and competition among plant functional types. *Global Biogeochemical Cycles*, **10**, 693–709.

Hilborn R. (2001) *Chaos and Nonlinear Dynamics. An Introduction for Scientists and Engineers*, 2nd edn. Oxford University Press.

Imbrie J. & Imbrie J.Z. (1980) Modelling the climatic response to orbital variations. *Science*, **207**, 943–953.

Imbrie J.J., Hays J.D., Martinson D.G., *et al.* (1984) The orbital theory of Pleistocene climate: Support from a revised chronology of the marine δO^{18} record. In: *Milankovitch and Climate*, Part I (Eds A. Berger, J. Imbrie, J. Hays, J. Kukla, and B. Saltzman), pp. 269–305. D. Reidel, Norwell, MA.

Johns T.C., Durman C.F., Banks H.T., *et al.* (2006) The new Hadley Centre climate model HadGEM1: Evaluation of coupled simulations. *Journal of Climate*, **19**, 1327–1353.

Jones, J.M. & Widmann, M. (2004) Reconstructing large-scale variability from palaeoclimatic evidence by means of Data Assimilation Through Upscaling and Nudging (DATUN). In: *The KIHZ Project: Towards a Synthesis of Holocene*

Proxy Data and Climate Models (Eds H. Fischer *et al.*), pp. 171–193, Springer-Verlag, Heidelberg.

Joos F., Gerber S., Prentice I.C., Otto-Bliesner B.L. & Valdes P.J. (2004) Transient simulations of Holocene atmospheric carbon dioxide and terrestrial carbon since the Last Glacial Maximum. *Global Biogeochemical Cycles*, **18**, GB2002, doi:10.1029/2003GB002156.

Joussaume S. & Jouzel J. (1993a) Paleoclimatic tracers: An investigation using an atmospheric general circulation model under ice age conditions. 1. Desert dust. *Journal of Geophysical Research*, **98**, 2767–2805.

Joussaume S. & Jouzel J. (1993b) Paleoclimatic tracers: An investigation using an atmospheric general circulation model under ice age conditions. 2. Water isotopes. *Journal of Geophysical Research*, **98**, 2807–2830.

Joussaume S. & Taylor K.E. (2000) The paleoclimate modelling intercomparison project. In: *Paleoclimate Modelling Intercomparison Projects* (Ed. P. Braconnot), pp. 7–24. Volume WCRP-111, Technical Document No. 1007, World Meteorological Organization, Geneva.

Kutzbach J. & Otto-Bliesner B.L. (1982) The sensitivity of the Afrian and Asian monsoon climate to orbital parameter changes for 9,000 years B.P. in a low-resolution general circulation model. *Journal of Atmospheric Science*, **39**, 1177–1188.

LeGrande A.N., Schmidt G.A., Shindell D.T., *et al.* (2006) Consistent simulations of multiple proxy responses to an abrupt climate change event. *Proceedings of the National Academy of Science USA*, **103**, 837–842.

Leung L.-Y. & North. G.R. (1990) Information theory and climate prediction. *Journal of Climate*, **3**, 5–14.

Liu Z., Kutzbach J. & Wu L. (2000) Modeling the climate shift of El-Niño variability in the Holocene. *Geophysical Research Letters*, **27**, 2265–2268.

Liu Z., Brady E. & Lynch-Stieglitz J. (2002) Global ocean response to orbital forcing in the Holocene. *Paleoceanography*, **18**, 1041, doi:10.1029/2002PA000819.

Lorenz E.N. (1963) Deterministic non-periodic flow. *Journal of Atmospheric Science*, **20**, 130–141.

Loutre M.F. & Berger A. (2003) Marine Isotope Stage 11 as an analog for the present interglacial. *Global Planetary Change*, **36**, 209–217.

Loutre M.F., Paillard D., Vimeux F. & Cortijo E. (2004) Does mean annual insolation have the potential to change the climate? *Earth and Planetary Science Letters*, **221**, 1–14.

Loutre M.F., Berger A., Crucifix M., Desprat S. & Sánchez-Goñi M.F. (2007) Interglacials as simulated by the LLN-2D NH and MoBidiC climate models. In: *The Climate of Past Interglacials* (Eds F. Sirocko, M. Claussen, M.F. Sánchez Goñi & T. Litt), pp. 547–582. Elsevier, Amsterdam.

Marchal O., Stocker T.F., Joos F., Indermuhle A., Blunier T. & Tschumi J. (1999) Modelling the concentration of atmospheric CO_2 during the Younger Dryas climate event. *Climate Dynamics*, **15**, 341–354.

Masson-Delmotte V., Dreyfus G., Braconnot P., *et al.* (2006) Past temperature reconstructions from deep ice cores: relevance for future climate change. *Climate of the Past*, **2**, 145–165.

Mitchell J.M.M. (1976) An overview of climatic variability and its causal mechanisms. *Quaternary Research*, **6**, 481–494.

Murphy J.M. (1998) An evaluation of statistical and dynamical techniques for downscaling local climate. *Journal of Climate*, **12**, 2256–2284.

Murphy J.M., Sexton D.M.H., Barnett D.N., *et al.* (2004) Quantification of modelling uncertainties in a large ensemble of climate change simulations. *Nature*, **430**, 768–772.

Oldfield F. (this volume) The role of people in the Holocene. In: *Global Warming and Natural Climate Variability: a Holocene Perspective* (Eds R.W. Battarbee & H.A. Binney), pp. 58–97. Blackwell, Oxford.

Otto-Bliesner B.L. (1999) El Niño/La Niña and Sahel precipitation during the middle Holocene. *Geophysical Research Letters*, **26**, 87–90.

Paillard D. (2001) Glacial cycles: toward a new paradigm. *Review of Geophysics*, **39**, 325–346.

Petit J.R., Jouzel J., Raynaud D., *et al.* (1999) Climate and atmospheric history of the past 420 000 years from the Vostok ice core, Antarctica. *Nature*, **399**, 429–436.

Petoukhov V., Ganopolski A., Brovkin V., *et al.* (2000) CLIMBER-2: a climate system model of intermediate complexity. Part I: model description and performance for present climate. *Climate Dynamics*, **16**, 1–17.

Raynaud D., Barnola J.-M., Souchez R., *et al.* (2005) The record for marine isotopic stage 11. *Nature*, **463**, 39–40.

Renssen H., Goosse H., Fichefet T. & Campin J.M. (2001) The 8.2 kyr BP event simulated by a global atmosphere-sea-ice-ocean model. *Geophysical Research Letters*, **28**, 1567–1570.

Renssen H., Goosse H. & Fichefet T. (2002) Modeling the effect of freshwater pulses on the early Holocene climate: The influence of high-frequency climate variability. *Paleoceanography*, **17**, doi:10.1029/2001PA000649.

Renssen H., Braconnot P., Tett S.F.B., von Storch H. & Weber S.L. (2004) Recent developments in Holocene climate modelling. In: *Past Climate Variability through Europe and Africa* (Eds R.W. Battarbee, F. Gasse & C.E. Stickley), pp. 495–514. Springer-Verlag, Berlin.

Renssen H., Goosse H. & Fichefet T. (2005) Contrasting trends in north Atlantic deep-water formation in the Labrador Sea and Nordic Seas during the Holocene. *Geophysical Research Letters*, **32**, L08711, doi:10.1029/2005GL022462.

Rignot E. & Thomas R.H. (2002) Mass balance of polar ice sheets. *Science*, **297**, 1502–1506.

Rodbell D.T., Seltzer G.O., Anderson D.M., Abbott M.B., Enfield D.B. & Newman J.H. (1999) A 15 000-year record of El-Niño-driven alluviation in southwestern Ecuador. *Science*, **283**, 516–520, doi:10.1126/science.283.5401.516.

Rougier J. (2006) Probabilistic inference for future climate using an ensemble of climate model evaluations. *Climate Change*, **81**, 247–264.

Ruddiman W.F. (2003) The anthropogenic greenhouse era began thousands of years ago. *Climate Change*, **61**, 261–293.

Ruddiman W.F. (2005) The early anthropogenic hypothesis a year later – an editorial reply. *Climate Change*, **69**, 427–434.

Ruddiman, W.F. (2006) On "The Holocene CO_2 rise: anthropogenic or natural?", *Eos (Transactions of the American Geophysical Union)*, **87**, 352–353.

Ruddiman, W.F. (2007) The early anthropogenic hypothesis: challenges and responses. *Review of Geophysics*, **45**, RG4001, doi:10.1029/2006RG000207.

Ruddiman W.F., Vavrus S.J. & Kutzbach J.E. (2005) A test of the overdue-glaciation hypothesis. *Quaternary Science Reviews*, **24**, 1–10.

Saltzman B. (1988) Modelling the slow climatic attractor. In: *Physically-based Modelling and Simulation of Climate and Climatic Change*, Part 2 (Ed. M.E. Schlesinger), pp. 737–754. NATO Advanced Science Institutes Series, Kluwer, Dordrecht.

Saltzman B. (2002) *Dynamical Paleoclimatology*. International Geophysics Series, Vol. 80, Academic Press, San Diego, 354 pp.

Saltzman B. & Maasch K.A. (1990) A first-order global model of late Cenozoic climate. *Transactions of the Royal Society Edinburgh Earth Science*, **81**, 315–325.

Schneider von Deimling T.S., Held H., Ganopolski A. & Rahmstorf S. (2006) Climate sensitivity estimated from ensemble simulations of glacial climate. *Climate Dynamics*, **27**, 149–163.

Siegenthaler U., Stocker T.F., Monnin E., *et al.* (2005) Stable carbon cycle–climate relationship during the Late Pleistocene. *Science*, **25**, 1313–1317.

Sirocko F., Claussen M., Sánchez Goñi M.F. & Litt T. (Eds) (2007) *The Climate of Past Interglacials*. Developments in Quaternary Science, Vol. 7, Elsevier, Amsterdam, 622 pp.

Solignac S., de Vernal A. & Hillaire-Marcel C. (2004) Holocene sea-surface conditions in the North Atlantic: contrasted trends and regimes in the western and eastern sectors (Labrador Sea vs. Iceland Basin). *Quaternary Research*, **23**, 319–334.

Stein U. & Alpert P. (1993) Factor separation in numerical simulations. *Journal of Atmospheric Science*, **50**, 2107–2115.

Stommel H. (1961) Thermohaline convection with two stable regimes of flow. *Tellus*, **13**, 224–230.

Tarasov L. & Peltier W.R. (1999) Impact of thermomechanical ice sheet coupling on a model of the 100 kyr ice age cycle. *Journal of Geophysical Research*, **104**, 9517–9545.

Teller J.T., Leverington D.W. & Mann J.D. (2002) Freshwater outbursts to the oceans from glacial Lake Agassiz and their role in climate change during the last deglaciation. *Quaternary Science Reviews*, **21**, 879–887.

Thomas E.R., Wolff E.W., Mulvaney R., *et al.* (2007) The 8.2 ka event from Greenland ice cores. *Quaternary Science Reviews*, **26**, 1–7.

Van der Schrier G. & Barkmeijer, J. (2005) Bjerknes' hypothesis on the coldness during AD 1790–1820 revisited. *Climate Dynamics*, **25**, 537–553.

Von Grafenstein U., Erlenkeuser H., Muller J., Jouzel J. & Johnsen S. (1998) The cold event 8200 years ago documented in oxygen isotope records of precipitation in Europe and Greenland. *Climate Dynamics*, **14**, 73–81.

Wang Y. & Mysak L.A. (2005) Response of the ocean, climate and terrestrial carbon cycle to Holocene freshwater discharge after 8 kyr BP. *Geophysical Research Letters*, **32**, L15705, doi:10.1029/2005GL023344.

Wang Y., Mysak L.A. & Roulet N.T. (2005a) Holocene climate and carbon cycle dynamics: Experiments with the "green" McGill Palaeoclimate Model. *Global Biogeochemical Cycles*, **19**, GB3022, doi:10.1029/2005GB002484.

Wang Y., Mysak L.A., Wang Z. & Brovkin V. (2005b) The greening of the McGill paleoclimate model; Part ii: Simulation of Holocene millennial-scale natural climate changes. *Climate Dynamics*, **24**, 481–496, doi:10.1007/s00382-004-0516-8.

Weber S.L. & Oerlemans J. (2003) Holocene glacier variability: three case studies using an intermediate-complexity model. *The Holocene*, **13**, 353–363.

Weber S.L., Crowley T.J. & van der Schrier G. (2004) Solar irradiance forcing of centennial climate variability during the Holocene. *Climate Dynamics*, **22**, 539–553, doi:10.1007/s00382-004-0396-y.

Weijer W. & Dijkstra H.A. (2003) Multiple oscillatory modes of the global ocean circulation. *Journal of Physical Oceanography*, **33**, 2197–2213.

Welander P. (1982) A simple heat–salt oscillator. *Dynamic Atmosphere and Oceans*, **6**, 233–242.

Welander P. (1986) Thermohaline effects in the ocean circulation and related simple models. In: *Large-scale Transport Processes in Oceans and Atmosphere* (Eds J. Willebrand & D.L.T. Anderson), pp. 163–200. Reidel, Dordrecht.

Wohlfahrt J., Harrison S.P. & Braconnot P. (2004). Synergistic feedbacks between ocean and vegetation on mid- and high-latitude climates during the mid-Holocene. *Climate Dynamics*, **22**, 223–238.

Wright Jr. H.E., Kutzbach J.E., Webb III T., Ruddiman W.F., Street-Perrot F.A. & Bartlein P.J (Eds) (1993) *Global Climates since the Last Glacial Maximum*. University of Minnesota Press, London, 569 pp.

5 The early to mid-Holocene thermal optimum in the North Atlantic

Eystein Jansen, Carin Andersson, Matthias Moros,
Kerim H. Nisancioglu, Birgitte F. Nyland
and Richard J. Telford

Keywords

Holocene, climate variability, ocean climate, North Atlantic, Nordic Seas

Introduction

The existence of an early to mid-Holocene thermal optimum is frequently referred to in the paleoclimate science literature. (Early Holocene is here loosely defined as 11–8 ka, and mid-Holocene is loosely defined as 8–4 ka, both in terms of calendar years BP.) The thermal optimum has been documented by a variety of paleoclimatic evidence, spanning many Northern Hemisphere high-latitude paleoclimatic archives, including ice cores, marine and limnic sediments, glacier records, and speleothems. A comprehensive overview is given in the *4th Assessment Report of the Intergovernmental Panel on Climate Change Working Group 1* (chapter 6, figure 6.9; Jansen *et al.* 2007). Based on alkenone data, Kim *et al.* (2004) found that the thermal optimum is most pronounced at high northern latitudes in the oceans, and that in low-latitude areas it is less pronounced or nonexistent. In some low-latitude areas, such as the Indian Ocean and parts of the tropical Atlantic, the period of the thermal optimum (8–6 ka) as recorded in the high-latitude North Atlantic is characterized by colder temperatures than in the late Holocene. Some of the sites providing evidence for a colder early to mid-Holocene are from upwelling regions; hence some of the cold anomalies may be related to enhanced wind-driven upwelling. There appears to be a poleward amplification of the thermal anomaly, as well as a clear tendency that the warmest temperature interval started earlier and lasted for a shorter period towards the Arctic than further south, where it often appeared as a warm phase between about 8 to 6 ka (Koç *et al.* 1994; Sarnthein *et al.* 2003; Duplessy *et al.* 2005; Hald *et al.* personal communication).

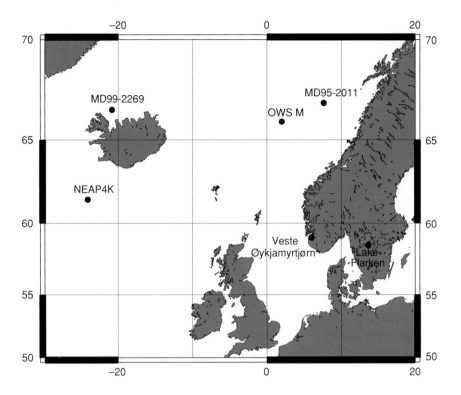

Figure 5.1 Map showing marine and terrestrial core sites.

Another feature is a delayed thermal maximum in the Labrador–Irminger Seas south of Iceland, where peak Holocene temperatures occurred at about 6 ka. This feature is probably related to the late demise of the Laurentide Ice Sheet (Andersen *et al.* 2004; Hall *et al.* 2004; Berner 2006). Many authors have attributed the thermal optimum to the combined effects of the demise of the continental ice sheets, removing their influence on temperature distribution, and higher summer solar irradiance than today due to orbital factors, especially the tilt of the Earth's axis, which provided a positive summertime radiative forcing anomaly in high-latitude regions in both hemispheres.

Figure 5.1 shows selected localities where examples of the thermal optimum have been demonstrated (Figure 5.2). Figure 5.3 summarizes the marine 6 ka thermal anomaly in the Mediterranean and the north-east Atlantic–Nordic Seas by using a data-set from alkenones only, thereby reducing errors when combining different proxy methods. This figure also shows the northward amplification of the temperature anomaly.

Despite these findings, however, there are a number of North Atlantic and Nordic Seas records where the thermal optimum is not apparent, and in some there is instead a tendency for a warming throughout the Holocene

In this chapter we analyze these contrasting records in the high-latitude ocean with the aim of attributing the contrasts to the forcing and/or to dynamics of the ocean–atmosphere system, and we evaluate the interplay between the long-term responses due to orbital forcing and shorter century to millennial scale variability.

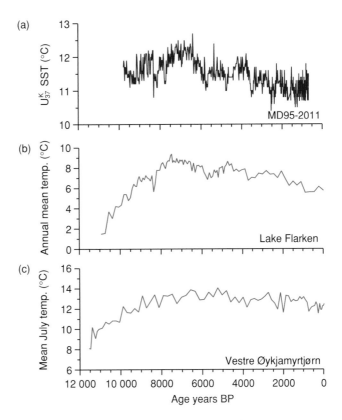

Figure 5.2 Examples of records displaying a mid-Holocene thermal maximum. (a) UK37 data from MD95-2011 (Calvo *et al.* 2002). (b) Pollen-based annual mean temperature reconstruction from Lake Flarken (Seppä *et al.* 2005). (c) Pollen-based reconstruction of mean July temperature from Vestre Øykjamyrtjørn (Folgefonna peninsula; Bjune *et al.* 2005).

Orbital forcing and climate model response

The effect of orbital forcing on sea-surface temperatures (SSTs) at 6 ka compared with pre-industrial conditions has been studied in the Paleoclimate Modeling Intercomparison Project PMIP2 with coupled ocean–atmosphere climate models (AOGCMs) (see Crucifix this volume, Figure 4.3). Similar simulations have also been performed with other models, including models of intermediate complexity (e.g. Liu *et al.* 2003; Rimbu *et al.* 2004; Renssen *et al.* 2005). A consistent feature of these simulations is the positive summertime SST anomaly in the North Atlantic–Nordic Seas area. The SST response of the MIT-EMIC (Michigan Institute of Technology – Earth System Model of Intermediate Complexity) to the 6 ka orbital forcing compared with the response of the model to pre-industrial forcing, i.e. the response solely due to changing orbital parameters, is shown in Figure 5.4. The response is strongest in high latitudes, with very little response in the tropics, and the most pronounced warming occurs near the Polar Front in the North Atlantic–Nordic Seas. This poleward amplification in the Atlantic sector is consistent with observational evidence (Figure 5.3; data from Kim *et al.* 2004; Kim and Schneider 2004a,b). The degree to which the models are able to separate surface and thermocline temperatures and the seasonal response is variable. The

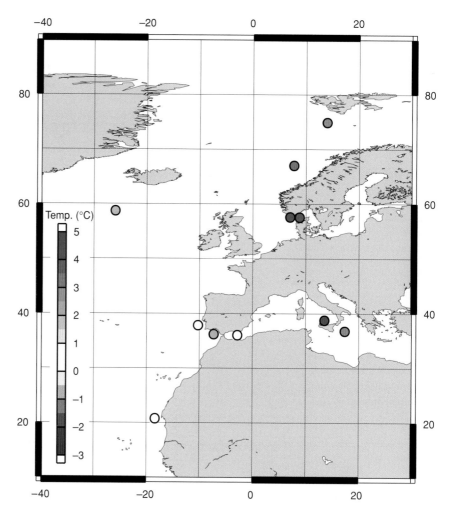

Figure 5.3 6−0 ka alkenone sea-surface temperature differences for selected North Atlantic and Mediterranean cores. (Data from Kim and Schneider 2004a,b.)

imprint of the seasonal forcing on the ocean sub-surface is clearly different between models and observations, as noted in the following section. This implies that for climate models to represent the true ocean sub-surface and thermocline variability in response to the forcing, a better representation of high-latitude surface ocean stratification is required.

Ocean records and responses to orbital forcing

Sea-surface-temperature reconstructions

Reconstructions of surface layer temperatures of the high-latitude North Atlantic and Nordic Seas through the Holocene have been obtained from various proxy methods. Transfer functions have been produced based on data from alkenones,

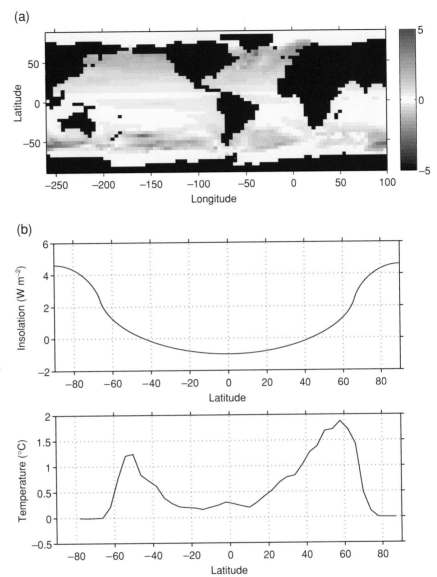

Figure 5.4 Annual mean response in a simulation with the Massachusetts Institute of Technology Earth System model of intermediate complexity (Dutkiewicz *et al.* 2005) using 6 ka boundary conditions. (a) 6 ka sea-surface temperature (SST) increase compared with pre-industrial (AD 1850) SSTs. (b) Annual mean insolation change relative to pre-industrial (upper panel), and annual mean SST change relative to pre-industrial.

coccolith species, diatom, and forminifers, as well as planktonic foraminiferal oxygen isotope records and some Mg/Ca records, for parts of the Holocene. Multiple proxies have been analyzed in some cores, enabling a direct comparison of the output from the different methods without the errors of temporal correlation. Core MD95-2011, obtained from beneath the main flow of warm Atlantic water towards the Arctic, is a good example (Figure 5.5). The Holocene temperature development reconstructed by the different proxy methods is very different, both in terms of the amplitude of millennial and shorter time-scale variability, and for overall Holocene trends. The proxies cluster into two distinct categories: those

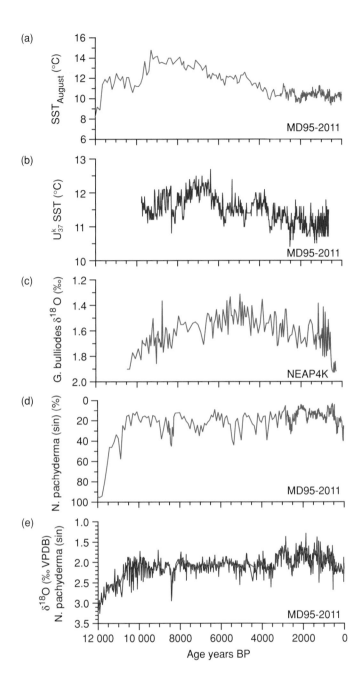

Figure 5.5 Comparison between sea-surface temperature reconstructions from: (a) alkenones (Calvo *et al.* 2002); (b) diatoms in MD95-2011 (Andersen *et al.* 2004; Koc personal communication); (c) planktonic stable oxygen isotope data from NEAP4K (Hall *et al.* 2004); (d) stable oxygen isotope data and (e) planktonic census data from MD95-2011 (Andersson *et al.* 2003; Risebrobakken *et al.* 2003).

with a distinct Holocene thermal maximum (diatom-based SSTs, alkenones); and foraminiferal-based proxies (transfer functions, species distribution, oxygen isotope data corrected for ice-volume effects), which do not show an early to mid-Holocene maximum. Instead, these proxies show the opposite, a warming trend towards the late Holocene and more high-amplitude changes at century–millennial time-scales after about 4 ka. The close correspondence between the independent foraminiferal-based proxies indicates that the general pattern of the foraminifer-based temperature series is robust. Likewise, there is no a priori reason to distrust the fidelity of the diatom- and alkenone-based reconstructions. They are consistent with summer season orbital forcing and consistent with other surface ocean data from the region sampled from different cores, from ice-core data, ice-core borehole data, and terrestrial records.

We must assume that both categories (i.e. with and without a pronounced early to mid-Holocene thermal optimum) of temperature histories are valid and reflect different aspects of the Holocene thermal evolution of the high-latitude ocean. To understand why the difference has developed, the seasonality of the forcing, the habitat of the biological proxy indicators, and the seasonality of the vertical structure of the upper layers of the high-latitude ocean must be considered. The main character of the orbital forcing of the early versus late Holocene is a strong high-latitude early to mid-Holocene positive thermal anomaly during the summer season, due to the combined effects of tilt and precession (see Crucifix this volume, Figure 4.4). A significant, yet smaller, negative anomaly occurred in the winter season, but is more confined to low and mid-latitudes.

The result is an enhanced seasonality in the forcing and in the expected ocean response as compared with the late Holocene. The discrepancy between the two categories of temperature series might be explained as a response to the seasonality of the forcing. Yet it remains to be argued why the foraminiferal-based records should respond to the winter-time or annual mean forcing and not the summer-time forcing, when we know that the primary foraminifer production season is in the spring and summer. It is, however, well known that different species of foraminifers have their main habitat at different depths in the ocean surface layer. In contrast to diatoms and alkenone producing algae, which photosynthesize and live in the euphotic zone of the upper 50 m of the ocean, planktonic foraminifers are zooplankton and are often found at the depths of the main food sources near the thermocline at the base of the surface layer. Risebrobakken *et al.* (2003) showed that both the left- and right-coiling varieties of *Neogloboquadrina pachyderma* share this pattern, i.e. showing a lack of evidence for the Holocene thermal optimum, yet there is a difference between their O-isotope values, with the left-coiling variety indicating colder temperatures than the right-coiling variety.

The seasonal forcing of surface layer temperatures provides strong contrasts between the upper 50 m and deeper layers. Temperature data from Ocean Weather Station M in the Norwegian Sea (Figure 5.6) shows that the summer solar insolation warms and stratifies the surface layer, imposing a strong heating in the upper 50 m, whereas the temperature of the thermocline is unrelated to the summer season, and instead reflects the winter-time ventilation of the whole surface layer when the summer season stratification breaks down. Thus at 100 m, summer temperatures are set by the winter-time situation and reflect this or the annual mean temperatures. As can been seen in the hydrographic data (Figure 5.6), a temperature proxy derived from organisms that thrive below approximately 50 m in these waters will not register any spring or summer-time warming, even though they have their largest standing stock during the summer season. Enhanced solar

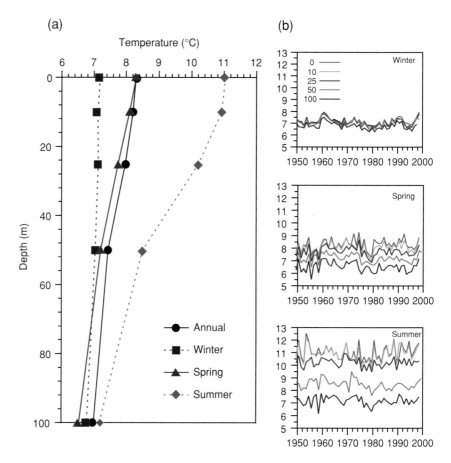

Figure 5.6 (a) Seasonal and annual water temperature averages for the last 50 years from Ocean Weather Ship Mike in the upper 100 m of the water column. The averages are based on the temperature records presented in (b). (From (Nyland *et al.* 2006.)

insolation, as was the case for the early to mid-Holocene, would exacerbate this feature, stabilize the surface layer in the warm season, and produce a stronger contrast between the euphotic zone proxies and those derived from thermocline species.

For the thermal maximum this implies that the strong solar insolation maximum of the early to mid-Holocene originating from summer season orbital forcing did not manifest itself below the euphotic zone, and was only captured by the temperature reconstructions derived from phytoplankton with an upper surface ocean habitat. The difference between the alkenone- and diatom-based SST reconstructions is most likely due to the calibration of the methodologies. The diatom transfer functions are calibrated to surface August temperatures, whereas alkenones are mainly produced in the spring season and calibrated to the mean annual temperatures of the uppermost surface layer. Those foraminiferal species (e.g. *Globigerina bulloides*) that normally calcify above the summer-time thermocline would be expected to show the same pattern as phytoplankton-based proxies. For foraminiferal-based transfer functions the higher degree of stratification will give a mixed response depending on the blend of species with surface or thermocline habitat preference. In the Nordic Seas, thermocline species dominate the foraminiferal assemblage and the resulting transfer function temperatures are consistent with the O-isotope-based records (Andersson *et al.* 2003; Risebrobakken *et al.* 2003). As a consequence of these findings, the wide use of *N. pachyderma*, as well as foraminiferal transfer functions in paleoceanographic studies from high-latitude areas, should be qualified with the information that the reconstruction is primarily that of temperature in the thermocline, and thus reflects winter-time or annual mean temperatures.

Dynamical responses or radiative forcing only?

So far we have seen that the seasonal character of orbital forcing explains the thermal maximum and also the temperature reconstructions that do not display any early to mid-Holocene optimum. The difference between the data-sets reflects seasonality anomalies. It has also been inferred that the early to mid-Holocene thermal maximum reflects enhanced advection of heat, either in the form of advection of **warmer** waters from the North Atlantic, by **stronger** advection of Atlantic waters towards the Arctic or by enhanced activity of the Westerlies (e.g. a mechanism akin to that described by Blindheim *et al.* 2000).

The MD95-2011 data rule out the existence of an advection of North Atlantic warm water anomalies to the north. The core location lies underneath the main flow path of Atlantic waters towards the Arctic. At present, the thickness of the Atlantic water layer in this area is 400–600 m, much thicker than the suggested depth distribution of the proxy temperature recorders from the site (Figure 5.5). The advective time constant of the present Norwegian Sea from south to north is in the order of several years (Skagseth 2004). Therefore, an advective temperature anomaly would be expected to override seasonal changes and be manifested down to several hundred meters water depth. The thermal maximum is, however, only manifest in the upper 50 m, which implies that it is attributable to the radiative forcing from the orbital configuration. This is consistent with the model results of Liu *et al.* (2003) who found a difference between the surface layer and thermocline waters in the model's response to orbital forcing. It remains to be tested if this is a robust feature of model behavior in the Nordic Seas.

Ruling out advective transport of warm water anomalies as an explanation, it is possible that oceanic heat fluxes towards the Arctic, larger than later in the Holocene, could have occurred. One possible case is that a larger heat flux could have been contained in stronger along-slope currents, similar to the positive mode of the North Atlantic Oscillation (NOA) observed in recent decades (Blindheim *et al.* 2000; Skagseth 2004). The available data cannot register such a situation with stronger flow of warm waters towards the Arctic and without any change in the mean temperature of the Atlantic water masses. There is, at present, no reliable kinetic flow proxy data-set from which to evaluate this aspect. Nevertheless, it seems likely that the polar amplification of the thermal maximum is mainly due to the polar amplification of orbital forcing and to sea-ice albedo feedbacks due to the summer insolation anomaly, as also noted from the PMIP2 models depicted in

Crucifix (this volume, Figure 4.3). This is also consistent with the shorter duration and early timing of the maximum temperatures near the present sea-ice margin, which implies a connection to the time of the strongest insolation anomaly along the sea-ice margin. Further to the south the maximum temperatures occurred later, after 8 ka (Figure 5.5), and continued at times when temperatures close to the present sea-ice margin had started to fall (Sarnthein *et al.* 2003; Kim *et al.* 2004; Hald *et al.* personal communication). Thus the polar amplification does not appear to be mainly due to advection of oceanic heat.

Holocene climate variability

Internal variability is an inherent part of the climate system. On sub-orbital time-scales internal variability of the ocean and ocean–atmosphere systems has been used to attribute the Atlantic Multi-decadal Oscillation (e.g. Sutton and Hodson 2005) to multi-century to millennial variability (e.g. Schulz *et al.* 2007), primarily suggesting an origin from low-frequency variations in the Atlantic Meridional Overturning Circulation.

Some studies have suggested that there may have existed different prevailing atmospheric patterns in the early to mid-Holocene than in recent times, indicating that atmospheric forcing might have influenced the ocean response and vice versa. Rimbu *et al.* (2004), comparing the alkenone data-set for the Northern Hemisphere of Kim *et al.* (2004) with climate model experiments, note that the SST anomaly pattern appears to be similar to the tripole pattern of the positive phase of the NAO, suggesting a prevailing situation with a stronger tendency for meridional ocean and atmosphere heat transport in the eastern sector of the high-latitude North Atlantic realm. This pattern of model response to 6 ka orbital forcing is, however, not robust when comparing the coupled AOGCMs used in the PMIP2 experiments. A PMIP2 model–model intercomparison shows that three of the nine models support a positive NAO-like atmospheric circulation in the mean state for the mid-Holocene as compared with the pre-industrial period without significant changes in simulated NAO variability (Gladstone *et al.* 2005). Some observational evidence comes from lake studies in northern Scandinavia, indicating a generally more maritime climate in the early to mid-Holocene than in the late Holocene (Hammarlund *et al.* 2002). A general feature of NAO-related responses in northern Europe is a strong increase in winter precipitation in years of positive NAO index, i.e. the pattern indicated by Rimbu *et al.* (2004). When combining several glacier mass balance records and calculating the winter precipitation contribution to Norwegian glaciers, Nesje (personal communication) (Figure 5.7) finds no strong early to mid-Holocene winter precipitation maximum, thus documenting lack of evidence for a general NAO positive state. Rather, the glacier-based precipitation records follow the pattern of cold intervals punctuating the overall trends and occurring at century to millennial time-scales. This is comparable to the pattern of warm and cold phases seen in many other records.

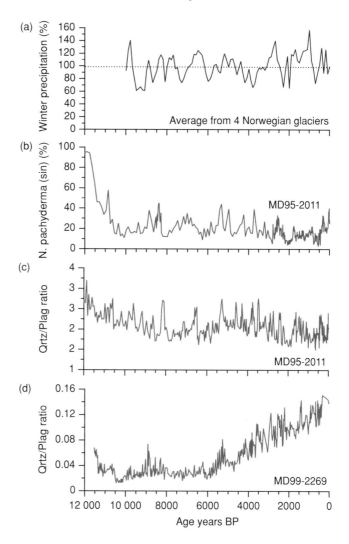

Figure 5.7 (a) Mean winter (October–April) precipitation in percent of 1961–90 mean (= 100%) calculated from Jostedalsbreen, Folgefonna, Spørteggbreen, and Bjørnbreen (compilation by Atle Nesje, personal communication). (b) Relative percent *Neogloboquadrina pachyderma* (sin) in MD95-2011 (Andersson *et al.* 2003; Risebrobakken *et al.* 2003). (c) Ice-rafting proxy data (quartz/plagioclase ratio) from cores MD95-2011 and (d) MD99-2269 (Moros *et al.* 2006).

A feature of many Holocene climate records is that variability on century to multi-century time-scales often displays somewhat lower amplitudes between about 8 and 6 ka, and a tendency for an increased amplitude of the variability towards present (see Figures 5.5 and 5.7). In a number of records there is strong variability in the early Holocene with the 8.2 ka event being probably the most pronounced. Since much of this variability and the 8.2 ka event may plausibly be linked with the final demise of the large Northern Hemisphere ice sheets and thus have a dynamical origin from Ice-Age processes, we focus here on the evolution after the 8.2 ka event. In Figure 5.5 it is apparent that the amplitude of multi-century to millennial scale variability is not stationary through the Holocene. Risebrobakken *et al.* (2003) showed that there is an increase in the amplitude of century to millennial scale variability after the mid-Holocene, and that there is very little persistence in the main frequencies of variability through the Holocene. The noticeable increased amplitude after about 4 ka in the Nordic Seas thermocline temperature (Figure 5.5) and sea-ice proxies (Figure 5.7) indicates increased climate variability affecting winter-time conditions. This pattern is also generally consistent with the onset of neoglaciation in Europe (Nesje *et al.* 2000) and an increase in Arctic sea-ice cover (Koç and Jansen 1994).

We do not have evidence of stronger external forcing throughout the pre-industrial Holocene. It is therefore likely that the nonstationary aspect of the variability indicates that the stronger century to millennial scale variability is excited by changes in boundary conditions, which amplify processes occurring at time-scales of more than 100 years. The amplitude of millennial to sub-millennial scale variability appears potentially linked with sea-ice extent, Northern Hemisphere snow cover, and ocean surface temperature, and indicates that increased sea-ice cover following the reduced summer insolation may have put in place amplification mechanisms leading to stronger ocean temperature variability. One plausible mechanism, drawing on the general trends of orbital forcing through the Holocene, is that the reduction in boreal summer insolation and a less pronounced summertime surface ocean stratification induced sea-ice/snow albedo feedbacks which drove the overturning circulation past a threshold into a more variable mode of operation. In the absence of a clear attribution of this variability to external forcings (e.g. solar, volcanic; e.g. Risebrobakken *et al.* 2003), it appears most likely that the century to millennial scale variability is primarily caused by the long time-scale internal dynamics of the climate system.

Moros *et al.* (2006) (Figure 5.7) published records of ice-rafted debris (IRD) in the form of quartz content of marine sediments, presumed to originate from melting sea-ice. The data show a strong increase in the drift-ice occurrence off East Greenland in the latest half of the Holocene. This is coherent with orbital forcing, which led to an increased sea-ice cover in the marginal ice zone. Koç and Jansen (1994) found a similar pattern based on sea-ice diatom records, although this study had a much lower temporal resolution. The IRD record from the central Nordic Seas basically follows the foraminiferal temperature data (Figure 5.7), indicating that sea-ice here is related to protrusions of colder waters to the south-east and probably related to changes in atmospheric dynamics. The IRD record from the central Nordic Seas is, in a number of aspects, similar to the IRD data of Bond *et al.* (2001), although the different methods and temporal resolution do not make a detailed correlation possible. Overall, the data of Bond *et al.* as well as the IRD data from the Nordic Seas indicate that there is not one single sea-ice response in the region and that the general sea-ice response appears as one following the orbital forcing, but punctuated by colder intervals at millennial to century scales with increased presence of sea-ice and colder thermocline temperatures. As first noted by Risebrobakken *et al.* (2003), there is not a single persistent variability in the high-latitude North Atlantic through the Holocene. The emergence of higher amplitude millennial scale variability in many records appears to be linked to a threshold reached after the thermal optimum, and caused by decreased summer insolation and seasonality. The winter-time precipitation records are in line with this evidence, indicating that the cold intervals in the ocean records are linked with a colder, less moist winter-time climate over Scandinavia.

Conclusions

The early to mid-Holocene thermal maximum is clearly documented in SST records in the high-latitude North Atlantic–Nordic Seas, with a polar amplification, and an earlier and shorter lasting extension in the areas near the modern sea-ice limit in the Nordic Seas than to the south. Further south, the maximum is most clearly expressed between 8 and 6 ka. In the Irminger Sea the maximum SSTs were delayed, probably due to downwind effects from the remnants of the Laurentide Ice Sheet.

The maximum is expressed by proxy records which sample the uppermost surface layer above the seasonal thermocline. Proxy data from recorders that sample the sub-surface do not document the maximum temperatures. This shows that the SST maximum is mainly forced by the summer insolation maximum and sub-surface proxy data that reflect annual mean or winter-time conditions are not influenced by the summer-time insolation forcing.

The origin of the thermal maximum is primarily from orbital forcing and not advection of SST anomalies from the south. It is difficult to attribute the thermal maximum to long-term persistent atmospheric patterns.

Superimposed on the SST trends is century to millennial scale climate variability. This variability does not occur as a stationary response to specific external forcing, but appears primarily to be induced and later amplified after the thermal maximum by changes in climate system internal dynamics.

Acknowledgments

The authors are grateful to Atle Nesje for compiling winter precipitation data and to Øyvind Lie and Trond Dokken for fruitful discussions and comments. This study is an outcome of the NORPAST-2 project funded by the Norwegian Research Council, and the MOTIF and PACLIVA projects funded by the European Union. This is publication A174 from the Bjerknes Centre.

References

Andersen C., Koç N., Jennings A. & Andrews J.T. (2004) Non-uniform response of the major surface currents in the Nordic Seas to insolation forcing: implications for the Holocene climate variability. *Paleoceanography*, **19**(2): Art. No. PA2003 APR 6.

Andersson C., Risebrobakken B., Jansen E. & Dahl S.O. (2003) Late Holocene surface ocean conditions of the Norwegian Sea (Voring Plateau). *Paleoceanography*, **18**(2): Art. No. 1044 JUN 4.

Berner K.S. (2006) *Variability of the two main branches of the North Atlantic Drift through the Holocene*. PhD thesis, University of Tromsø.

Bjune A.E., Bakke J., Nesje A. & Birks H.J.B. (2005) Holocene mean July temperature and winter precipitation in western Norway inferred from palynological and glaciological lake-sediment proxies. *The Holocene*, **15**, 177–189.

Blindheim J., Borovkov V., Hansen B., Malmberg S.A., Turrell W.R. & Østerhus S. (2000) Upper layer cooling and freshening in the Norwegian Sea in relation to atmospheric forcing. *Deep-Sea Research Part I–Oceanographic Research Papers*, **47**, 655–680.

Bond G., Kromer B., Beer J., *et al.* (2001) Persistent solar influence on north Atlantic climate during the Holocene. *Science*, **294**, 2130–2136.

Calvo E., Grimalt J.O. & Jansen E. (2002) High resolution Uk37 sea surface temperature reconstruction in the Norwegian Sea during the Holocene. *Quaternary Science Reviews*, **21**, 1385–1394.

Crucifix M. (this volume) Modelling the climate of the Holocene. In: *Global Warming and Natural Climate Variability: a Holocene Perspective* (Eds R.W. Battarbee & H.A. Binney), pp. 98–122. Blackwell, Oxford.

Duplessy J.C., Cortijo E., Ivanova E., *et al.* (2005) Paleoceanography of the Barents Sea during the Holocene. *Paleoceanography*, **20** (4): Art. No. PA4004 OCT 7.

Dutkiewicz S., Sokolov A., Scott J. & Stone P. (2005) *A Three-dimensional Ocean–Sea-ice–Carbon Cycle Model and its Coupling to a Two-dimensional Atmospheric Model: Uses in Climate Change Studies*. MIT Joint Program on the Science and Policy of Global Change, Report 122. http://web.mit.edu/globalchange/www/MITJPSPGC_Rpt122.pdf.

Gladstone R.M., Ross I., Valdes P.J., *et al.* (2005) Mid-Holocene NAO: a PMIP2 model intercomparison. *Geophysical Research Letters*, **32**(16): Art. No. L16707 AUG.

Hall I.R., Bianchi G.G. & Evans J.R. (2004) Centennial to millennial scale Holocene climate-deep water linkage in the North Atlantic. *Quaternary Science Reviews*, **23**, 1529–1536.

Hammarlund D., Barnekow L., Birks H.J.B., Buchardt B. & Edwards T.W.D. (2002) Holocene changes in atmospheric circulation recorded in the oxygen-isotope stratigraphy of lacustrine carbonates from northern Sweden. *The Holocene*, **12**, 339–351.

Jansen E., Overpeck J., Briffa K.R., *et al.* (2007) Palaeoclimate. In: *Climate Change 2007: the Physical Science Basis. Contribution of Working Group I to the Fourth Assessment Report of the Intergovernmental Panel on Climate Change* (Eds S. Solomon, D. Qin, M. Manning, *et al.*), Cambridge University Press, Cambridge.

Kim J.-H. & Schneider R.R. (2004a) *GHOST Global Database for Alkenone-derived 6ka Sea-surface Temperatures*. http://www.pangaea.de/Projects/GHOST/.

Kim J.-H. & Schneider R.R. (2004b) *GHOST Global Database for Alkenone-derived Holocene Sea-surface Temperature Records*. http://www.pangaea.de/Projects/GHOST/.

Kim J.-H., Rimbu N., Lorenz S.J., *et al.* (2004) North Pacific and North Atlantic sea-surface temperature variability during the Holocene. *Quaternary Science Reviews*, **23**, 2141–2154.

Koç N. & Jansen E. (1994) Response of the high-latitude Northern-Hemisphere to orbital climate forcing – Evidence from the Nordic Seas. *Geology*, **22**, 523–526.

Liu Z., Brady E. & Lynch-Stieglitz J. (2003) Global ocean response to orbital forcing in the Holocene. *Paleoceanography*, **18**(2): Art. No. 1041 MAY 22.

Moros M., Andrews J.T., Eberl D.D. & Jansen E. (2006) Holocene history of drift ice in the northern North Atlantic: evidence for different spatial and temporal modes. *Paleoceanography*, **21**(2): Art. No. PA2017 JUN 8.

Nesje A., Dahl S.O., Andersson C. & Matthews J.A. (2000) The lacustrine sedimentary sequence in Sygneskardvatnet, western Norway: a continuous, high-resolution record of the Jostedalsbreen ice cap during the Holocene. *Quaternary Science Reviews*, **19**, 1047–1065.

Nyland B.F., Jansen E., Elderfield H. & Andersson C. (2006) *Neogloboquadrina pachyderma* (dex. and sin.) Mg/Ca and delta O-18 records from the Norwegian Sea. *Geochemistry Geophysics Geosystems*, **7**: Art. No. Q10P17 OCT 12.

Renssen H., Goosse H., Fichefet T., Brovkin V., Driesschaert E. & Wolk F. (2005) Simulating the Holocene climate evolution at northern high latitudes using a coupled atmosphere–sea ice–ocean–vegetation model. *Climate Dynamics*, **24**, 23–43.

Rimbu N., Lohmann G., Lorenz S.J., Kim J.H. & Schneider R.R. (2004) Holocene climate variability as derived from alkenone sea surface temperature and coupled ocean–atmosphere model experiments. *Climate Dynamics*, **23**, 215–227.

Risebrobakken B., Jansen E., Andersson C., Mjelde E. & Hevroy K. (2003) A high-resolution study of Holocene paleoclimatic and paleoceanographic changes in the Nordic Seas. *Paleoceanography*, **18**(1): Art. No. 1017 MAR 26.

Sarnthein M., Van Kreveld S., Erlenkeuser H., *et al.* (2003) Centennial-to-millennial-scale periodicities of Holocene climate and sediment injections off the western Barents shelf, 75 degrees N. *Boreas*, **32**, 447–461.

Schulz M., Prange M. & Klocker A. (2007) Low-frequency oscillations of the Atlantic Ocean meridional overturning circulation in a coupled climate model. *Climate of the Past*, **3**, 97–107.

Seppä H., Hammarlund D. & Antonsson K. (2005) Low-frequency and high-frequency changes in temperature and effective humidity during the Holocene in south-central Sweden: implications for atmospheric and oceanic forcings of climate. *Climate Dynamics*, **25**, 285–297.

Skagseth Ø. (2004) Monthly to annual variability of the Norwegian Atlantic slope current: connection between the northern North Atlantic and the Norwegian Sea. *Deep-Sea Research Part I-Oceanographic Research Papers*, **51**, 349–366.

Sutton R.W. & Hodson D.L.R. (2005) Atlantic Ocean Forcing of North American and European Summer Climate. *Science*, **309**.

6 Holocene climate change and the evidence for solar and other forcings

Jürg Beer and Bas van Geel

Keywords

Climate change, solar forcing, orbital forcing, volcanic forcing, 850 BC climate event, socio-economic impacts

Introduction

Future climate change may have considerable effects on the hydrologic cycle and temperature, with significant consequences for sea level, food production, world economy, health, and biodiversity. How and why does the natural climate system vary on decadal to millennial time-scales? Do we sufficiently understand natural climate change, and what is the relative importance of natural processes versus human activities in explaining the global warming of the last few decades? How much of the recent warming is induced by natural (solar, volcanic) rather than anthropogenic forcing?

In addition to historical documents produced by people observing environmental change, nature itself provides a wealth of information stored in various archives such as ice cores, sediments, tree rings, and speleothems. A growing proportion of this stored information can be deciphered thanks to impressive developments in analytic techniques (cf. Birks, this volume). An increasingly clear view of the past environment is emerging. We are still far away, however, from having a complete picture of all the physical, chemical, and biological processes involved. The main problem is identifying all the relevant processes and quantifying the forcing factors, and the corresponding response of the numerous interlinked components of the climate system (atmosphere, hydrosphere, cryosphere, biosphere). Therefore, any primary change will trigger a chain of secondary reactions, some of which amplify the primary change (positive feedback). These positive feedback mechanisms occur on very different time-scales, ranging from minutes (e.g. the ozone chemistry in the stratosphere) to millennia (e.g. the formation of continental ice sheets).

In this chapter we give an overview of the mechanisms causing natural climate change on decadal to millennial time-scales. Proxy climate records provide the basis for both a better understanding of the processes involved, and for testing climate models. Based on selected literature from high-resolution Holocene records of climate change, we evaluate the role of the various potential forcing factors.

Forcing mechanisms

The climate system generally can be described in terms of energy transport. The main energy source is the Sun, which emits electromagnetic radiation with wavelengths covering the full spectrum and peaking in the visible part (400–750 nm). On its way through the atmosphere to the Earth's surface, part of this radiation is reflected, scattered, or absorbed. The absorbed energy (~70 percent) is ultimately re-emitted into space with much longer wavelengths. Since the incoming solar radiation covers only half the globe (day side) and peaks at low latitudes, there are permanent energy gradients on the Earth's surface, which the climate system tries to eliminate by transporting energy through the atmosphere and the ocean (thermohaline circulation).

All the radiative processes depend strongly on the composition of the atmosphere (gases, aerosols, dust, clouds), which provide another level of complexity, as these components are strongly connected to chemical, thermal, and dynamical changes taking place in the atmosphere on vastly differing time-scales.

Any change in these complex processes can force the climate to change. We distinguish between external (orbital and solar) and internal (volcanic and ocean circulation) forcings which we will address in the following sections.

Orbital forcing

The amount of solar radiation arriving at a given point on the top of the atmosphere depends on the solar luminosity (the total amount of radiation emitted by the Sun) and the relative position of Sun and Earth in space (distance, direction of the Earth's rotational axis relative to the ecliptic). Whereas the former depends on processes taking place within the Sun, the latter is determined by the distortion of the Earth's orbit by the gravitational forces of the other planets (mainly Jupiter and Saturn) (Laskar *et al.* 2004). Orbital forcing affects three parameters with typical periodicities: the eccentricity (the deviation of the orbit from a circle, with periods of ~400 and ~100 kyr), the obliquity (the tilt angle of the Earth's axis with a period of ~40 kyr), and the precession of the Earth's axis (~20 kyr period). Whereas the eccentricity leads to changes in the mean annual distance between Sun and Earth, and therefore to changes in the total incoming radiation, the other two parameters affect only the relative distribution of the solar radiation, with implications for the energy transport within the climate system. At the time-scale of the Holocene,

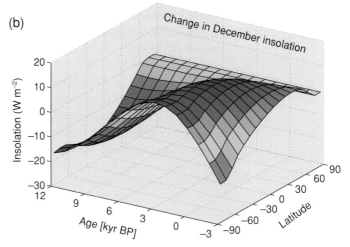

Figure 6.1 Changes in orbital forcing during the Holocene (12 000 years BP to 3000 years in the future) for the months June (a) and December (b). The figures show clear trends depending on latitude. In June there is a strong decreasing trend in the north (a). In December, the insolation on the Southern Hemisphere first increases and then reaches its maximum between 2000 and 4000 cal. years BP before decreasing again (b).

orbital forcing is slow. Nevertheless, the change in insolation over the past 12 000 years is considerable and cannot be neglected (cf. Jansen *et al.*, this volume). Figure 6.1a shows the change in insolation in June for the Holocene period as a function of latitude. At high latitude the change in insolation amounts to 54 W m^{-2}. In Figure 6.1b the change in insolation is shown for December. In this case, the insolation increases from the beginning of the Holocene and reaches a maximum between 4000 and 2000 years BP. The maximum change amounts to 37 W m^{-2}. Orbital forcing is the only forcing that is fully understood, and can be calculated not only for the past but also for the future several million years (Berger and Loutre 2004).

Solar activity and solar forcing

Since the Sun drives the climate system, solar variability is obviously a serious candidate for forcing climate change. This is especially so in view of the fact that the

Sun is a variable star showing considerable cyclic changes in its magnetic activity, as expressed, for example, in the sunspot number.

In the past, many attempts have been made to ascertain whether the solar radiation arriving at the top of the atmosphere at a distance of 1 astronomical unit from the Sun is constant (denoted by the solar constant or total solar irradiance, TSI). All the early attempts, however, failed due to the above-mentioned radiative atmospheric processes. Only after it became possible to mount radiometers on satellites outside the atmosphere was it discovered that the solar constant fluctuates in phase with the magnetic activity of the Sun. It is possible to divide the measured total solar irradiance (TSI) into three components: a background component, a darkening component controlled by the sunspots, and a brightening component related to the faculae, which overcompensate the negative effect of the sunspots and lead to a positive correlation between TSI and solar activity (Fröhlich and Lean 2004).

The instrumental data from almost the past three decades show that the change of the TSI over an 11-year Schwabe cycle is about 0.1 percent, corresponding to 0.25 W m^{-2} (the global mean value at the Earth's surface), an estimate that is quite small compared with the 3.7 W m^{-2} estimated for a doubling of CO_2. The variability of the solar radiation, however, is strongly wavelength dependent and reaches values of more than 100 percent in the UV part of the spectrum. Such large changes in the spectral solar irradiance (SSI) strongly influence the photochemistry in the upper atmosphere, and in particular the ozone concentration. Model calculations show that through dynamical coupling SSI changes can cause shifts in the tropospheric circulation systems and therefore change the climate (Haigh and Blackburn 2006).

From a climate perspective, changes in forcings on decadal and shorter time-scales are less important, because many processes within the climate system occur on much longer time-scales (e.g. the thermohaline circulation, build up of ice-sheets). Therefore, the crucial questions are whether changes of TSI and SSI occur on centennial and millennial time-scales, and how large these changes are. These questions are still being debated. They can be answered in two steps: (i) how variable is the Sun's magnetic activity? (ii) how is this magnetic activity related to TSI and SSI? As we will show, the magnetic variability is indeed larger on longer time-scales. From the solar physics perspective, however, it is not yet clear if this is also the case for the TSI and the SSI. On the other hand, paleoclimate reconstructions provide growing evidence for larger changes in solar forcing than has been experienced during the past 30 years (Bond *et al.* 2001; Neff *et al.* 2001; Wang *et al.* 2005; Haltia-Hovi *et al.* 2007).

The longest historical record of solar activity is the sunspot record, which goes back to 1610 when the telescope was invented. It shows the well-known 11-year Schwabe cycle superimposed on a generally increasing trend from 1610 to the present, which is interrupted by distinct periods of low solar activity (the Maunder minimum of 1645–1715, and the Dalton minimum of 1795–1820).

To extend this record of solar activity beyond the era of direct observations we have to rely on indirect proxy data. Such data can be derived from measurements of the cosmogenic radionuclides ^{10}Be and ^{14}C in natural archives such as ice cores and tree rings (Beer *et al.* 1990; Stuiver *et al.* 1991; Muscheler *et al.* 2004). Cosmogenic radionuclides are produced continuously in the atmosphere as a result of the interaction of galactic cosmic rays with nitrogen and oxygen (Masarik and Beer 1999). The higher the cosmic ray intensity, the larger the production rate, and vice versa. The cosmic ray intensity is modulated by two magnetic effects: the solar activity and the geomagnetic field. Depending on the magnetic activity, the Sun emits plasma (solar wind) carrying magnetic fields into the heliosphere, which acts as a shield and reduces the cosmic ray intensity. The second shielding effect is due to the geomagnetic field, which prevents cosmic rays with too low an energy from penetrating the atmosphere and producing cosmogenic radionuclides. After production, the fate of the cosmogenic radionuclides depends on their geochemical properties. ^{10}Be becomes attached to aerosols and is removed from the atmosphere mainly by wet deposition within 1–2 years. Ice cores are excellent archives to measure ^{10}Be and to reconstruct its production rate in the past. On the other hand, ^{14}C forms ^{14}CO$_2$ and exchanges between atmosphere, biosphere, and ocean. As a consequence of the large size of these reservoirs and the long residence times (ocean: 1–2 kyr), the amplitude of an observed ^{14}C change in the atmosphere is considerably smaller than the corresponding change in the production rate and the ^{14}C change is delayed. The ideal archive for the reconstruction of past ^{14}C changes are tree rings.

Although the physics of the production processes in the atmosphere are well understood, the transport from the atmosphere into the archives is rather complex and has not yet been fully elucidated (Beer *et al.* 2002). Comparisons between ^{10}Be and ^{14}C records show that during the Holocene the production signal was dominant and system effects generally can be neglected in a first-order approach. This means that both ^{10}Be and ^{14}C provide an independent record of the cosmic ray intensity of the past. To extract the solar component from this signal the geomagnetic effect has to be removed. This can be achieved by taking into account the changes in the geomagnetic field intensity derived from archeomagnetic and paleomagnetic measurements. The result is a record of the solar modulation function Φ (Figure 6.2) (Vonmoos *et al.* 2006). $\Phi = 0$ means a completely quiet Sun, a condition that has probably never occurred. $\Phi = 1000$ MeV corresponds to an active Sun as typical for a solar cycle maximum during recent times. The Φ-record is characterized by a long-term trend superimposed on which are short-term fluctuations. These short-term fluctuations can be divided into cyclic and episodic features. The cyclic features show periodicities around 11 years (Schwabe cycle), 80 years (Gleissberg cycle), 205 years (DeVries or Suess cycle), and 2200 years (Halstatt cycle) (Stuiver and Braziunas 1993).

Most episodic features are strong negative spikes corresponding to so-called grand solar minima, periods when the Sun was very quiet as during the Maunder minimum. It should be noted that the record does not cover the past 300 years BP. The mean Φ value of the past five decades is about 700 MeV. This means that we are presently in a period of high solar activity, and that in the past there were periods with considerably lower solar activity. Similar conclusions were obtained by Solanki *et al.* (2004). Whether these large changes of activity are also reflected in

Figure 6.2 Reconstruction of the solar modulation function Φ covering the period 9300–340 cal. years BP. Φ depends on the intensity of the solar open magnetic field, which reduces the cosmic ray flux and therefore the production rate of cosmogenic radionuclides. This record is based on ^{10}Be data from the GRIP ice core and shows short-term as well as long-term fluctuations. The average solar activity of the past few decades corresponds to a value of 700 MeV. The data have been low-pass filtered with cut-off frequencies of 1/(100 years) and 1/(500 years).

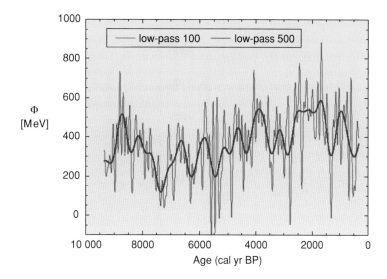

the TSI and SSI is not known. Therefore we prefer not to speculate about the corresponding forcing in W m^{-2}. There is clear evidence for larger changes from paleoclimatic records, as we will show below. Most authors have used the Δ^{14}C record as an indicator of past solar activity. However, Δ^{14}C reflects the deviation of the atmospheric ^{14}C content relative to a standard value, so it also contains a long-term geomagnetic and a system component which perturbs the solar signal to some extent. Consequently, the ^{10}Be flux and the ^{14}C production rate are better proxies, although they still contain a geomagnetic component.

In Figure 6.3, the x-axis is based on tree ring counting. At 850 BC in Figure 6.3a, a rapid decrease of the radiocarbon age occurs followed by a plateau. The decrease corresponds to the Δ^{14}C rise in Figure 6.3b (red data points). The blue curve in Figure 6.3b depicts the changes in the radiocarbon production that are needed to cause the observed Δ^{14}C fluctuations. Note the difference in the amplitudes (ca. 40 percent in production, ca. 2 percent in Δ^{14}C) and the lag of the Δ^{14}C peak by about 20 years relative to the production. These system effects are caused by the large ^{14}C reservoirs (ocean, atmosphere, biosphere) and the exchange between them.

Volcanic forcing and ice cores

Volcanic eruptions are the main cause of strong short-term (annual) climate forcing. The injection of large amounts of gases (SO_2, CO_2, H_2O, N_2) and dust into the atmosphere perturbs the radiative balance via enhanced absorption and scattering of solar radiation leading to a warming in the upper atmosphere and a cooling in the lower atmosphere (Figure 6.4). Due to the short tropospheric lifetime (1–3 weeks) the injection of gases must reach the stratosphere (lifetime: 1–3 years) to become globally active. The most important substance is SO_2, which oxidizes quickly to H_2SO_4. Beside volcanic eruptions, there are other comparatively stable

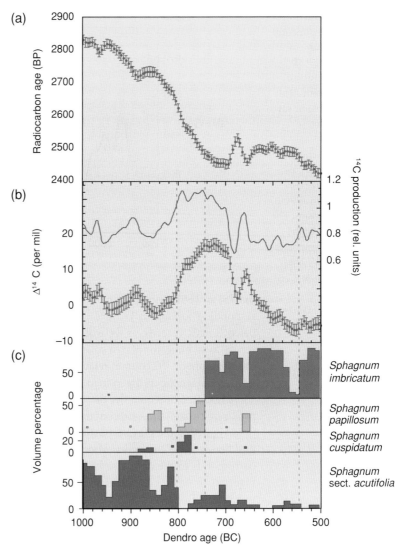

Figure 6.3 (a) The ^{14}C calibration curve for the period 1000–500 BC. The horizontal axis shows the calendar age. The vertical axis shows the ^{14}C age in BP (BP: before 1950). The calibration curve is based on mainly decadal ^{14}C measurements of exactly dated (dendrochronology) wood. Fluctuations of the curve are caused by changes of the cosmic ray intensity due to fluctuations in solar activity. (b) Δ^{14}C is the deviation of the atmospheric ^{14}C/^{12}C ratio from a standard value in ‰ as calculated from the data of the calibration curve. The blue curve depicts the changes in the radiocarbon production that are needed to cause the observed Δ^{14}C data. The Δ^{14}C curve provides information about the ^{14}C production rate and the carbon system in the past. The interval ca. 850–750 BC represents the phase of extreme climate at the start of the Sub-atlantic. The "^{14}C-clock" accelerates during that period (300 ^{14}C "years" in about 100 calendar years). The Δ^{14}C curve shows a sharp rise of atmospheric radiocarbon, related to a temporary decline in solar activity. (c) Changes in species composition of *Sphagnum* during the Sub-boreal–Sub-atlantic transition in Engbertsdijksveen, a raised bog (after van Geel *et al.* 1996). *Sphagnum* species of the section *acutifolia* prefer relatively dry and warm climatic conditions. *Sphagnum cuspidatum* and *S. papillosum* prefer a relatively high water table. *Sphagnum imbricatum* prefers high air humidity and cool conditions. A major change started ca. 850 BC, when solar activity showed a temporary decline (a fast increase of Δ^{14}C).

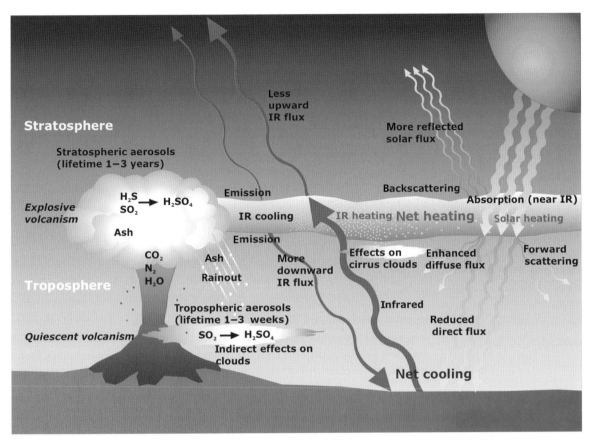

Figure 6.4 Schematic overview of the main atmospheric processes related to volcanic forcing (Fischer 2003, 2005). (From Robock (2000), plate 1.)

sources of SO_2 such as the marine biota producing dimethylsulfide, and noneruptive volcanic emissions. Figure 6.4 is a schematic overview of the main processes involved in volcanic forcing (Fischer 2003, 2005).

Ice cores are an important source of information on volcanic eruptions. Figure 6.5 shows a record of volcanic sulfate derived from the GISP2 ice core (Zielinski and Mershon 1997). Using additional information about the volcanoes responsible for the emissions, Crowley (2000a) estimated the corresponding forcing for the past 1000 years (inset in Figure 6.5). The very sharp peaks indicate that the sulfate is rapidly removed from the atmosphere. Bay *et al.* (2004) linked volcanic ash layers in Siple Dome (Antarctica) with the onset of millennium-scale cooling recorded in a Greenland ice core and interpreted the results as evidence for a causal connection between volcanism and millennial climate change. The high volcanic activity during the main deglaciation of the early Holocene suggests that ice-sheet unloading and/or sea-level rise was responsible for increased volcanism during that period. One of the difficulties in quantifying the volcanic forcing is that without additional information it is impossible to distinguish between regional tropospheric and global stratospheric eruptions. Whereas one single eruption can

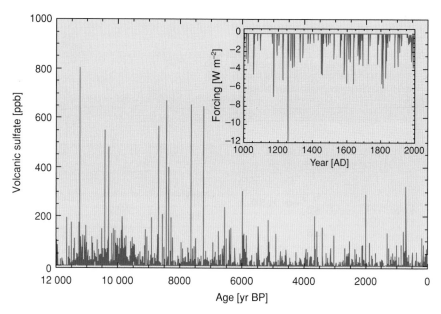

Figure 6.5 A record of volcanic sulfate concentrations in ppb derived from H_2O_4 measurements on the GISP2 ice core (Zielinski and Mershon 1997) covering the past 12 000 calendar years. The inset shows an estimate of the sulfate-induced forcing in W m^{-2} by Crowley (2000a). Note that the forcing can be very strong, but does not last long (1–2 years).

change weather patterns significantly over a 1–2 year period, a series of eruptions may have a long-term climatic effect. Castellano *et al.* (2004) compared millennial scale volcanic frequencies against δD in the EPICA-Dome-C ice core; they found no clear evidence for a close relationship between climatic change and volcanism.

Oscillations in the ocean system

An abrupt cold period occurred around 8200 calendar years BP in the North Atlantic area. It lasted for ca. 300 years and in Greenland ice-core records it is characterized by a reduction in temperature greater than 1°C, a decrease in ice accumulation rate, increasing wind speeds, and a drop in atmospheric methane levels (Wiersma and Renssen 2006, and references therein). A slowing down of the thermohaline circulation as a result of a freshwater perturbation has been proposed as the cause of the event. The slowdown resulted in a decrease of the northward heat transport in the North Atlantic Ocean, leading to pronounced cooling. The proglacial Laurentide Lakes in front of the Laurentide ice sheet were most probably the source of the freshwater pulse (Clarke *et al.* 2003). Model–data comparisons by Goosse *et al.* (2002) and Wiersma and Renssen (2006) confirm the catastrophic drainage of Laurentide Lakes as a forcing mechanism.

Reactions of the climate system to forcings

It is well known that the climate system shows variability on vastly different time-scales. It is not yet clear as to what extent this variability is caused by internal processes within the climate system, and which role the different forcing factors play. The results of climate model experiments indicate significant internal variability, as well as a significant sensitivity to solar, volcanic, and greenhouse gas forcing (see Goosse *et al.*, this volume; Claussen, this volume). During the Holocene, greenhouse gases did not vary greatly. Volcanic aerosols remain in the atmosphere for only a few years. Therefore, as the focus here is on climate change on decadal to millennial time-scales, we are principally concerned with solar forcing, although we need to recognize that multiple volcanic eruptions can cause climatic effects also on decadal time-scales (Crowley 2000b). There is a rapidly growing number of examples pointing to the Sun as a major forcing factor during the Holocene (Beer *et al.* 2000, 2006). Here we discuss evidence for solar-induced climate change based on records from various natural archives.

Peat deposits and climate change

Holocene peat deposits, especially the rainwater-fed raised bogs such as those in north-west Europe, are natural archives of climate change (cf. also Verschuren and Charman, this volume). Climate-related changes in precipitation and temperature are reflected in the changing species composition of the peat-forming vegetation (Figure 6.3c). Plant remains can be identified and, by using ecological information about peat-forming species, changes in species composition of sequences of peat samples can be interpreted as evidence for changing local hydrologic conditions, often linked to climate change. The degree of decomposition of the peat-forming plants is also related to former climatic conditions (peat decomposes more under drier conditions; wetter climatic conditions allow plant remains to be preserved better; Figure 6.6). Blytt and Sernander sub-divided the Holocene in alternating periods, which were supposed to represent differing climatic conditions (Blytt, 1882; Sernander 1910) (Figure 6.7; Birks, this volume). Later researchers found the Blytt–Sernander sub-division to be too simple, but the so-called Sub-boreal–Sub-atlantic transition of the Blytt–Sernander scheme, which occurred around 850 BC, is a consistently observed abrupt and intense climate shift. In north-west Europe, there is strong peat-stratigraphic and archaeological evidence that the climate changed from relatively dry and warm to cooler, wetter conditions (Figures 6.3 and 6.6; van Geel *et al.* 1998). The radiocarbon method is generally used for dating of Holocene climate-induced shifts in peat deposits. Calibration of a single radiocarbon date usually yields an irregular probability distribution of calendar ages, quite often over a long time-interval. This is problematic in paleoclimatological studies, especially when a precise temporal comparison between different climate proxies is required. Closely spaced sequences of (uncalibrated) ^{14}C dates from peat deposits, however, display wiggles that can be fitted to the wiggles in the radiocarbon calibration curve. The practice of dating peat samples using ^{14}C "wiggle-match dating" has greatly improved the precision of radiocarbon chronologies since its application by van Geel and Mook (1989). By wiggle-matching ^{14}C measurements, high precision calendar age chronologies for peat sequences can be generated (Blaauw *et al.* 2003), which show that mire surface wetness increased together with rapid increases of atmospheric production of ^{14}C during the early Holocene, the Sub-boreal–Sub-atlantic transition (Figure 6.3b: a sharp increase of ^{14}C production

Figure 6.6 Sampling a profile in the former raised bog area of Bargerveen (Drenthe, The Netherlands). The Sub-boreal–Sub-atlantic transition is indicated by an arrow. (Photograph by Bas van Geel. Left, D.G. van Smeerdijk; right, W.A. Casparie.)

and evidence for wetter conditions in Figure 6.3c), and the Little Ice Age (Wolf, Spörer, Maunder, and Dalton minima of solar activity). Peat records show that this phenomenon occurred in The Netherlands (Kilian *et al.* 1995; van Geel *et al.* 1998; van der Plicht *et al.* 2004), the Czech Republic (Speranza *et al.* 2003), and the UK and Denmark (Mauquoy *et al.* 2002). The production of radiocarbon is regulated by solar activity, and therefore periods of increased mire surface wetness have been interpreted as evidence for solar forcing of climate change (the effects of sudden declines in solar activity).

Lake sediments, glacier variations, and solar activity

Denton and Karlén (1973) linked the radiocarbon record with geologic data such as the extension of glaciers and made important conclusions about the solar

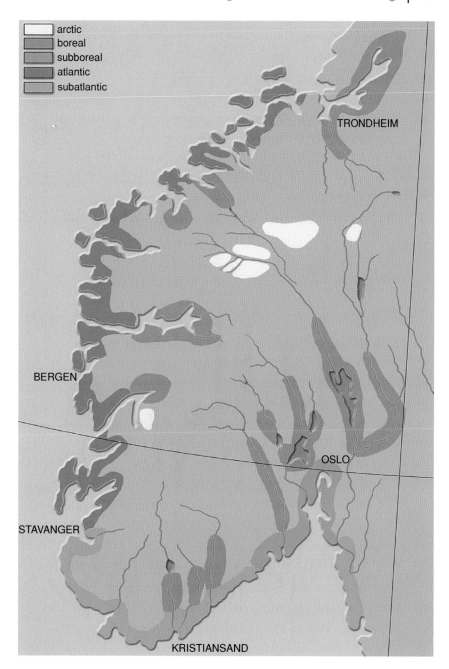

arctic
boreal
subboreal
atlantic
subatlantic

TRONDHEIM

BERGEN

OSLO

STAVANGER

KRISTIANSAND

Figure 6.7 Map of southern Norway (after Blytt 1882) showing areas characterized by specific vegetation types related to differences in temperature and precipitation. This pattern was the modern reference for Axel Blytt in his work subdividing the Holocene.

forcing of climate change. Magny (2004, 2007) published a long record of Holocene climate-related water table changes in lakes in south-eastern France and adjacent Switzerland. The lake-level fluctuations closely correspond to the atmospheric ^{14}C fluctuations and therefore also to the history of solar activity (Figure 6.8). Lake levels were low during periods of high solar activity (low values of Δ^{14}C and Φ), whereas high lake-stands occurred when solar activity was low. Holzhauser *et al.*

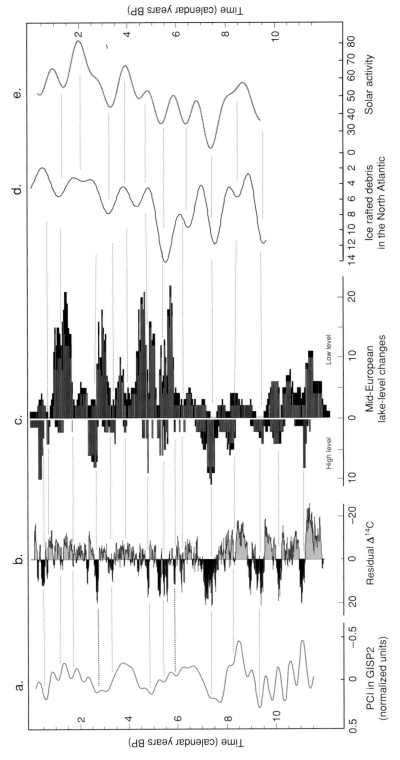

Figure 6.8 Comparison of various climate parameters (redrawn after Magny 2004) with the detrended $\Delta^{14}C$ and the solar modulation function Φ. The records show the following parameters: (a) the Polar Circulation Index (PCI) at GISP2 (Mayewski *et al.* 1997); (b) the detrended deviation of the atmospheric $^{14}C/^{12}C$ ratio from a standard value as an indicator of solar activity; (c) lake-level changes after Magny (2007); (d) the amount of ice-rafted debris found in sediment cores as a measure of the southward drift of icebergs in the North Atlantic (after Bond *et al.* 2001); (e) the solar modulation function of Figure 6.2.

(2005) compared glacier and lake-level fluctuations in west-central Europe over the past 3500 years and demonstrated synchroneity between glacier advances, periods of higher lake levels, and maxima of atmospheric radiocarbon. Lake-sediment records have also yielded strong evidence for relationships between solar forced changes to cooler, wetter climatic conditions and the economy of prehistoric people. Based on changes in proportions of wild to domesticated animals and archaeobotanical data in excavated lake-side settlements Schibler *et al.* (1997) and Arbogast *et al.* (2006) concluded that people shifted their caloric intake from domesticated cereals to wild sources of food in response to adverse climatic conditions. There are numerous other examples of sediment and lake-level data that point to a solar link (Verschuren *et al.* 2000; Patterson *et al.* 2004; Baker *et al.* 2005; Stager *et al.* 2005; Morrill *et al.* 2006; Wu *et al.* 2006).

Marine sediments and stalagmites

Precise radiocarbon dating of marine sediments is problematic, because the $^{14}C/^{12}C$ ratio in the ocean differs from the one in the atmosphere (reservoir effect). Nevertheless, Bond *et al.* (2001) showed that more ice-rafted debris in the North Atlantic Ocean was transported to the south during periods of relatively low solar activity (low Φ values). The agreement between cosmogenic isotope fluctuations and ice-rafted debris points to a dominant influence of solar activity changes on the North Atlantic climate (Figure 6.8).

Neff *et al.* (2001) studied the climate archive of stalagmites in Oman. The oxygen isotope record was interpreted as a proxy for fluctuations in monsoon rainfall. After some adjustments of the U–Th chronology, the oxygen isotope record was linked to the $\Delta^{14}C$ record, suggesting that changes of monsoon intensity are driven by solar activity fluctuations. $\delta^{18}O$ in stalagmites from the Dongge cave in southern China also show clear evidence for a solar signal in the monsoon variability on decadal to centennial time-scales (Wang *et al.* 2005). Mangini *et al.* (2005) reconstructed temperature changes during the past 2000 years based on a stalagmite from a cave in the Central Alps. Based on a high correlation of that record with $\Delta^{14}C$, the conclusion was made that solar variability was a major driver of climate in central Europe during the past two millennia.

Tree rings and climate

Localized site tree-ring studies mainly reflect the complex ecological processes that operate on small scales in forest ecosystems. Tree-ring density and tree-ring width data enhance our understanding of past temperatures and other climate changes. Extensive sets of tree-ring data can be used for the reconstruction of regional and even hemispheric-scale temperature changes (Esper *et al.* 2002; Briffa *et al.* 2004; D'Arrigo *et al.* 2006). Tree rings are excellent proxies to study short-term effects such as cooling induced by large volcanic eruptions (Briffa *et al.* 1998).

The reconstruction of long-term climate changes is, however, more ambiguous because of the necessity to remove tree-age-related (biological) trends that do not reflect climate signals. It is particularly difficult to retain long-term climate information on time-scales longer than the individual length of the tree-ring series combined in a mean chronology (Cook 1995). Methods to overcome this limitation and to reconstruct the full spectrum of climate variability include "Regional Curve Standardization" (Esper *et al.* 2003) and "Age Banding" (Briffa *et al.* 2001).

The 850 BC event: an example of climate change with socio-economic impacts

As pointed out by Oldfield (this volume) climate change can have important impacts on human society. The two fundamental characteristics of a change are its magnitude and rate. Although the former is obvious, the latter is often not taken seriously enough. Although every change causes problems of adaptation it is much easier to deal with if it occurs slowly.

The climate shift around 850 cal. BC (Sub-boreal–Sub-atlantic transition), one of the most important climate shifts during the Holocene, with evidence for climatic tele-connections is such an example. It had strong socio-economic impacts in areas that were marginal from a hydrologic point of view. In lowland regions in The Netherlands, the climate shift caused a sudden, considerable rise of the groundwater table so that arable land was transformed into wetland, where peat growth started. Farming communities living in such areas were forced to migrate because they could no longer produce sufficient food (van Geel *et al.* 1996). In Figure 6.9, the landscape development in the northern Netherlands is shown for three successive phases. At the start of the second phase, coincident with an abrupt decline of solar activity, the atmospheric circulation changed, leading to cooler and wetter climate conditions. The rise of the water table forced the farmers to migrate to well-drained areas in the northern Netherlands where salt marshes offered them new fertile land. Magny (2004) showed that over a period of several millennia the presence of lakeside villages in south-eastern France and adjacent Switzerland was strongly linked with lake levels and solar activity. No lakeside villages occurred after 850 cal. BC.

In north-west and central Europe, the climatic shift around 850 cal. BC preceded a considerable rise in palynological indicators of human impact on the landscape. Van Geel and Berglund (2000) suggested a causal link between the climate shift around 850 BC and the evidence for a subsequent increase in human population density. They postulated that the climatic crisis in the first instance caused an environmental and social crisis. A collapse of societies resulted in a weakening of the position of dominating groups, which brought about a change in the social structure of farming communities. This facilitated the introduction of a new technological complex, which again created further social change combined with a leap forward in production, food consumption, and population density.

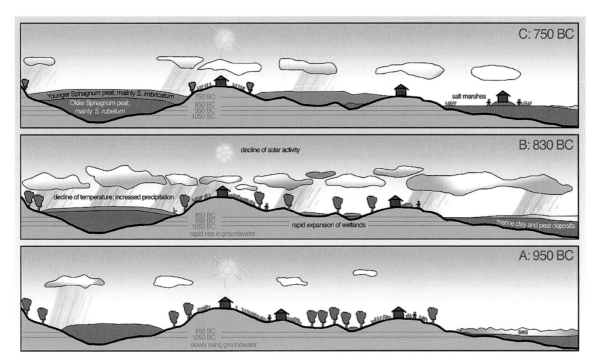

Figure 6.9 The effects of climate change during the period ca. 950–750 BC in the northern Netherlands. From phase A to phase B, solar activity abruptly declined (the Sub-boreal–Sub-atlantic transition). Atmospheric circulation changes occurred and the Westerlies became more intense. Precipitation in north-west Europe suddenly increased, while temperatures declined. As a consequence, a sudden rise in the water table occurred. Fenlands extended, often at the expense of arable land. Species composition in raised bogs changed as a consequence of cooler, wetter climatic conditions (compare Figure 6.3c). Farmers in marginal areas could no longer produce enough food and they had to migrate to well-drained areas. The thermal contraction of ocean water caused a temporary pause or slowing of the rise in sea level that had been occurring since the end of the last Ice Age. Salt marshes occurred for the first time along the coast of the northern Netherlands. These fertile areas were ideal for farmers who had lost their habitat as a consequence of rising groundwater tables. Artificial earth mounds were constructed for protection.

In south-central Siberia, archaeological evidence suggests an acceleration of cultural development and a sudden increase in density and geographic distribution of the nomadic Scythian population after 850 BC. Van Geel *et al.* (2004a) hypothesized a relationship with an abrupt climatic shift towards increased humidity (equatorward relocation of mid-latitude storm tracks). The hypothesis is supported by pollen-analytic evidence. Areas that initially may have been hostile semi-deserts changed into attractive steppe landscapes with a high biomass production and carrying capacity. Newly available steppe areas could be utilized by herbivores, making them attractive for nomadic tribes. The Central Asian horse-riding Scythian culture expanded, and an increased population density was a stimulus for westward migration towards south-eastern Europe.

There is strong evidence for climate change in the Central African rain forest belt around 850 BC (van Geel *et al.* 1998). Palynological studies point to a drastic change in the vegetation cover (from predominantly rain forest to a more open savannah landscape) as a consequence of an aridity crisis. The change of climate and vegetation was also important for prehistoric humans. A population of farmers migrated from the south into the area. The contrast between this change to dryness in central west Africa and the contemporary increase of precipitation in the temperate zones fits well with the hypothesis that, after a decline of solar activity, there was a decrease in the latitudinal extent of the Hadley Cell circulation and consequently the monsoon decreased in intensity, while the mid-latitude storm tracks in the temperate zones were enhanced and moved in the direction of the Equator (van Geel and Renssen 1998). A dryness crisis caused by a weak monsoon intensity in north-west India after 850 BC also supports this hypothesis (van Geel *et al.* 2004b). Furthermore, massive glacier advance in the south-central Andes of Chile, probably resulting from an equatorward relocation of mid-latitude storm tracks (like in the Northern Hemisphere), forms part of a wealth of evidence for worldwide climate change around 850 BC (van Geel *et al.* 2000).

Evidence from paleodata indicates that the climate shift around 850 BC occurred suddenly, probably within a decade. The end of the event, however, seems to have been gradual (a time-transgressive passing of thresholds), so that, given present knowledge, it is not yet possible to pinpoint an end of the event. Changing climatic conditions at 850 BC may have been similar to climatic cooling shifts during the Little Ice Age (Mauquoy *et al.* 2002).

Discussion

The attribution of climate change to potential causes is a key issue in climate research. It is very important to be able to separate climate effects due to human activities from the response due to natural climate factors and internal climate variability. Climate feedbacks dominate the hydrologic cycle due to the large amounts of water on our planet and its strong impact on the energy balance (van Dorland 2006). Climate change caused by freshwater flux may have been restricted to the early Holocene (Clarke *et al.* 2003; Wiersma and Renssen 2006).

Since the atmospheric greenhouse gas concentration did not change much before the start of the fossil fuel combustion by people, greenhouse gas forcing was not an important forcing on climate during the Holocene, with the exception of the past few decades. In spite of the fact that single volcanic eruptions are of short duration the combination of many eruptions can have significant climatic effects on annual to decadal time-scales (Crowley 2000b; Robock 2000; Ammann *et al.* 2003).

On longer time-scales there is an increasing amount of evidence for solar forcing of climate change (van Geel *et al.* 1999; Renssen *et al.* 2000; Bond *et al.* 2001; Mauquoy *et al.* 2002; Hu *et al.* 2003; van der Plicht *et al.* 2004; van Geel *et al.* 2004a,b; Maasch *et al.* 2005; Versteegh 2005; Xiao *et al.* 2006). Climate proxy data indicate that variations in solar activity are an important driving force of variations in climate. The apparently large sensitivity of the climate system to small changes in solar activity indicates that amplification by feedback processes in the climate system has to be taken into account. Two mechanisms have been proposed for this: cosmic ray-cloud correlations and UV-ozone atmospheric circulation changes.

Variation in solar UV radiation controls stratospheric ozone production and may trigger climate change. Haigh (1994, 1999) performed simulations with climate models to study the relation between the 11-year solar activity cycle, ozone production, and climate change. A chemical-atmospheric model showed that a 1 percent increase in UV radiation at the maximum of a solar cycle generated 1–2 percent more ozone in the stratosphere. This increased amount of ozone used as input in a climate model resulted in the warming of the lower stratosphere by the absorption of more UV radiation. In addition, stratospheric winds were strengthened, and the tropospheric westerly jet streams were displaced poleward. The position of these jets determines the latitudinal extent of the Hadley Cell circulation and, therefore, the poleward shift of the jets resulted in a similar displacement of the descending parts of the Hadley Cells. The change in circulation ultimately caused a poleward relocation of mid-latitude storm tracks. The opposite effect to that described by Haigh may have played a role during the climate shift of 850 BC. Reduced solar activity, as indicated by the observed strong increase of atmospheric ^{14}C and ^{10}Be, may have resulted in a decrease in stratospheric ozone. A decrease in latitudinal extent of the Hadley Cell circulation with equatorward relocation of mid-latitude storm tracks would consequently follow (Haigh 1994, 1999; van Geel *et al.* 1998).

The second, and more controversial, potential amplification factor involves changes in cosmic ray flux (as a consequence of a fluctuating solar wind) and its relationship with global cloud cover. According to Pudovkin and Raspopov (1992), and Raspopov *et al.* (1997), ionization by cosmic rays positively affects aerosol formation and cloud nucleation. Svensmark and Friis-Christensen (1997) found indications for the possible importance of this process. They recorded a correlation between the variation in cosmic ray flux and observed global cloud cover over the solar cycle from 1983 to 1995. An increase in global cloud cover was thought to cause global cooling. The increase in cloudiness and accompanying cooling corresponds with reconstructed wetter and cooler conditions at mid-latitude in Europe at around 850 BC, when there was an abrupt and steep increase of cosmogenic radiocarbon in the atmosphere, but according to van Geel *et al.* (2001) the UV-ozone amplification mechanism is more probable than the cosmic–ray–cloud hypothesis. Wagner *et al.* (2001) also did not find evidence for a cosmic ray climate connection.

Conclusions

The past few decades were the warmest during the past millennium (IPCC 2001; Osborn and Briffa 2006, and references therein) owing to greenhouse gas-induced warming. The temperature rise during the first half of the last century, however, is clearly linked to increased solar activity (Lockwood *et al.* 1999). Reconstruction of the solar variability reveals that, at present, the Sun is in a very active phase, but shows no clear indication of a trend since 1980 when the satellite-based measurements of the total solar irradiance (TSI) began. Based on paleorecords, and considering the fact that amplification mechanisms for changing solar activity are not well understood – and therefore cannot yet be sufficiently quantified in climate models – we conclude that the solar forcing of climate change may be more important than has been suggested by the Intergovernmental Panel on Climate Change (IPCC 2001). If the Little Ice Age and the subsequent warming were mainly driven by changes in solar activity this component of natural forcing plays an important role in estimating the future climate warming induced by greenhouse gases due to human activities. A future major decline of solar activity as predicted by de Jager (2005) may bring further progress in our understanding of present and future climate change.

Acknowledgments

The authors thank Jan van Arkel (IBED-UvA) for help with the illustrations and Rick Battarbee and two anonymous reviewers for valuable comments. Investigations by BvG and colleagues were supported by INTAS and the Research Council for Earth and Life Sciences (ALW) with financial aid from The Netherlands Organization for Scientific Research (NWO). The work of JB is supported by the Swiss National Science Foundation. Dan Yeloff kindly corrected the English.

References

Ammann C.M., Meehl G.A., Washington W.M. & Zender C.S. (2003) A monthly and latitudinally varying volcanic forcing dataset in simulations of 20th century climate. *Geophysical Research Letters*, **30**(12), art. no.-1657.

Arbogast R., Jacomet S., Magny M. & Schibler J. (2006) The significance of climate fluctuations for lake levels changes and shifts in subsistence economy during the Late Neolithic (4300–2400 B.C.) in Central Europe. *Vegetation History and Archaeobotany*, **15**, 403–418.

Baker P.A., Fritz S.C., Garland J. & Ekdahl E. (2005) Holocene hydrologic variation at Lake Titicaca, Bolivia/Peru, and its relationship to North Atlantic climate variation. *Journal of Quaternary Science*, **20**(7–8), 655–662.

Bay R.C., Bramall N. & Price P.B. (2004) Bipolar correlation of volcanism with millennia climate change. *Proceedings of the National Academy of Sciences*, **101**, 6341–6345.

Beer J., Blinov A., Bonani G., *et al.* (1990) Use of ^{10}Be in polar ice to trace the 11-year cycle of solar activity. *Nature*, **347**(6289), 164–166.

Beer J., Mende W. & Stellmacher R. (2000) The role of the Sun in climate forcing. *Quaternary Science Reviews*, **19**(1–5), 403–415.

Beer J., Muscheler R., Wagner G., *et al.* (2002) Cosmogenic nuclides during isotope stages 2 and 3. *Quaternary Science Reviews*, **21**, 1129–1139.

Beer J., Vonmoos M. & Muscheler R. (2006) Solar variability over the past several millennia. *Space Science Reviews*, doi:10.1007/s11214-006-9047-4.

Berger A. & Loutre M.F. (2004) Astronomical theory of palaeoclimates. *Comptes Rendus Geoscience*, **336**(7–8), 701–709.

Birks J. (this volume) Holocene climate research – progress, paradigms, and problems. In: *Global Warming and Natural Climate Variability: a Holocene Perspective* (Eds R.W. Battarbee & H.A. Binney), pp. 7–57. Blackwell, Oxford.

Blaauw M., Heuvelink G.B.M., Mauquoy D., van der Plicht J. & van Geel B. (2003) A numerical approach to C-14 wiggle-match dating of organic deposits: best fits and confidence intervals. *Quaternary Science Reviews*, **22**(14), 1485–1500.

Blytt A. (1882) Die Theorie der wechselnden kontinentalen und insularen Klimate. In: *Botanische Jahrbücher für Systematik, Pflanzengeschichte und Pflanzengeographie 2*, pp. 1–50.

Bond G., Kromer B., Beer J., *et al.* (2001) Persistent solar influence on North Atlantic climate during the Holocene. *Science*, **294**(5549), 2130–2136.

Briffa K.R., Jones P.D., Schweingruber F.H. & Osborn T.J. (1998) Influence of volcanic eruptions on northern hemisphere summer temperature over the past 600 years. *Nature*, **393**(4), 450–455.

Briffa K.R., Osborn T.J., Schweingruber F.H., *et al.* (2001) Low-frequency temperature variations from a northern tree ring density network. *Journal of Geophysical Research–Atmospheres*, **106**(D3), 2929–2941.

Briffa P.D., Osborn T.J. & Schweingruber F.H. (2004) Large-scale temperature inferences from tree rings: a review. *Global and Planetary Change*, **40**, 11–26.

Castellano E., Becagli S., Jouzel J., *et al.* (2004) Volcanic eruption frequency in the last 45 ky as recorded in EPICA-Dome-C ice core (East Antarctica) and its relationship to climate changes. *Global and Planetary Change*, **42**, 195–205.

Clarke G., Leverington D., Teller J. & Dyke A. (2003) Superlakes, megafloods, and abrupt climate change. *Science*, **301**, 922–923.

Claussen M. (this volume) Holocene rapid land-cover changes – evidence and theory. In: *Global Warming and Natural Climate Variability: a Holocene Perspective* (Eds R.W. Battarbee & H.A. Binney), pp. 232–253. Blackwell, Oxford.

Cook E.R. (1995) Temperature histories from tree-rings and corals. *Climate Dynamics*, **11**(4), 211–22.

Crowley T.J. (2000a) *Causes of Climate Change Over the Past 1000 Years*. IGBP PAGES/World Data Center for Paleoclimatology Data Contribution Series #2000-045. International Geosphere–Biosphere Program, Bern.

Crowley T.J. (2000b) Causes of climate change over the past 1000 years. *Science*, **289**, 270–277.

Crucifix M. (this volume) Modelling the climate of the Holocene. In: *Global Warming and Natural Climate Variability: a Holocene Perspective* (Eds R.W. Battarbee & H.A. Binney), pp. 98–122. Blackwell, Oxford.

D'Arrigo R., Wilson R. & Jacoby G. (2006) On the long-term context for late twentieth century warming. *Journal of Geophysical Research-Atmospheres*, **111**(D3).

De Jager C. (2005) Solar forcing of climate. 1: solar variability. *Space Science Reviews*, **120**, 197–241.

Denton G.H. & Karlén W. (1973) Holocene climatic variations – their pattern and possible cause. *Quaternary Research*, **3**(2), 155–205.

Esper J., Cook E.R. & Schweingruber F.H. (2002) Low-frequency signals in long tree-ring chronologies for reconstructing past temperature variability. *Science*, **295**, 2250–2253.

Esper J., Cook E.R., Krusic P.J., Peters K. & Schweingruber F.H. (2003) Tests of the RCS method for preserving low-frequency variability in long tree-ring chronologies. *Tree-Ring Research*, **59**(2), 81–98.

Fischer E.M. (2003) *Regional and seasonal impact of volcanic eruptions on European climate over the last centuries*. Diploma, University of Bern, Bern.

Fischer E.M. (2005) Climate response to tropical eruptions. *PAGES Newsletter* **13**(3), 8–10.

Fröhlich C. & Lean J. (2004) Solar radiative output and its variability: evidence and mechanisms. *Astronomy and Astrophysics Review*, **12**(4), 273–320.

Goosse H., Renssen H., Selten F.M., Haarsma R.J. & Opsteegh J.D. (2002) Potential causes of abrupt climate events: a numerical study with a three-dimensional climate model. *Geophysical Research Letters*, **29**(18), 1860, doi:10.1029/2002 GL014993.

Goosse H., Mann M.E. & Renssen H. (this volume) Climate of the past millennium: combining proxy data and model simulations. In: *Global Warming and Natural Climate Variability: a Holocene Perspective* (Eds R.W. Battarbee & H.A. Binney), pp. 163–188. Blackwell, Oxford.

Haigh J.D. (1994) The role of stratospheric ozone in modulating the solar radiative forcing of climate. *Nature*, **370**, 544–546.

Haigh J.D. (1999) Modelling the impact of solar variability on climate. *Journal of Atmospheric And Solar Terrestrial Physics*, **61**, 63–72.

Haigh J. & Blackburn M. (2006) Solar influences on dynamical coupling between the stratosphere and troposphere *Space Science Reviews*, **125**(1–4), 331–3444.

Haltia-Hovi E., Saarinen T. & Kukkonen M. (2007) A 2000-year record of solar forcing on varved lake sediment in eastern Finland. *Quaternary Science Reviews*, **26**, 678–689.

Holzhauser H., Magny M. & Zumbuhl H.J. (2005) Glacier and lake-level variations in west-central Europe over the last 3500 years. *The Holocene*, **15**(6), 789–801.

Hu F.S., Kaufman D., Yoneji S., *et al.* (2003) Cyclic variation and solar forcing of Holocene climate in the Alaskan subarctic. *Science*, **301**(5641), 1890–1893.

IPCC Working Group 1 (2001) *Climate Change 2001: The Scientific Basis*. Contribution of Working Group I to the Third Assessment Report of the Intergovernmental Panel on Climate Change, Cambridge University Press, Cambridge, 881 pp.

Jansen E., Andersson C., Moros M., Nisancioglu K.H., Nyland B.F. & Telford R.J. (this volume) The early to mid-Holocene thermal optimum in the North

Atlantic. In: *Global Warming and Natural Climate Variability: a Holocene Perspective* (Eds R.W. Battarbee & H.A. Binney), pp. 123–137. Blackwell, Oxford.

Kilian M.R., Van der Plicht J. & Van Geel B. (1995) Dating raised bogs: new aspects of AMS 14C wiggle matching, a reservoir effect and climatic change. *Quaternary Science Reviews*, **14**, 959–966.

Laskar J., Robutel P., Joutel F., Gastineau M., Correia A.C.M. & Levrard B. (2004) A long-term numerical solution for the insolation quantities of the Earth. *Astronomy and Astrophysics*, **428**(1), 261–285.

Lockwood M., Stamper R. & Wild M.N. (1999) A doubling of the Sun's coronal magnetic field during the past 100 years. *Nature*, **399**, 437–39.

Maasch K.A., Mayewski P.A., Rohling E.J., *et al.* (2005) A 2000-year context for modern climate change. *Geografiska Annaler*, **87A**, 7–15.

Magny M. (2004) Holocene climate variability as reflected by mid-European lake-level fluctuations and its probable impact on prehistoric human settlements. *Quaternary International*, **113**, 65–79.

Magny M. (2007) Lake level studies – West-Central Europe. In: *Encyclopedia of Quaternary Studies*, Vol. 2, pp. 1389–99. Elsevier, Amsterdam.

Mangini A., Spotl C. & Verdes P. (2005) Reconstruction of temperature in the Central Alps during the past 2000 yr from a delta O-18 stalagmite record. *Earth and Planetary Science Letters*, **235**(3–4), 741–51.

Masarik J. & Beer J. (1999) Simulation of particle fluxes and cosmogenic nuclide production in the Earth's atmosphere. *Journal of Geophysical Research*, **104**(D10), 12,099–12,111.

Mauquoy D., Van Geel B., Blaauw M. & van der Pflicht J. (2002) Evidence from northwest European bogs shows "Little Ice Age" climatic changes driven by variations in solar activity. *The Holocene*, **12**(1), 1–6.

Mayewski P.A., Meeker L.D., Twickler M.S., *et al.* (1997) Major features and forcing of high-latitude northern hemisphere atmospheric circulation using a 110 000 year-long glaciochemical series. *Journal of Geophysical Research*, **102**(C12), 26345–26366.

Morrill C., Overpeck J.T., Cole J.E., Liu K.B., Shen C.M. & Tang L.Y. (2006) Holocene variations in the Asian monsoon inferred from the geochemistry of lake sediments in central Tibet. *Quaternary Research*, **65**(2), 232–243.

Muscheler R., Beer J., Wagner G., *et al.* (2004) Changes in the carbon cycle during the last deglaciation as indicated by the comparison of ^{10}Be and ^{14}C records. *Earth and Planetary Science Letters*, **219**(3–4), 325–340.

Neff U., Burns S., Mangini A., Mudelsee M., Fleitmann D. & Matter A. (2001) Strong coherence between solar variability and the monsoon in Oman between 9 and 6 kyr ago. *Nature*, **411**(6835), 290–293.

Oldfield F. (this volume) The role of people in the Holocene. In: *Global Warming and Natural Climate Variability: a Holocene Perspective* (Eds R.W. Battarbee & H.A. Binney), pp. 58–97. Blackwell, Oxford.

Osborn T.J. & Briffa K.R. (2006) The spatial extent of 20th-century warmth in the context of the past 1200 years. *Science*, **311**, 841–844.

Patterson R.T., Prokoph A. & Chang A. (2004) Late Holocene sedimentary response to solar and cosmic ray activity influenced climate variability in the NE Pacific. *Sedimentary Geology*, **172**(1–2), 67–84.

Pudovkin M.I. & Raspopov O.M. (1992) The mechanism of action of solar activity on the state of the lower atmosphere and meteorological parameters (a review). *Geomagnetism and Aeronomy*, **32**, 593–608.

Raspopov O.M., Shumilov O.I., Kasatkina E.A., Jacoby G. & Dergachev V.A. (1997) *The Cosmic Ray Influence on Cloudy and Aerosol Layers of the Earth and Connection of Solar Cycle Lengths to Global Surface Temperature.* Preprint 1694, Physical-Technical Institute of Russian Academy of Sciences, St.-Petersburg, 20 pp.

Renssen H., van Geel B., van der Plicht J. & Magny M. (2000) Reduced solar activity as a trigger for the start of the Younger Dryas? *Quaternary International*, **68–71**, 373–383.

Robock A. (2000) Volcanic eruptions and climate. *Reviews of Geophysics*, **38**(2), 191–219.

Schibler J., Jacomet S., Hüster-Plogmann H. & Brombacher C. (1997) Economic crash in the 37th and 36th centuries cal. BC in Neolithic lake shore sites in Switzerland. *Anthropozoologica*, **25/26**, 553–570.

Sernander R. (1910) *Die schwedischen Torfmoore als Zeugen postglazialer Klimaschwankungen.* Die Veränderungen des Klimas seit dem Maximum der Letzten Eiszeit, Stockholm.

Solanki S.K., Usoskin I.G., Kromer B., Schussler M. & Beer J. (2004) Unusual activity of the Sun during recent decades compared to the previous 11 000 years. *Nature*, **431**(7012), 1084–1087.

Speranza A., van Geel B. & van der Plicht J. (2003) Evidence for solar forcing of climate change at ca. 850 cal BC from a Czech peat sequence. *Global and Planetary Change*, **35**(1–2), 51–65.

Stager J.C., Ryves D., Cumming B.F., Meeker L.D. & Beer J. (2005) Solar variability and the levels of Lake Victoria, East Africa, during the last millennium. *Journal of Paleolimnology*, **33**(2), 243–251.

Stuiver M. & Braziunas T.F. (1993) Sun, ocean, climate and atmospheric $^{14}CO_2$, an evaluation of causal and spectral relationships. *The Holocene*, **3**, 289–305.

Stuiver M., Braziunas T.F., Becker B. & Kromer B. (1991) Climatic, solar, oceanic and geomagnetic influences on Late-Glacial and Holocene atmospheric $^{14}C/^{12}C$ change. *Quaternary Research*, **35**, 1–24.

Svensmark H. & Friis-Christensen E. (1997) Variation of cosmic ray flux and global cloud coverage – a missing link in solar-climate relationships. *Journal of Atmospheric and Terrestrial Physics*, **59**(11), 1225–1232.

Van der Plicht J., van Geel B., Bohncke S.J.P., *et al.* (2004) The Preboreal climate reversal and a subsequent solar-forced climate shift. *Journal of Quaternary Science*, **19**(3), 263–269.

Van Dorland R. (Ed.) (2006) *Scientific Assessment of Solar Induced Climate Change.* Netherlands Environmental Assessment Agency, Bilthoven, 154 pp.

Van Geel B. & Berglund B.E. (2000) A causal link between a climatic deterioration around 850 cal BC and a subsequent rise in human population density in NW-Europe? *Terra Nostra*, **7**, 126–130.

Van Geel B. & Mook W.G. (1989) High-resolution ^{14}C dating of organic deposits using natural atmospheric 14C variaitions. *Radiocarbon*, **31**, 151–156.

Van Geel B. & Renssen H. (1998) Abrupt climate change around 2650 BP in North-West Europe: evidence for climatic teleconnections and a tentative explanation. In: *Water, Environment and Society in Times of Climatic Change* (Eds A.S. Issar & N. Brown), pp. 21–41. Kluwer Academic Publishers, Dordrecht.

Van Geel B., Buurman J. & Waterbolk H.T. (1996) Archaeological and palaeo-ecological indications of an abrupt climate change in The Netherlands, and evidence for climatological teleconnections around 2650 BP. *Journal of Quaternary Science*, **11**(6), 451–460.

Van Geel B., van der Plicht J., Kilian M.R., *et al.* (1998) The sharp rise of Δ^{14}C ca. 800 cal BC: possible causes, related climatic teleconnections and the impact on human environments. *Radiocarbon*, **40**(No. 1), 535–550.

Van Geel B., Raspopov O.M., Renssen H., van der Plicht J., Dergachev V.A. & Meijer H.A.J. (1999) The role of solar forcing upon climate change. *Quaternary Science Reviews*, **18**(3), 331–338.

Van Geel B., Heusser C.J., Renssen H. & Schuurmans C.J.E. (2000) Climatic change in Chile at around 2700 BP and global evidence for solar forcing: a hypothesis. *The Holocene*, **10**(5), 659–664.

Van Geel B., Renssen H. & van der Plicht J. (2001) Evidence from the past: solar forcing of climate change by way of cosmic rays and/or by solar UV? In: *Workshop on Ion–Aerosol–Cloud Interactions* (Ed. J. Kirkby), Vol. CERN 2001–007, pp. 24–29. Conseil Européen pour la Recherche Nucléaire (European Council for Nuclear Research), Geneva.

Van Geel B., Bokovenko N.A., Burova N.D., *et al.* (2004a) Climate change and the expansion of the Scythian culture after 850 BC: a hypothesis. *Journal of Archaeological Science*, **31**(12), 1735–1742.

Van Geel B., Shinde V. & Yasuda Y. (2004b) Solar forcing of climate change and a monsoon-related cultural shift in western India around 800 cal. yrs BC. In: *Monsoon and Civilization* (Eds Y. Yasuda & V. Shinde), pp. 275–279. Roli Books, New Delhi.

Verschuren D. & Charman D. (this volume) Latitudinal linkages in late-Holocene moisture-balance variation. In: *Global Warming and Natural Climate Variability: a Holocene Perspective* (Eds R.W. Battarbee & H.A. Binney), pp. 189–231. Blackwell, Oxford.

Verschuren D., Laird K.R. & Cumming B.F. (2000) Rainfall and drought in equatorial east Africa during the past 1100 years. *Nature*, **403**(27 January 2000), 410–414.

Versteegh G.J.M. (2005) Solar forcing of climate: evidence from the past. *Space Science Reviews*, **120**, 243–286.

Vonmoos M., Beer J. & Muscheler R. (2006) Large variations in Holocene solar activity – constraints from [10]Be in the GRIP ice core. *Journal of Geophysical Research D*, **111**, doi:10.1029/2005JA011500.

Wagner G., Livingstone D.M., Masarik J., Muscheler R. & Beer J. (2001) Some results relevant to the discussion of a possible link between cosmic rays and the Earth's climate, *Journal of Geophysical Research*, **106**(D4), 3381–3388.

Wang Y.J., Cheng H., Edwards R.L., *et al.* (2005) The Holocene Asian monsoon: Links to solar changes and North Atlantic climate. *Science*, **308**(5723), 854–857.

Wiersma A.P. & Renssen H. (2006) Model-data comparison for the 8.2 ka BP event: confirmation of a forcing mechanism by catastrophic drainage of Laurentide Lakes. *Quaternary Science Reviews*, **25**, 62–88.

Wu Y.H., Lucke A., Jin Z.D., *et al.* (2006) Holocene climate development on the central Tibetan plateau: a sedimentary record from Cuoe Lake. *Palaeogeography, Palaeoclimatology, Palaeoecology*, **234**(2–4), 328–340.

Xiao S., Li A., Liu J.P., *et al.* (2006) Coherence between solar activity and the East Asian winter monsoon variability in the past 8000 years from Yangtze River-derived mud in the East China Sea. *Palaeogeography, Palaeoclimatology, Palaeoecology*, **237**, 293–304.

Zielinski G.A. & Mershon G.R. (1997) Paleoenvironmental implications of the insoluble microparticle record in the GISP2 (Greenland) ice core during the rapidly changing climate of the Pleistocene-Holocene transition. *Geological Society of America Bulletin*, **109**(5), 547–559.

7 Climate of the past millennium: combining proxy data and model simulations

Hugues Goosse, Michael E. Mann, and Hans Renssen

Keywords

Climate reconstructions, past millennium, model results, proxy data, forced climate variability, internal climate variability, NAO, ENSO, data assimilation

Introduction

By studying the climate of the past millennium in great detail, it is possible to analyze the impact of human activities during the past few centuries compared with natural driving forces, such as volcanic eruptions and solar variations, and internal variations related only to the complex dynamics of the climate system itself. An optimal analysis is obtained when proxy data and model results are combined. Proxy data provide the "ground truth" (e.g. Mann *et al.* 1998; Jones and Mann 2004; Luterbacher *et al.* 2004; Moberg *et al.* 2005), whereas models can be used to interpret the observed changes, in particular to analyze the mechanisms responsible for the observed response of the climate system to changes in external forcing (e.g. Crowley 2000; Shindell *et al.* 2001; Stott *et al.* 2001; Bertrand *et al.* 2002; Bauer *et al.* 2003; Goosse *et al.* 2005b). The majority of past studies have been heavily biased towards one or other of these two particular ways of analyzing past changes, while only briefly discussing (if at all) the other. Our goal here is to show that very interesting insights and synergies could be gained from a combined analysis of proxy records and model results.

Model results and proxy data can be complementary in several ways. First, proxy data can be used to evaluate the validity of model results (e.g. Shindell *et al.* 2001; Bertrand *et al.* 2002; Bauer *et al.* 2003; Goosse *et al.* 2005a). This is a necessary step before any analysis of model results, but it is not a trivial task, as any discrepancy between model results and data could be due not only to model deficiencies but also to several other processes. The uncertainties in the forcing history could have an impact on model results. Proxy data record a climatic signal but also include a component of nonclimatic noise that is difficult to estimate. Furthermore, proxy records could be influenced by regional or local changes whereas model results provide information at a spatial scale of hundreds of kilometers. Second, model results could be used to provide information on the climatic signal that is recorded by the proxy data, disentangling the contribution of various climatic variables. For instance, changes in ground surface temperature are influenced by variations in surface temperature, snow cover, and changes in surface vegetation. Model results include an estimate of variations of all such variables and could thus help in the interpretation of the recorded variations in proxy data (e.g. González-Rouco *et al.* 2003; Mann and Schmidt 2003). Additionally, models provide a physically consistent and complete set of data that could be used to test the method applied to reconstruct past changes from proxy data with incomplete temporal and spatial coverage (e.g. Rutherford *et al.* 2003; von Storch *et al.* 2004; Mann *et al.* 2005b). Finally, evidence deduced from proxy data and models can be combined to describe and analyze the causes of past changes. All these elements will be illustrated below.

Reviews have extensively described recent work on the past millennium climate (e.g. Jones and Mann 2004), and we therefore provide only some general information about proxy records, reconstructions, and model simulation below to give an introduction to the subject. In subsequent sections, reconstruction methods are tested using synthetic time-series derived from model results; reconstructions and model results are used to detect the role of various forcings at the hemispheric scale and changes at regional scale are discussed; and some considerations are presented about promising techniques that could allow the direct assimilation of proxy records in models, providing a single estimate of past changes and of the mechanisms responsible for those changes based on all the available information.

Proxy data and proxy reconstructions

The lack of widespread instrumental climate records prior to the mid-19th century requires that we turn to "proxy data" in our attempts to reconstruct how the climate has changed over past centuries. When, as is typically the case in studies of the past millennium, our interest is primarily in annually resolved climate variations, we must turn to "high-resolution" climate proxy data, such as tree rings, corals, ice cores, and historical documentary records. Mann *et al.* (1998) assembled a network of 415 annually resolved proxy data (predominantly dendroclimatic), for use in reconstructing temperature patterns over the past 1000 years. More recently, Mann *et al.* (submitted) have assembled a much larger network of 1209 annual and decadally resolved proxy data consisting of tree rings (including 105 gridded maximum latewood density or "MXD" tree-ring series – see Briffa *et al.* 2001), corals and sclerosponge series, ice cores, lake sediments, speleothems, historical

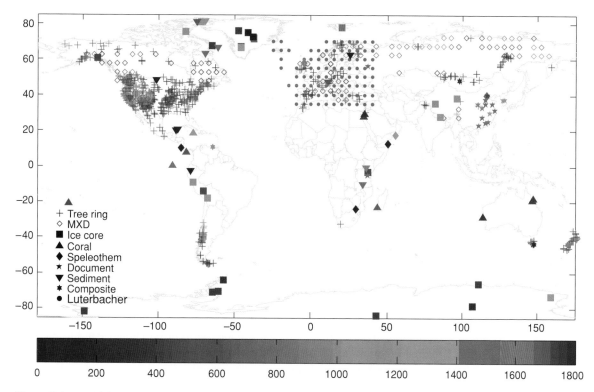

Figure 7.1 Spatial distribution of high-resolution climate proxy data. Nine different proxy types are denoted with different symbols. Starting date of proxy record is indicated by color scale.

documentary series, and a set of 89 gridded European surface temperature reconstructions back to AD 1500 that are based on a composite of proxy, historical, and early instrumental data (Luterbacher *et al.* 2004). The spatial distribution of the proxy data is shown in Figure 7.1.

Most previous proxy data studies have focused on hemispheric or global mean temperature (an example of one reasonably representative reconstruction is shown later in Figure 7.3), although some studies have also attempted to reconstruct the underlying spatial patterns of past surface temperature changes at global (e.g. Mann *et al.* 1998) and regional (e.g. Luterbacher *et al.* 2004) scales, and other fields such as regional sea-level pressure, and North American drought. Other studies have focused on the reconstruction of particular climate indices such as the North Atlantic Oscillation (NAO) and related Arctic Oscillation (AO), the Pacific Decadal Oscillation (PDO), and Atlantic Multi-decadal Oscillation (AMO), and indices such as the Southern Oscillation Index (SOI) and Niño3 index, which attempt to describe the variability in the El Niño–Southern Oscillation (ENSO) phenomenon. An extensive review is provided by Jones and Mann (2004).

Most reconstructions of hemispheric or global mean temperatures or of climate indices have employed the "Composite Plus Scale" (CPS) methodology, where a selection of natural climate "proxies", such as tree rings, ice cores, or corals, are first standardized, then composited to form a regional or hemispheric mean temperature series (cf. Jones and Mann 2004). More recently, methodological adjustments have been applied to the CPS method, including the use of proxies selected specifically for their retention of low-frequency variability (Esper *et al.* 2002; Mann and Jones 2003) or the inclusion of low-resolution (decadal or centennial-scale) proxies that might be well-suited to reconstructing low-frequency climate variability (Moberg *et al.* 2005). In the latter case, one must take care during the standardizing and compositing process to ensure that the low-frequency component is not artificially inflated relative to the higher-frequency component (Mann *et al.* 2005b).

An alternative is the climate field reconstruction (CFR) approach, which combines information from multiple proxy records in reconstructing the underlying spatial patterns of past climate change (e.g. Mann *et al.* 1998; Luterbacher *et al.* 2004; Rutherford *et al.* 2005). In this case, hemispheric or global means, as well as any climate indices of interest, are computed directly from the reconstructions of the underlying spatial field. A key advantage of the CFR approach is that the spatial reconstructed patterns can be directly compared with model-predicted patterns of climate responses to forcing (Shindell *et al.* 2001, 2004; Waple *et al.* 2002). Climate field reconstruction methods typically make use of both local and nonlocal information by relating predictors (i.e. the long-term proxy climate data) to the temporal variations in the large-scale patterns of the spatial field of interest. Such an approach takes greater advantage of the potential climate information in the proxy data-set. Two good examples are the close link between drought-sensitive tree rings from the western USA and surface temperature patterns in the Pacific Ocean, and ice-core records from Greenland which are closely tied to the behavior of the NAO, which influences cold-season temperature patterns over North America and Eurasia. These relations can be exploited in climate reconstructions. Climate field reconstruction approaches depend more on assumptions about the stationarity of relationships between proxy indicators and large-scale climate patterns than simpler methods such as CPS. Investigations using synthetic proxy data ("pseudo-proxies"), however, suggest that these assumptions are likely to hold well for the range of variability inferred over the past one or two millennia. These investigations, which are described in more detail below, demonstrate that CFR methods are quite skilful in independent tests of their fidelity.

Models and climate forcing

Numerous simulations devoted to the study of the past millennium climate have been performed using Energy Balance Models (EBMs; e.g. Free and Robock 1999; Crowley 2000; Hegerl *et al.* 2003), general circulation atmospheric models coupled to a simplified ocean model (e.g. Shindell *et al.* 2001; Waple *et al.* 2002), Earth Models of Intermediate Complexity (EMIC; e.g. Bertrand *et al.* 2002; Bauer *et al.* 2003; Gerber *et al.* 2003, Goosse *et al.* 2005a), as well as comprehensive

Atmosphere–Ocean General Circulation Models (AOGCMs; e.g. Cubasch *et al.* 1997; González-Rouco *et al.* 2003, 2006; Widmann and Tett 2003). Such types of model have also been used to study other periods of the Holocene (cf. Crucifix, this volume; Claussen, this volume).

To reproduce the observed changes, climate models have to be driven by reconstructions of past variations in external forcings. These forcings can be grouped into two categories: natural and anthropogenic. The main radiative perturbations of purely natural origin over the past millennium are related to changes in solar irradiance (cf. Beer and van Geel, this volume) and the release of aerosols into the atmosphere during explosive volcanic eruptions. Orbital forcing, which is dominant on longer time-scales (cf. Crucifix, this volume), plays a weaker role during the past millennium. The largest radiative perturbation induced by human activity is related to the increase in greenhouse gas concentrations in the atmosphere. The increase in the atmospheric aerosol load is associated with a significant radiative forcing, particularly close to the main industrialized regions (e.g. Boucher and Pham 2002). These two forcings are mainly restricted to the past 250 years, with a strong amplification during the past 50 years. The rate of land-use change has also increased during the latter period, although in some regions its impact is significant throughout the past millennium (Ramankutty and Foley 1999) and earlier (cf. Oldfield, this volume).

Figure 7.2 shows typical reconstructions of external forcings during the past millennium. The uncertainties in these forcings are relatively large, except for the one associated with the increase in greenhouse gas concentrations. The majority of the forcings have to be reconstructed from indirect sources such as the amount of sulfates in ice cores for volcanic activities or of cosmogenic isotopes for solar irradiance (cf. Beer and van Geel, this volume). In addition, the magnitude of the negative aerosol forcing depends on numerous complex phenomena and the estimates of the present-day aerosol forcing range from nearly zero to more than 2 W m^{-2} (e.g. Andreae *et al.* 2005). This illustrates that the time evolution of the past radiative forcing, as well as its magnitude, is not well constrained.

Some model simulations include only a small number of the forcings mentioned above (generally one), the goal being to understand precisely the mechanism behind the response to a particular forcing (e.g. Shindell *et al.* 2001, 2004; Waple

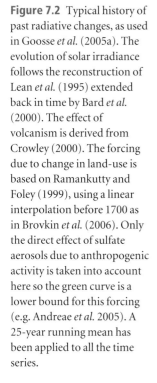

Figure 7.2 Typical history of past radiative changes, as used in Goosse *et al.* (2005a). The evolution of solar irradiance follows the reconstruction of Lean *et al.* (1995) extended back in time by Bard *et al.* (2000). The effect of volcanism is derived from Crowley (2000). The forcing due to change in land-use is based on Ramankutty and Foley (1999), using a linear interpolation before 1700 as in Brovkin *et al.* (2006). Only the direct effect of sulfate aerosols due to anthropogenic activity is taken into account here so the green curve is a lower bound for this forcing (e.g. Andreae *et al.* 2005). A 25-year running mean has been applied to all the time series.

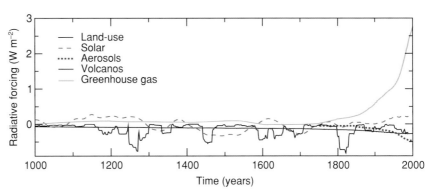

Figure 7.3 Annual mean temperature anomaly averaged over the Northern Hemisphere in simulations performed with two AOGCMs (CCSM, Ammann *et al.*, cited in Jones and Mann (2004) in green, and ECHO-G, González-Rouco *et al.* 2003, 2006) and in one three-dimensional Earth model of intermediate complexity (ECBILT–CLIO–VECODE, Goosse *et al.* (2005a) in blue). The two simulations performed in ECHO-G (ERIK and ERIK2, dashed red and red respectively) differ only in their initial conditions, ERIK using one that is probably too warm. The reconstruction of Jones and Mann (2004) is in magenta, with the reconstruction plus and minus two standard deviations in grey. The time series are grouped in 10-year averages. The reference period is 1500–1850. This reference period was chosen to eliminate possible problems related to initial conditions and to the strong differences of specified anthropogenic forcing in the 20th century.

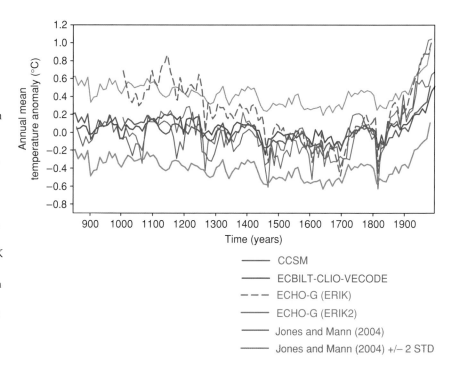

et al. 2002; Goosse and Renssen 2004; Mann *et al.* 2005a). Other simulations that try to reproduce the observed climate evolution tend to include all the important forcings. They all take into account at least the solar, volcanic, and greenhouse gas forcings, and some simulations include a more exhaustive list of forcings, accounting for the effect of land-use change, anthropogenic aerosols, and orbital forcing.

Figure 7.3 displays the annual mean temperature averaged over the Northern Hemisphere in such simulations using various three-dimensional climate models. The model results are in the range of the uncertainty associated with the reconstructions, showing generally colder temperatures during the 16–19th centuries and a larger warming during the 20th century. Nevertheless, differences between the models can be noticed, attributable to different factors. Firstly, the simulations use different reconstructions of past forcing changes. For solar radiations, despite the relatively large range in the reconstructions, recent model simulations tend to use similar reconstructions. The selection of the reconstruction of past solar irradiance has thus a weak impact on the differences shown in Figure 7.3 (Goosse *et al.* 2005b; Osborn *et al.* 2006). Models that do not include the anthropogenic aerosol forcing (such as the simulations using the ECHO-G model) simulate a much larger warming during the 20th century than the models that take into account this important forcing.

Secondly, the choice of initial conditions can play a role. This is illustrated in Figure 7.3 by the two simulations performed with the ECHO-G model that use the same forcing but in one case start from warm conditions (simulation ERIK) in year 1000 AD and in the other case from colder conditions (simulation ERIK2). In

ERIK, because of the selected warm initial state, the temperature drifts during several centuries towards colder temperatures until it becomes close to ERIK2, and the other simulations, around 1500. It is not possible to define precisely from observations the conditions that should be used to start a model in 1000 AD, but longer simulations starting in 850 AD or covering the past two millennia do not show warm conditions similar to the ones used in ERIK. The initial conditions used in ERIK2 and the subsequent evolution of temperature appear more reasonable than the ones of ERIK, although the lack of tropospheric aerosols in ERIK2 is still problematic for the realism of the simulation during the 20th century or the evaluation of the net changes in temperature from the pre-industrial to modern intervals.

Finally, different models use different grids and methods for the numerical resolution of the equations as well as different representations of some physical processes governing the evolution of the climate system, with a potential impact on the response of the model to a radiative perturbation. The behavior of models is often summarized by a few important characteristics, but this is a strong simplification of the complex behavior of three-dimensional models. Model sensitivity is classically defined by the change in global mean temperature between a controlled experiment using present-day conditions and an equilibrium experiment in which the atmospheric CO_2 concentration is doubled. The efficiency of the heat uptake by the ocean also plays an important role in the model response. In simple models, this could be related to bulk parameters such as an effective heat diffusion in the ocean (e.g. Hegerl *et al.* 2006; Osborn *et al.* 2006), whereas in more sophisticated models, the link with model parameters is much more complex. In Figure 7.3, such differences in model characteristics are likely responsible for the weaker temperature changes in ECBILT-CLIO-VECODE than in ECHO-G and CCSM, with the former model having relatively weak climate sensitivity.

At the hemispheric scale, precise model–data comparisons are difficult because of the uncertainties associated with the reconstructions, the forcing, and model formulation. As a consequence, an agreement between model and data is usually not a stringent test of the model results. At the regional scale, an additional difficulty is related to the large role played by the internal variability of the climate system, whereas at the hemispheric scale, the evolution of the system is largely imposed by the changes in the external forcing. As discussed later, the characteristics of some well-known modes of climate variability such as ENSO or NAO could be influenced by the forcing. Nevertheless, a large fraction of the variability of these modes is purely chaotic, related to internal dynamics, and cannot be related to any change in external conditions.

At regional scales, the internal variability of the system could easily mask the forced response. This is probably the main reason why the so-called Medieval Warm Period (covering roughly the period AD 900–1200) is not a synchronous phenomenon in all regions of the globe (e.g. Hughes and Diaz 1994; Bradley *et al.* 2003; Goosse *et al.* 2005a). The net effect of all forcings probably tended to induce warmer conditions during the period, but, depending on the sign of the anomalies associated with internal variability, the maximum temperatures occurred at different locations at different times.

Even if perfect models and perfect forcing time-series were available, each model experiment would simulate a possible evolution of the internal variability but not necessarily the one followed in the real world. Very precise initial conditions would only allow having an agreement between model and observations during a few weeks at most, a classic problem in weather forecasting (Lorenz 1963). In complex models (such as AOGCMs and some EMICs) that include a representation of the internal variability, it is possible to compare the statistics of one simulation with the observations or to try to extract the response to a forcing in the simulation and in available reconstructions. A precise comparison between temperature changes simulated in a model with a reconstruction in a particular region, for a particular year (or decade), is usually useless because it is generally impossible to state if any difference is due to uncertainties in the reconstruction, model, or forcing deficiencies, or simply to a different realization associated with the internal variability of the system.

In this framework, ensembles of simulations are particularly useful. To perform such an ensemble, slightly different initial conditions are selected. After a few simulated days, the internal variability evolves in a totally different way in the various members. If the ensemble is large enough, a reasonable sampling of the model variability could be obtained. In order to have an agreement between model and a reconstruction, the reconstruction must correspond to a reasonable member of the ensemble (Figure 7.4). As the members of the ensemble provide independent samples of the internal variability, averaging over all the members of the ensemble tends to filter out the internal variability, leaving only the forced response of the system. This provides a clear advantage compared with individual simulations of complex models that always provide a mixture between the forced response and the internal variability.

Figure 7.4 Anomaly of summer mean temperature in Fennoscandia averaged over 25 simulations that differ only in their initial conditions (black). The mean over the ensemble of 25 simulations plus and minus two standard deviations of the ensemble are in grey. The reconstruction of Briffa *et al.* (1992) is in green. As the reconstruction is within the range of the various simulations performed with the model, we can consider that there is no disagreement between model results and the reconstruction. The reference period is 1500–1850. The time series are grouped in 10-year averages.

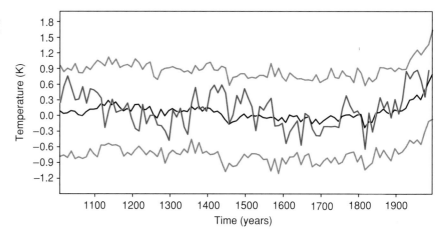

Using model-derived pseudo-proxies to test reconstruction methods

As discussed above there are primarily two distinct methods that have been used in proxy-based climate reconstruction: the CPS method, and the CFR method. Experiments using synthetic proxies or "pseudo-proxies" derived from climate model simulations have been used to test the performance of both the CPS (Mann *et al.* 2005b) and CFR methods (Mann and Rutherford 2002; Rutherford *et al.* 2003; Zorita *et al.* 2003; von Storch *et al.* 2004; Mann *et al.* 2005b; Wahl *et al.* 2006). Since our primary interest here is on regional and spatial patterns of change, we will focus on the CFR approach, which provides spatial patterns as well as hemispheric mean reconstructions.

Most of the studies described above have found that CFR-based reconstructions for past centuries are likely to be skilful, given the spatial distributions and estimated signal-to-noise characteristics of available proxy records. Von Storch *et al.* (2004) argue that proxy-based CFR reconstructions are likely to underestimate low-frequency variability, based on experiments using pseudo-proxy networks derived from a millennial simulation of the GKSS ECHO-G coupled model. Wahl *et al.* (2006) have challenged the von Storch *et al.* (2004) calculations, arguing that they suffer from an artefact of an originally undisclosed procedure in which data were detrended by von Storch *et al.* (2004) prior to calibration. Wahl *et al.* (2006) note that such a procedure a priori removes the primary pattern of low-frequency variability from the surface temperature data.

Von Storch *et al.* (2004) considered only the more primitive EOF-based CFR method of Mann *et al.* (1998), and not the RegEM CFR approach used in recent work by Mann and collaborators (e.g. Mann and Rutherford 2002; Rutherford *et al.* 2003, 2005). Mann *et al.* (2005b) tested the "RegEM" CFR method using a simulation of the climate of the past millennium using the NCAR CSM 1.4 coupled ocean–atmosphere model driven by estimated long-term natural and anthropogenic radiative forcing histories. The simulated Northern Hemisphere mean surface temperature history indicates modestly greater variability than most other simulations (cf. Figure 7.3), providing a challenging test for CFR methods. "Pseudo-proxy" data were constructed from the model surface temperature field to have signal-to-noise attributes similar to actual proxy data networks used in reconstructing past climates (e.g. as in Mann *et al.* 1998). Application of the RegEM method to the calibration of the surface temperature field against the pseudo-proxy networks produced skilful reconstructions of the actual surface temperature history, underscoring that CFR methods are likely to yield reliable reconstructions of past variations given realistic assumptions regarding the statistical attributes of actual proxy data networks.

Zorita *et al.* (2007) have, in turn, argued that the skilful results demonstrated by Mann *et al.* (2005b) may represent an artefact of the use of a long combined 19th–20th century (1856–1980) rather than a short, 20th century only (1900–1980) calibration interval. They also suggest that the skilful results demonstrated by Mann *et al.* (2005b) might be peculiar to the specific (NCAR CSM 1.4) model simulation used and not possible in other simulations such as the GKSS ECHO-G simulation used in their analyses. Finally, Zorita *et al.* (2007) and von Storch *et al.* (2006) argue that the results presented by Mann *et al.* (2005b) may represent an artefact of an inappropriate model for the proxy noise, and that poorer results would be achieved if proxy noise was assumed to be spectrally "red" rather than, as assumed in previous studies such as Mann *et al.* 2005b and von Storch *et al.* (2004), "white".

We investigate these issues based on more recent experiments by Mann *et al.* (2007a,b) wherein the RegEM approach was applied to surface temperature reconstructions using pseudo-proxy networks diagnosed from two different model simulations: the NCAR CSM 1.4 coupled simulation used by Mann *et al.* 2005b, and the GKSS ECHO-G "ERIK" simulation used by von Storch *et al.* (2004) and Zorita *et al.* (2007). The pseudo-proxy networks, as in Mann *et al.* 2005b, von Storch *et al.* (2004), and Zorita *et al.* (2007), have the spatial distribution of the full Mann *et al.* (1998) proxy network. Reconstructions were performed based on the "short" (1900–1980) calibration interval, and a **lower** signal-to-noise ratio ("SNR") than used by Zorita *et al.* (2007; the proxy signal-to-noise amplitude ratio was 0.4 whereas Zorita *et al.* (2007) used SNR = 0.5). The proxy noise was specified to be spectrally "red" as in Zorita *et al.* (2007), using the average noise autocorrelation coefficient $\rho = 0.32$ estimated from the actual Mann *et al.* (1998) network (see Mann *et al.* (2007a) for details). The resulting RegEM reconstructions are observed to track closely the actual model temperature histories for both the NCAR and GKSS simulations, and the Northern Hemisphere mean reconstructions lie entirely within the self-consistently estimated uncertainties of the true respective model histories (Figure 7.5). The most pronounced feature in the simulations – the cold temperatures of the 15th–19th centuries associated with anomalous negative radiative forcing by a combination of solar irradiance reduction and active explosive volcanic aerosol forcing – is well captured in both simulations, even with the short calibration interval and "red" proxy noise employed.

As discussed earlier the anomalous initial warmth and much of the subsequent long-term cooling trend in the GKSS "ERIK" simulation is an artefact of the model initialization procedures used. Mann *et al.* 2005b speculated that the unphysical drift resulting from this initialization might degrade CFR performance in tests using the "ERIK" simulation, since the drift pattern might not be captured over the modern training intervals used for calibration. The above test using the ERIK simulation (i.e. Figure 7.5b), however, demonstrates that this drift does not in fact pose an obstacle for the RegEM CFR method. In conclusion, the RegEM CFR approach should provide reliable reconstructions of past surface temperature histories over the past millennium within estimated uncertainties, given available proxy data networks.

Figure 7.5 Reconstruction of Northern Hemisphere (NH) mean temperature based on RegEM CFR reconstructions using "pseudo-proxy" networks taken from (a) NCAR CSM 1.4 and (b) GKSS ECHO-G "ERIK" simulations. In both cases, the pseudo-proxy network locations correspond to the 104 unique locations used by Mann *et al.* (1998), a proxy signal-to-noise ratio SNR = 0.4, red proxy noise with noise autocorrelation ρ = 0.32, and a 1900–1980 calibration interval is used. Self-consistent uncertainties in the reconstructions are estimated from the unresolved residual variance during an 1856–1899 "validation" interval, and are indicated by shading (95% uncertainty region). Actual model NH series is shown for comparison (black). All series are decadally smoothed. (From Mann *et al.* 2007a, ©American Meteorological Society.)

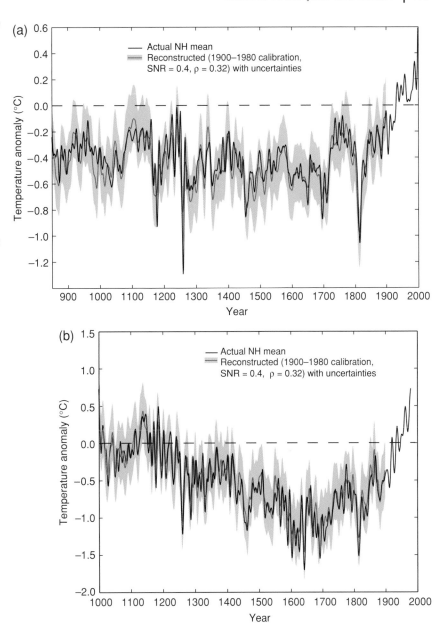

Using proxy reconstructions to constrain model parameters and detect the role of the various forcings

The time evolution during the past millennium of the annual mean temperature averaged over the Northern Hemisphere generally closely follows variations in the external forcing. This is probably the main reason for the success of relatively

Figure 7.6 Annual mean temperature anomaly averaged over the Northern Hemisphere in simulations performed with ECBILT–CLIO–VECODE using only one forcing at a time: greenhouse gas forcing (red), aerosol forcing (green), land-use changes (dark blue), volcanic forcing (orange), and solar forcing (light blue). The curves shown are the average over an ensemble of 10 simulations performed with the model using one of the forcings described in Figure 7.2. A 25-year running mean has been applied to all the time series. The reference period is 1500–1850.

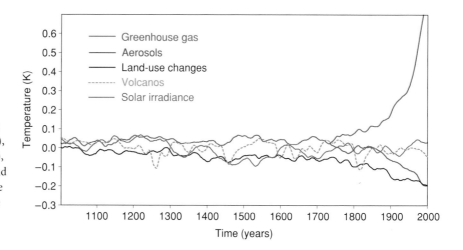

simple models in simulating this variable. At this spatial scale, the response of the climate system to the total forcing is, in good approximation, equal to the sum of the responses to all the individual forcings. It is thus possible to disentangle the role of the various forcings by performing simulations in which only one forcing is applied at a time (e.g. Crowley 2000; Bertrand *et al.* 2002; Bauer *et al.* 2003; see also Figure 7.6).

Such simulations have allowed that, as expected, a higher than average solar irradiance is associated with higher temperature and vice versa. As a consequence, the solar forcing alone, as reconstructed in Figure 7.2, induces similarly high temperatures during the 12th, 13th, and 20th centuries, whereas lower temperatures are simulated during the 16th and the period covering the late 18th to early 19th centuries. The large volcanic forcing is responsible for the simulated cold periods after large volcanic eruptions, in particular those of 1258, 1452, 1600, 1641, and 1815. The land–use changes are associated with a long-term cooling trend and the simulated cooling due to aerosols is mainly restricted to the 20th century. Greenhouse gas forcing mainly induces a large warming during the past 150 years, and if the forcing is excluded it is not possible to simulate the large temperature increase observed during the 20th century.

The temperature response averaged over the Southern Hemisphere is more complex than in the Northern Hemisphere. Between 50°S and 70°S, the Earth's surface is nearly exclusively covered by oceans, implying that the effective heat capacity at these latitudes is much larger than in the Northern Hemisphere, inducing a damped and delayed response to the forcing and thus a less clear imprint of the volcanic signal than in the Northern Hemisphere (Figure 7.7). Furthermore, the Southern Ocean experiences large-scale upwelling of relatively old water masses that acquire their characteristics decades to centuries before they reach the surface in the Southern Ocean. As a consequence, the surface temperature at one particular time in the Southern Ocean is not only influenced by the radiative forcing at that time but also by the previous history of forcing changes, which has left its imprint on the characteristics of the older water masses that upwell there. This

Figure 7.7 Annual mean temperature anomaly averaged over the Northern Hemisphere (red) and over the latitude band 50–70°S in simulations performed with ECBILT–CLIO–VECODE driven by natural and anthropogenic forcings. The time series are grouped by 25-year average and divided by their standard deviation to facilitate the comparison. The reference period is 1500–1850.

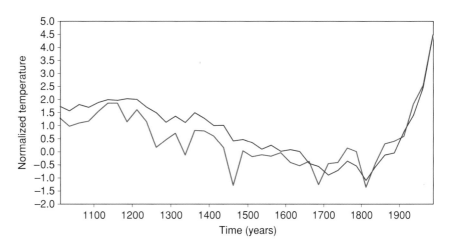

results in a much more complex link between the forcing and the response than in the Northern Hemisphere. For instance, this has led to the hypothesis that the Medieval Warm Period was delayed in the Southern Ocean by 50 to 200 years compared with the Northern Hemisphere (Goosse *et al.* 2004).

The previous discussion is mainly based on model results. The data are only used to show that the models are providing results in agreement with the reconstructions if all the forcings are included and that the conclusions deduced from model simulations appear reasonable. The results could thus be model-dependent and strongly influenced by uncertainties in the reconstruction of one particular forcing. It is possible, however, to combine directly the model results and observations in order to detect the influence of various forcings on Northern Hemispheric temperature during the past millennium (Hegerl *et al.* 2003), using techniques similar to the ones applied for the 20th century, and to distinguish the anthropogenic and natural forcings (e.g. Tett *et al.* 1999; Stott *et al.* 2001). The first step is to obtain, from model simulations, the signal associated with a forcing, i.e. the time evolution of the temperature change when only one forcing is applied. This is usually called the "fingerprint" of the forcing. A multiple regression is then used to obtain the linear combination of those fingerprints that provide the best agreement with the reconstructions. In other words, the model provides the shape of the response to a forcing whereas it is the fitting of the linear combination of the fingerprints with observations that gives the amplitude of the response to this forcing. The conclusions of such analyses are relatively insensitive to the magnitude of the response of the model to a particular forcing or to uncertainties in the amplitude of the forcing. Indeed, an error in the amplitude of the forcing of a factor of two would simply induce a change by a factor of two in the regression coefficient, if the response of the system could be considered as linear (e.g. Tett *et al.* 1999).

Using an EBM to obtain the fingerprints of the externally forced signals and various reconstructions of Northern Hemisphere mean temperature, Hegerl *et al.* (2003) were able to detect clearly the influence of volcanic forcing on temperature changes during the past millennium. Furthermore, in their model, the time-scale of the response is similar to the one deduced from the reconstructions. The influence of anthropogenic forcings during the late 20th century is also detected, in agreement with studies focusing on the 20th century. Solar forcing has been marginally detected only over some periods and in some records, thus they conclude that the role of this forcing has probably been weak for the multi-decadal variability of the temperature averaged over the Northern Hemisphere.

It is possible to combine model results and data to constrain the magnitude of the changes during the past millennium as well as the value of some important characteristics of the climate system such as the climate sensitivity. Using a coupled physical–biogeochemical climate model, driven by different reconstructions of past changes in solar irradiance and volcanic forcing, Gerber *et al.* (2003) compared the simulated CO_2 and temperature evolution with available observations. In their analysis, for large changes in solar irradiance and hemispheric mean temperature, they find large discrepancies between observed CO_2 concentration and that simulated by the model. They argue that, between 1100 and 1700, changes in multi-decadal Northern Hemispheric mean surface temperature larger than 1°C are inconsistent with CO_2 records and thus not realistic. Following a similar objective to Gerber *et al.* (2003), Hegerl *et al.* (2006) recently performed a large ensemble of simulations with an EBM over the past millennium varying two important model parameters, namely the equilibrium climate sensitivity to a doubling of CO_2 and the effective ocean diffusivity, and using different scaling of the reconstructions of past radiative forcing. By comparing the results of this ensemble with several reconstructions of surface temperature in the Northern Hemisphere at (nearly) hemispheric scale, using a technique based on a Bayesian approach, they derive a probability density function (PDF) for climate sensitivity. Such methods, however, could be strongly dependent on prior assumptions made in the process of the PDF estimation (e.g. Frame *et al.* 2005) and additional work is still required before being able to constrain efficiently the model parameters using all the available information from the data.

Analyzing regional changes using observations and models

At regional scales, the temperature and precipitation anomalies are strongly influenced by the dominant mode of variability of the climate system, such as the North Atlantic Oscillation (NAO) or El Niño–Southern Oscillation (ENSO). Any change in these modes of variability, for instance, a shift from a positive to a negative phase, could be purely related to internal dynamics or caused by changes in the forcing. The analysis of proxy data could allow determination of the magnitude of the changes and the potential correlation with the reconstructions of past forcings. Alternatively, if a model is able to simulate reasonably well the observed changes, this could provide a reasonable hypothesis for the mechanism responsible for the change. Despite the complementary information from model and data, determining

Figure 7.8 Reconstructed (top) and simulated (bottom: GCM, general circulation model) annual (ANN) average surface temperature differences between 1660–80 and 1770–90. The reconstructed surface temperatures are based on a multi-proxy estimate using tree rings, ice cores, corals, and historical data. Model results are based on the sum of the response in two simulations: one incorporating reconstructed solar irradiance changes during this period and one using volcanic forcing scaled to changes over this time. (From Schmidt *et al.* 2004. Copyright Elsevier (2004).)

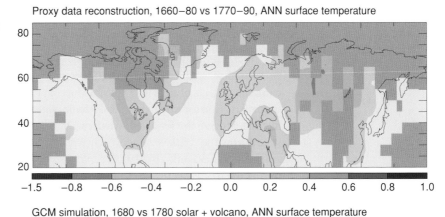

Proxy data reconstruction, 1660–80 vs 1770–90, ANN surface temperature

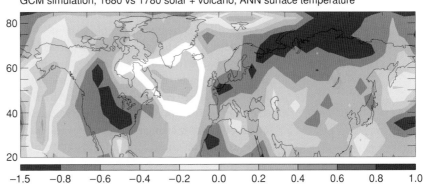

GCM simulation, 1680 vs 1780 solar + volcano, ANN surface temperature

the cause of a change in one mode of variability or in the probability to have one phase of a mode is nearly impossible, currently, as internal variability alone could explain the observed changes in those modes.

The observed substantial cooling in large parts of Europe during the late 17th and early 18th centuries appears to be related to long-term variations in the NAO pattern of climate variability (Figure 7.8), which are associated with variations over time in the pattern of the Northern Hemisphere jet stream. In model simulations, these changes are associated with a large-scale dynamical response of the climate system to natural radiative forcing by explosive volcanic activity (Robock 2000; Schmidt *et al.* 2004; Shindell *et al.* 2004) and solar output (Shindell *et al.* 2001; Schmidt *et al.* 2004) that interacts with the NAO pattern of atmospheric circulation. This has led to the suggestion that the moderate apparent lowering of solar irradiance during the 17th century leads to only moderate decreases in hemispheric mean temperature, but a tendency for strong annual mean cooling in some regions, such as Europe.

Recent work suggests that the El Niño–Southern Oscillation may also be an important component of the response of the climate to forcing over the past 1000 years. Model simulations using simple models of the coupled tropical ocean–atmosphere to study the response of ENSO to solar and volcanic forcing (Mann *et al.* 2005a) indicate a counterintuitive tendency towards El Niño (warm eastern and central tropical Pacific) conditions in response to negative radiative forcing (past explosive tropical volcanic eruptions or decreases in solar irradiance) and a tendency for La Niña-like conditions in response to positive radiative forcing (i.e., increases in solar irradiance). This prediction matches available evidence from tropical Pacific coral records (Cobb *et al.* 2003). The response of ENSO to the combined effects of solar and volcanic forcing in past centuries provides a potential explanation for why the tropical Pacific appears to have been in a cold La Niña-like state during the so-called "Medieval Warm Period" and a warm El Niño-like state during the "Little Ice Age" (Adams *et al.* 2003).

Changes in oceanic circulation could also have an impact on regional changes during the past millennium. Unfortunately, there are not enough high-frequency observations in the ocean to derive a clear and comprehensive view of oceanic changes during this period. The available information is mainly derived from some local observations and model results. A thorough model–data comparison has not been performed yet but that could certainly be very useful and would help to fill some of the gaps in our knowledge. In particular, when driven by a reduction of the solar irradiance (or more generally in case of a moderate cooling), models tend to simulate an intensification of the meridional overturning circulation in the Atlantic (Cubasch *et al.* 1997; Goosse and Renssen 2004; Weber *et al.* 2004). This is due to an increase in the density of the surface water in the North Atlantic, caused by cooling and by changes in the freshwater budget, inducing more vigorous deep water formation there. The intensification of the oceanic circulation implies a larger heat transport towards the high latitudes and is thus associated with a negative feedback in those regions of the initial cooling. For a moderate warming, the opposite mechanism is noticed, presenting some similarities with what is expected for the 21st century (e.g. Gregory *et al.* 2005). A reduction of the solar forcing might also modify the probability of having years with significantly reduced inflow of warm Atlantic waters at high latitudes (Goosse and Renssen 2004). This leads to a higher probability of having very cold conditions in the Nordic Seas and over Scandinavia, resulting overall in a very strong positive feedback in the northern North Atlantic and surrounding regions during those cold periods. There is also evidence for the potentially important response of the Atlantic meridional overturning to low-frequency changes in the NAO (Delworth and Dixon 2000), which have in turn been simulated as a possible response to long-term solar and volcanic forcing in past centuries (Shindell *et al.* 2001, 2004; Schmidt *et al.* 2004).

Assimilation of observations

Classically, the information deduced from the analysis of observations and from model simulations is obtained from two different, clearly separated streams that are then combined to provide additional insight, in order to give more strength to the conclusions, or to propose a new interpretation. In the sections above, we have shown examples where a stronger coupling between modeling and data analysis could be very instructive. This is precisely the goal of the assimilation of observations or data assimilation, which is defined by Talagrand (1997) "as the process through which all the information is used to estimate as accurately as possible the state of the atmospheric or oceanic flow. The available information essentially consists of the observations proper, and of the physical laws which govern the evolution of the flow. The latter are available in practice under the form of a numerical model."

Data assimilation techniques have been used successfully in various domains of meteorology and oceanography during the past decades (for a review, see for instance Talagrand 1997; Kalnay 2003, Brasseur 2006). One of the initial and most important applications of such techniques is in weather forecasting, to determine the initial conditions at a time t_0 from which subsequent evolution of the system at time $t > t_0$ will be deduced. This step, which is called the analysis, combines some background information (also denoted as first guess or prior information) with available observations. Following the notation and formulation of Kalnay (2003), observed variables are noted y^0, while x^b is the background field. It is generally obtained from a previous forecast using a model simulation starting from an analysis at time $t - 1$ $(t - 1 < t_0)$ until time t_0. Both y^0 and x^b are three-dimensional fields. Variables y^0 and x^b are not usually obtained at the same location and some interpolation is needed. Furthermore, observations could provide a different variable than the one needed by the model. Some transformation of model results (or alternatively of observations) is then needed before they can be compared with data. This is represented by $\mathbf{H}(x^b)$, where $\mathbf{H}$ is called the observation operator. The difference between the first guess obtained from the model forecast and observations, $y - \mathbf{H}(x^b)$, is called observational increment or innovation. The analysis x^a is then obtained by adding this innovation to the model forecast, using some weight $\mathbf{W}$ that is determined by accounting for uncertainties in model results as well as in observations. This could be mathematically represented by

$$x^a = x^b + \mathbf{W}(y^0 - \mathbf{H}(x^b)) \qquad (1)$$

The sequence of obtaining a first guess from model results and then updating using data to provide the analysis can be repeated from time to time in an "analysis cycle". Such a technique is often called sequential assimilation (e.g. Talagrand 1997; Kalnay 2003).

In weather forecasting, model and data assimilation schemes (for instance the formulation of $\mathbf{W}$ and $\mathbf{H}$) are improved regularly, providing potential inconsistencies between different periods. Reanalyses of the past 50 years, using the same procedure during all periods covered (e.g. Kalnay *et al.* 1996), have provided a very valuable and widely used comprehensive set of atmospheric fields.

Such a procedure could, in theory, be applied to any period and thus also for the past millennia. However, several difficulties arise, a major one being the amount of data. Hundreds of thousands of atmospheric and surface observations are used for the analyses in weather forecasting, covering all areas of the world, and providing a very detailed knowledge of the atmospheric state, whereas only a few tens up to hundreds of proxy records are available for the past millennium, mainly localized on the mid-latitude continents. Secondly, analysis in weather forecasting is performed several times a day, typically every 6 hours. By contrast, proxy data provide at best seasonal resolution. It is therefore not possible to constrain the day-to-day variability of the climate system. Thirdly, proxy data contain climatic as well as nonclimatic signals and the variables measured, such as tree-ring width, wood density or varve thickness, are usually not a simple function of climate. Determining precisely the form of the observation operator **H** is thus a difficult task. Finally, sophisticated data assimilation techniques are expensive, in particular if they have to be applied on relatively long periods such as the past millennium. The central processing unit (CPU) time requirements could be a limiting factor for the AOGCMs, and the technique developed for data assimilation exercises covering the past 20 to 50 years for the atmosphere and the ocean cannot be transferred to the analysis of climate variations during the past millennium without significant adaptation.

Because of the difficulties, only a few attempts of data assimilation have been published for the past millennium (von Storch *et al.* 2000; Jones and Widmann 2003; van der Schrier and Barkmeijer 2005, Goosse *et al.* 2006). The first technique proposed was called DATUN, for Data Assimilation Through Upscaling and Nudging (von Storch *et al.* 2000). In this technique, the first step, upscaling, is to derive from proxy data the intensity of some large-scale patterns, for instance the one of the Northern Annular Mode (also known as the Arctic Oscillation and is linked to the NAO, see previous section). The goal is to increase the signal-to-noise ratio, as proxy data are usually more suitable for reconstructing large-scale fields than small-scale ones. In the second step, a simple nudging technique is applied to force the model to remain close to the pattern obtained by upscaling. In DATUN, the nudging is performed directly on the pattern space.

The goal of van der Schrier and Barkmeijer (2005) is also to bring model results close to a pattern deduced from observations. In the example they studied, the target for the model surface atmospheric circulation in winter is a reconstruction for the period 1790–1820. This is not achieved through nudging as in DATUN, but by adding small perturbations to model variables that optimally lead the atmospheric model to the target pattern. These perturbations are called "forcing singular vectors" and are computed using the adjoint of the atmospheric model. To apply this technique, the adjoint model must be available, which induces a limitation of the applications. As the forcing singular vectors have small amplitudes, they affect only the large-scale variability while the synoptic variability evolves freely. Van der Schrier and Barkmeijer (2005) argue that this provides an advantage compared with nudging techniques, which tend to suppress the variability in the nudged pattern. Unfortunately, no systematic comparison of the two techniques is presently available.

In the technique described by Goosse *et al.* (2006), it is not necessary to obtain first a large-scale pattern from observations. The proxy data are directly compared with model results of an ensemble of simulations, using a simple quadratic law to compute a cost function *CF*:

$$CF_k(t) = \sqrt{\sum_{i=1}^{n} w_i (F_{\text{obs}}(t) - F_{\text{mod}}^{k}(t))^2} \qquad (2)$$

where $CF_k(t)$ is the value of the cost function for the experiment k – a member of the ensemble – for a particular period t, and n is the number of reconstructions or proxy data used in the model–data comparison. F_{obs} is the reconstruction of a variable F, based on observations, in a particular location for the period t. F_{obs}^{k} is the value of the corresponding variable F simulated in experiment k in the model and w_i is a weight factor characterizing the statistics and reliability of proxy data.

For the technique of Goosse et al. (2006), a relatively large number of simulations (i.e. at least 30) is required. $CF_k(t)$ is first evaluated for all the simulations and all the time periods. The minimum for each t over all k experiments is then selected and the corresponding experiment corresponds to the "best" one for this period. Such a "best" model estimate for the whole millennium is obtained by grouping the best simulations for all the periods t, obtaining what is called the best pseudo-simulation. The technique selects among the ensemble of simulations the one that is the closest to the available reconstructions, with the cost function measuring the misfit between model results and proxy records. From this cost function, it is then also possible to assess the quality of the reconstruction for all the periods investigated. An example of the application of this technique is illustrated in Figure 7.9, showing that it is possible to find, for nearly all the periods, an element of the ensemble that is close to all the available proxies, at least in the case tested by Goosse et al. (2006) using only a few temperature-sensitive proxies that covered (nearly) the whole past millennium. We must be aware, however, that this technique implicitly assumes that the ensemble members cover a sufficiently wide range of internal variability compared with the observed one, in order to find a good analog of the observed situations among the simulations. Problems could thus occur in regions where this assumption is not valid.

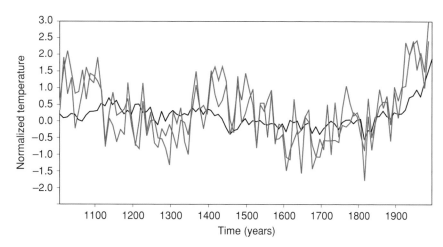

Figure 7.9 Anomaly of summer mean temperature in Fennoscandia. The average over an ensemble of 125 simulations performed with ECBILT–CLIO–VECODE is in black. The reconstruction of Briffa et al. (1992) is in green. The red line is the best pseudo-simulation obtained by using 12 proxy records in the Northern Hemisphere extratropics (Goosse et al. 2006). The time series are grouped by 10-year average and divided by their standard deviation. The reference period is 1500–1850.

A main goal of the data assimilation techniques is to provide a reconstruction of the state of the climate system that is consistent with proxy records, the physical laws described by the model, and forcing reconstructions. These techniques can be considered as a physically based way to interpolate the information provided by data that have an incomplete spatial and temporal coverage. This reconstruction of past climate variations includes the forced response of the system as well as the random variations associated with internal variability. As mentioned before, models alone could not estimate the history of the latter component if model evolution is not constrained by observations.

Although observations provide only regional- and seasonal-scale constraints, the estimates obtained by these techniques could be used to analyze high-frequency changes. For instance, van der Schrier and Barkmeijer (2005) analyze, in their model, the influence of the assimilation of a pattern of winter-mean sea-level pressure corresponding to the period 1790–1820 on the number of cyclones in the Atlantic and found an enhanced cyclonic activity both south and north of the modern storm track position. The model results also include information on variables that are not assimilated, such as oceanic currents. This has allowed van der Schrier and Barkmeijer (2005) to show that changes in oceanic surface circulation and temperature in the North Atlantic have played a role during the coldest periods observed in Europe in winter. This kind of information on variables that are not directly accessible using proxy records provides a very powerful tool to analyze past climate variations. Indeed, it is possible to propose mechanisms that are consistent with proxy evidence as well as with the physics governing the atmosphere, the ocean, and sea ice. In a following step, the proposed hypotheses could be tested using new data or the results of the data assimilation exercise could be used to select data locations that would give essential information and a strong constraint on the mechanisms involved.

Conclusions

In this chapter, we have discussed some recent developments that aim to optimize the combination of proxy data and climate model results, with a focus on the climate of the past millennium.

One development is to evaluate temperature reconstructions based on proxy records by using independent "pseudo-proxy records" that are derived from model experiments. An important application of this method is the validation of the reliability of the statistical CFR method that has been used for several temperature reconstructions of the past millennium and improved since its introduction.

Another recent development is to run climate models to compute the climatic "fingerprint" associated with a particular forcing, and subsequently to apply multiple regression to find the combination of fingerprints that gives the best match with proxy-based temperature reconstructions. This technique has provided a detailed analysis of the impact of different forcings on the climate of the past millennium, including the important effect of anthropogenic forcings on the late 20th century climate. Model and proxy data could also be used to constrain model parameters as well as important characteristics of the climate system.

Progress is also made by the application of data assimilation techniques that use both proxy data and models' results in order to obtain a climate state that is consistent with the proxy records and model physics. Once such a climate state is obtained, aspects of this state that are not registered in proxies can be analyzed in detail, thus providing additional information on the climate of the past millennium. In addition, this climate state can be used to find key locations for data collection. At the moment, the number of suitable proxy data and the computation time involved still limit the application of data assimilation methods for the past millennium.

The examples displayed here have shown that the combined analysis of proxy and model results is certainly an interesting way to improve our understanding of past climate changes. The examples also show that several difficulties remain that need intense collaboration between scientists from different backgrounds to solve them. Any improvement in the representation of past changes by models or in the number and quality of proxy records would certainly be very beneficial. The method used to combine proxy data and models, however, needs also to be refined and different methods could be required if one is interested in reconstructions of past changes or in understanding the causes of those changes. In particular, estimating precisely the impact on the results of the analysis of the various hypotheses or prior assumptions involved in the method, sometimes hidden in complex formulations that could be assessed by only a few specialists, is certainly an important task.

Acknowledgments

We would like to thank all the scientists who sent us or made available their datasets and results. H. Renssen is sponsored by the Netherlands Organization for Scientific Research (NWO). M.E. Mann gratefully acknowledges support from the ATM program of the National Science Foundation (Grant ATM-0542356). H. Goosse is a Research Associate with the Fonds National de la Recherche Scientifique (Belgium) and is supported by the Belgian Federal Science Policy Office.

References

Adams J.B., Mann M.E. & Ammann C.M. (2003) Proxy evidence for an El Niño-like response to volcanic forcing. *Nature*, **426**, 274–278.

Andreae M.O., Jones C.D. & Cox P.M. (2005) Strong present-day aerosol cooling implies a hot future. *Nature*, **435**, 1187–1190.

Bard E., Raisbeck G., Yiou F. & Jouzel J. (2000) Solar irradiance during the last 1200 years based on cosmogenic nuclides. *Tellus*, **52B**, 985–992.

Bauer E., Claussen M., Brovkin V. & Huenerbein A. (2003) Assessing climate forcings of the Earth system for the past millennium. *Geophysical Research Letters*, **30**(6), 1276 doi: 10:1029/2002GL016639.

Beer, J. & van Geel B. (this volume) Holocene climate change and the evidence for solar and other forcings. In: *Global Warming and Natural Climate Variability: a Holocene Perspective* (Eds R.W. Battarbee & H.A. Binney), pp. 138–162. Blackwell, Oxford.

Bertrand C., Loutre M.F., Crucifix M. & Berger A. (2002) Climate of the last millennium: a sensitivity study. *Tellus*, **54A**, 221–244.

Boucher O. & Pham M. (2002) History of sulfate aerosol radiative forcings. *Geophysical Research Letters*, **29**(9), 1308, doi: 10.1029/2001GL014048.

Bradley R.S., Hughes M.K. & Diaz H.F. (2003) Climate in Medieval time. *Science*, **302**, 404–405

Brasseur P. (2006) Ocean data assimilation using sequential methods based on the Kalman filter. In: *Ocean Weather Forecasting* (Eds E.P. Chasignet and J. Verron), pp. 271–316. Springer-Verlag, Berlin.

Briffa K.R., Jones P.D., Bartholin T.S., *et al.* (1992) Fennoscandian summers from AD 500: temperature changes on short and long timescales. *Climate Dynamics*, **7**, 111–119

Briffa K.R., Osborn T.J., Schweingruber F.H., *et al.* (2001) Low-frequency temperature variations from a northern tree-ring density network. *Journal of Geophysical Research*, **106**, 2929–2941.

Brovkin V., Claussen M., Driesschaert E., *et al.* (2006) Biogeophysical effects of historical land cover changes simulated by six Earth system models of intermediate complexity. *Climate Dynamics*, **26**, 587–600.

Claussen M. (this volume) Holocene rapid land-cover changes – evidence and theory. In: *Global Warming and Natural Climate Variability: a Holocene Perspective* (Eds R.W. Battarbee & H.A. Binney), pp. 232–253. Blackwell, Oxford.

Cobb K.M., Charles C.D., Cheng H. & Edwards R.L. (2003) El Niño–Southern Oscillation and tropical Pacific climate during the last millennium. *Nature*, **424**, 271–276.

Crowley T.J. (2000) Causes of climate change over the past 1000 years. *Science*, **289**, 270–277.

Crucifix M. (this volume) Modelling the climate of the Holocene. In: *Global Warming and Natural Climate Variability: a Holocene Perspective* (Eds R.W. Battarbee & H.A. Binney), pp. 98–122. Blackwell, Oxford.

Cubasch U., Voss R., Hegerl G.C., Waszkewit J. & Crowley T.J. (1997) Simulation of the influence of solar radiation variations on the global climate with an ocean–atmosphere general circulation model. *Climate Dynamics*, **13**, 757–767.

Delworth T.L. & Dixon K.W. (2000) Implications of the recent trend in the Arctic/North Atlantic Oscillation for the North Atlantic thermohaline circulation. *Journal of Climate*, **13**, 3721–3727.

Esper J., Cook E.R. & Schweingruber F.H. (2002) Low-frequency signals in long tree-line chronologies for reconstructing past temperature variability. *Science*, **295**, 2250–2253.

Frame D.J., Booth B.B.B., Kettleborough J.A., *et al.* (2005) Constraining climate forecasts: the role of prior assumptions. *Geophysical Research Letters*, **32**, L09702, doi: 10.1029/2004GL022241.

Free M. & Robock A. (1999) Global warming in the context of the Little Ice Age. *Journal of Geophysical Research*, **104**(D16), 19057–19070.

Gerber S., Joos F., Brugger P., *et al.* (2003) Constraining temperature variations over the last millennium by comparing simulated and observed atmospheric CO_2. *Climate Dynamics*, **20**, 281–299.

González-Rouco F., von Storch H. & Zorita E. (2003) Deep soil temperature as proxy for surface air-temperature in a coupled model simulation of the last thousand years. *Geophysical Research Letters*, **30**(21), 2116.

González-Rouco J.F., Beltrami H., Zorita E. & von Storch H. (2006) Simulation and inversion of borehole temperature profiles in surrogate climates: spatial distribution and surface coupling. *Geophysical Research Letters*, **33**, L01703, doi: 10.1029/2005GL024693.

Goosse H. & Renssen H. (2004) Exciting natural modes of variability by solar and volcanic forcing: idealized and realistic experiments. *Climate Dynamics*, **23**, 153–163.

Goosse H., Masson-Delmotte V., Renssen H., *et al.* (2004) A late medieval warm period in the Southern Ocean as delayed response to external forcing? *Geophysical Research Letters*, **31**(6), L06203, doi: 10.1029/2003GL019140.

Goosse H., Renssen H., Timmermann A. & Bradley R.S. (2005a) Internal and forced climate variability during the last millennium: a model–data comparison using ensemble simulations. *Quaternary Science Reviews*, **24**, 1345–1360.

Goosse H., Crowley T., Zorita E., Ammann C., Renssen H. & Driesschaert E. (2005b) Modelling the climate of the last millennium: What causes the differences between simulations? *Geophysical Research Letters*, **32**, L06710, doi: 10.1029/2005GL22368.

Goosse H., Renssen H., Timmermann A., Bradley R.S & Mann M.E. (2006) Using paleoclimate proxy-data to select optimal realisations in an ensemble of simulations of the climate of the past millennium. *Climate Dynamics*, **27**, 165–184,

Gregory J., Dixon K.W., Stouffer R.J., *et al.* (2005) A model intercomparison of changes in the Atlantic thermohaline circulation in response to increasing atmospheric CO_2 concentration. *Geophysical Research Letters*, **32**, L12703, doi: 10.1029/2005GL023209.

Hegerl G.C., Crowley T.J., Baum S.K., Kim K-Y & Hyde W.T. (2003) Detection of volcanic, solar and greenhouse gas signals in paleo-reconstructions of Northern Hemispheric temperature. *Geophysical Reserch Letters*, **30**(5), 1242, doi: 10.1029/2002GL016635

Hegerl G.C., Crowley T.C., Hyde W.T. & Frame D.J. (2006) Climate sensitivity constrained by temperature reconstructions over the past seven centuries. *Nature*, **440**, 1029–1032.

Hughes M.K. & Diaz H.F. (1994) Was there a "Medieval Warm Period", and if so, where and when? *Climatic Change*, **26**, 109–142.

Jones J. & Widmann M. (2003) Reconstructing large-scale variability from paleoclimatic evidence by means of Data Assimilation Through Upscaling and Nudging (DATUN). In: *The KIHZ Project: Towards a Synthesis of Holocene Proxy Data and Climate Models* (Eds Fischer H., Lohmann G., Flser G., Miller H., von Storch H. and Negendank J.F.). Springer-Verlag, Berlin, pp. 171–193.

Jones P.D. & Mann M.E. (2004) Climate over past millennia. *Reviews of Geophysics*, **42**, RG2002, doi: 10.1029/2003RG000143.

Kalnay E. (2003) *Atmospheric Modelling, Data Assimilation and Predictability*. Cambridge University Press, Cambridge, 341 pp.

Kalnay E., Kanamitsu M., Kistler R., *et al.* (1996) The NCEP/NCAR 40-year reanalysis project. *Bulletin of the American Meteorological Society*, **77**, 437–471.

Lean J., Beer J. & Bradley R. (1995) Reconstruction of solar irradiance since 1610: implications for climate change. *Geophysical Research Letters*, **22**, 1591–1594.

Lorenz E.D. (1963) Deterministic nonperiodic flow. *Journal of the Atmospheric Sciences*, **29**, 130–141.

Luterbacher J., Dietrich D., Xoplaki E., Grosjean M. & Wanner H. (2004) European seasonal and annual temperature variability, trends, and extremes since 1500. *Science*, **303**(5663), 1499–1503.

Mann M.E. & Jones P.D. (2003) Global surface temperatures over the past two millennia. *Geophysical Research Letters*, **30**, 1820, doi: 10.1029/2003GL017814.

Mann M.E. & Rutherford S. (2002) Climate reconstruction using "pseudoproxies". *Geophysical Research Letters*, **29**, 1501, doi: 10.1029/2001GL014554.

Mann M.E. & Schmidt G.A. (2003) Ground vs. surface air temperature trends: Implications for borehole surface temperature reconstructions *Geophysical Research Letters*, **30**(12), 1607.

Mann M.E., Bradley R.S. & Hughes M.K. (1998) Global-scale temperature patterns and climate forcing over the past six centuries. *Nature*, **392**, 779–787.

Mann M.E., Cane M.A., Zebiak S.E. & Clement A. (2005a) Volcanic and solar forcing of the tropical Pacific over the past 1000 years. *Journal of Climate*, **18**, 447–456.

Mann M.E., Rutherford S., Wahl E. & Ammann C. (2005b) Testing the fidelity of methods used in proxy-based reconstructions of past climate. *Journal of Climate*, **18**, 4097–4107.

Mann M.E., Rutherford S., Wahl E & Ammann C. (2007a) Robustness of proxy-based climate field reconstruction methods. *Journal of Geophysical Research*, **112**, D12109, doi:10.1029/2006JD008272.

Mann M.E., Rutherford S., Wahl E & Ammann C. (2007b) Reply to Comment by Zorita *et al.* on Mann, Rutherford, Wahl and Ammann (2005). *Journal of Climate*, **20**, 3699–3703.

Moberg A., Sonechkin D.M., Holmgren K., Datsenko N.M. & Karlen W. (2005) Highly variable Northern Hemisphere temperatures reconstructed from low and high-resolution proxy data. *Nature*, **433**, 613–617.

Oldfield F. (this volume) The role of people in the Holocene. In: *Global Warming and Natural Climate Variability: a Holocene Perspective* (Eds R.W. Battarbee & H.A. Binney), pp. 58–97. Blackwell, Oxford.

Osborn T.J., Raper S.C.B. & Briffa K.R. (2006) Simulated climate change during the last 1000 years: comparing the ECHO-G general circulation model with the MAGICC simple climate model. *Climate Dynamics*, **27**, 185–197.

Ramankutty N. & Foley J.A. (1999) Estimating historical changes in global land cover: croplands from 1700 to 1992. *Global Biogeochemical Cycles*, **13**(4), 997–1027.

Robock A. (2000) Volcanic eruptions and climate. *Reviews of Geophysics*, **38**, 191–219.

Rutherford S., Mann M.E., Delworth T.L. & Stouffer R. (2003) Climate field reconstruction under stationary and nonstationary forcing. *Journal of Climate*, **16**, 462–479.

Rutherford S., Mann M.E., Osborn T.J., *et al.* (2005) Proxy-based Northern Hemisphere surface temperature reconstructions: sensitivity to methodology, predictor network, target season and target domain. *Journal of Climate*, **18**, 2308–2329.

Schmidt G.A., Shindell D.T., Miller R.L., Mann M.E. & Rind D. (2004) General Circulation Modeling of Holocene climate variability. *Quaternary Science Reviews*, **23**, 2167–2181.

Shindell D., Schmidt G.A., Mann M.E., Rind D. & Waple A. (2001) Solar forcing of regional climate change during the Maunder Minimum. *Science*, **294**, 2149–2152.

Shindell D.T., Schmidt G.A., Mann M.E. & Faluvegi G. (2004) Dynamic winter climate response to large tropical volcanic eruptions since 1600. *Journal of Geophysical Research*, **109**, D05104, doi: 10.1029/2003JD004151.

Stott P.A., Tett S.F.B., Jones G.S., Ingram W.J. & Mitchell J.F.B. (2001) Attribution of twentieth century temperature change to natural and anthropogenic causes. *Climate Dynamics*, **17**, 1–21.

Talagrand O. (1997) Assimilation of observations, an introduction. *Journal of the Meteorological Society of Japan*, **75**(1B), 191–209.

Tett S.F.B., Stott P.A., Allen M.R., Ingram W.J. & Mitchell J.F.B. (1999) Causes of twentieth-century temperature change near the Earth's surface. *Nature*, **399**, 569–572.

Van der Schrier G. & Barkmeijer J. (2005) Bjerknes' hypothesis on the coldness during AD 1790–1820 revisited. *Climate Dynamics*, **24**, 335–371.

Von Storch H, Cubasch U., González-Rouco F., *et al.* (2000) Combining paleoclimatic evidence and GCMs by means of data assimilation through upscaling and nudging (DATUN). *Proceedings of the 11th Symposium on Global Change Studies*, American Meteorological Society, Long Beach, CA.

Von Storch H, Zorita E., Jones J.M., Dimitriev Y., González-Rouco F. & Tett S.F.B. (2004) Reconstructing past climate from noisy data. *Science*, **306**, 679–682.

Von Storch H., Zorita E., Jones J.M., González-Rouco F. & Tett S.F.B. (2006) Response to Comment on 'Reconstructing Past Climate from Noisy Data'. *Science*, **312**, 529c.

Wahl E.R., Ritson D.M. & Ammann C.M. (2006) Comment on 'Reconstructing Past Climate from Noisy Data'. *Science*, **312**, 529b.

Waple A.M., Mann M.E. & Bradley R.S. (2002) Long-term patterns of solar irradiance forcing in model experiments and proxy based surface temperature reconstructions. *Climate Dynamics*, **18**, 563–578.

Weber S.L., Crowley T.J. & van der Schrier G. (2004) Solar irradiance forcing of centennial climate variability during the Holocene. *Climate Dynamics*, **22**(5), 539–553

Widmann M. & Tett S.F.B. (2003) Simulating the climate of the last millennium. *PAGES News*, **11**, 21–23.

Zorita E., González-Rouco F. & Legutke S. (2003) Testing the Mann *et al.* (1998) approach to paleoclimate reconstructions in the context of a 1000-yr control simulation with the ECHO-G Coupled Climate Model. *Journal of Climate*, **16**, 1378–1390.

Zorita E., González-Rouco F. & von Storch H. (2007) Reply to Comment on Mann, Rutherford, Wahl and Ammann (2005). *Journal of Climate*, **20**, 3693–3698.

8 Latitudinal linkages in late Holocene moisture-balance variation

Dirk Verschuren and Dan J. Charman

Keywords

Africa, dendroclimatology, Europe, Holocene, Indian Ocean monsoon, lake-level records, Little Ice Age, Medieval Warm Period, PEPIII transect, paleohydrology, peat surface-wetness records, rainfall and drought, speleothem records

Introduction

Projected changes in the amount and distribution of global precipitation over the next century will almost certainly have greater impacts on the quality of human life and natural ecosystem functioning than temperature change (IPCC 2007). Most paleoclimate research to date, however, has concentrated on global and regional histories of temperature change. This is partly because most high-quality continental climate-proxy archives are located in regions with positive water balance and above-average temperature sensitivity. More fundamentally, reconstructing past precipitation variation independent of coincident temperature changes is a difficult task. Although temperature and precipitation co-vary on large spatial and long time-scales (Trenberth and Shea 2005), breakdown of this co-variation at smaller spatial and shorter time-scales is often a key to understanding the climate mechanisms involved. Improving our understanding of regional climate and hydrologic variability at decade to century time-scales thus represents a significant challenge for both the paleoclimate data and paleoclimate modeling communities. External climate forcing at this time-scale (solar variability, volcanic eruptions) is relatively modest and expressed through both direct and cumulative effects, therefore delays in the response of individual climate-system components and their associated climate archives to this forcing become more important than they are over longer time-scales. Consequently, uncertainty exists over the extent to which observed decadal to century-scale changes truly reflect abrupt events or result from the cumulative effect of interannual variation. A satisfactory answer to this question is critical to evaluate the exact relationship between climate variability and water-resource availability at the landscape scale that has often determined the success or failure of societies in the past (Weiss and Bradley 2001), and will do so in the future (IPCC 2007). A first step in providing this answer can be made through syntheses of paleohydrologic data at the continental scale, in an attempt to reconstruct large-scale spatial patterns of variability at decadal to centennial time-scales. In this contribution we aim to (i) review initial attempts to integrate proxy records of Holocene continental paleohydrology across regions; and (ii) juxtapose high- and low-latitude patterns of hydrologic change through time. Geographically, we focus on the PAGES–PANASH Europe–Africa transect (Bradley *et al.* 1995), which includes continental regions of Europe, adjacent Eurasia, and Africa bounded on the west by the Atlantic Ocean and extending east into monsoonal Asia (Figure 8.1).

Challenges to proxy data integration

Integration of paleohydrologic data across sites

Most traditional hydrologic climate-proxy indicators tend to reflect "effective moisture availability", rather than rainfall, with poorly constrained temperature effects on evaporation. Rainfall variation is also spatially more heterogeneous and temporally more pronounced than temperature variation, because the atmospheric circulation patterns ultimately determining rainfall location and intensity depend on ocean currents, pressure contrasts between oceans and adjacent land masses, and local land topography. To draw a coherent picture of latitudinal linkages and the associated climate dynamics for rainfall variability during past millennia thus requires appropriate geographic coverage of well-dated, high-quality hydrologic proxy records. This in turn presents two other challenges: (i) what is the proper spatial scale to integrate paleodata, when the main aim of such integration is to capture and highlight characteristic regional climate patterns for comparison with those of other regions?; (ii) how to compare climate-proxy time series between two or more regions where the selected proxy indicators are influenced to a different degree by temperature and precipitation variations?

High-quality paleodata are often generated from natural climate archives that are common in certain regions but have limited distribution or poor archival quality in other regions. Consequently, meeting the second challenge (if not also the first, cf. Moberg *et al.* 2005) requires integration of climate-proxy time series across archives that have a different resolution and chronologic control. As with multi-proxy studies on a single archive, comparison of proxy records between archives from within the same region can provide unique insight into the seasonal characteristics of past climate change and into how well we understand individual proxies as indicators of climate change. It also provides an independent test of the quality of natural climate archives as recorders of climate change in a manner that any number of proxy indicators from the same archive can not.

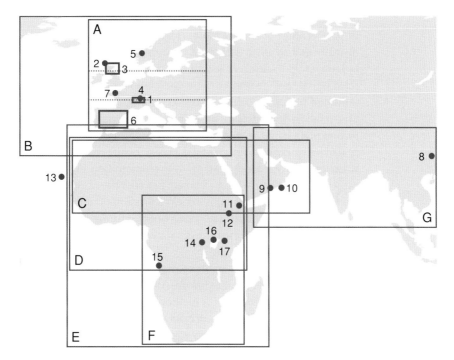

Figure 8.1 Western Eurasia and Africa with main study sites and regions discussed in the text: lakes in the Jura, French pre-Alps and Swiss Plateau (1: Magny 2004); Uamh an Tartair Cave, Scotland (2: Proctor *et al.* 2002); northern Britain peatlands (3: Charman *et al.* 2006); Great Aletsch Glacier, Switzerland (4: Holzhauser *et al.* 2005); Bjørnbreen Glacier, Norway (5: Matthews *et al.* 2005); fluvial chronology of Spain (6: Thorndycraft and Benito 2006); oak tree-rings, western France (7: Masson-Delmotte *et al.* 2005); Dongge Cave, China (8: Wang *et al.* 2005); Qunf Cave, Oman (9: Fleitmann *et al.* 2003); marine sediments, western Arabian Sea (10: Gupta *et al.* 2003); Lakes Abhé and Ziway-Shala, Ethiopia (11 and 12: references in Gasse 2000); marine sediments, eastern tropical Atlantic Ocean (13: de Menocal *et al.* 2000); Lake Edward, DR Congo/Uganda (14: Russell and Johnson 2005b); agricultural drought index, coastal Angola (15: Miller 1982); Pilkington Bay, Lake Victoria (16: Stager *et al.* 2005); Lake Naivasha, Kenya (17: Verschuren *et al.* 2000). Continental-scale syntheses of Holocene and/or historical paleohydrology are available for Europe (A: Yu and Harrison 1995; B: Pauling *et al.* 2006), the arid–sub-arid belt of North Africa and the Arabian Peninsula (C: Hoelzmann *et al.* 2004), north and intertropical Africa (D: Street and Grove 1976), the entire African continent (E: Gasse 2000), sub-Saharan Africa (F: Verschuren 2004) and the Indian and East Asian monsoon domains (G: Fleitmann *et al.* 2007a).

Calibration of paleohydrologic proxy indicators in space and time

Any exercise in paleodata integration across sites and archives necessarily assumes that the relative magnitude of reconstructed proxy climate signals has been calibrated locally. This is accomplished either by time series regression against instrumental climate data (e.g. precipitation) or derived measures of hydrologic change (e.g. effective moisture, Palmer drought index) or through a spatial transfer function relating variation in the proxy indicator to modern geographic variation in the climate variable. Time-series calibration (as is customary in dendroclimatology) is generally considered superior to the spatial approach (customary in palynology or paleoecology of aquatic biota). This is not necessarily the case, however, when instrumental weather data from the immediate vicinity of the proxy record are lacking, or of insufficient length to capture local climate variability at the time-scale of interest and to confirm the stationarity (constancy through time) of indicator response to this variability (Jones et al. 1998). Lack of suitably long precipitation records can sometimes be overcome by a regional climate field reconstruction, i.e. a multi-variate statistical calibration of proxy data against spatial networks of instrumental data (e.g. Zhang et al. 2004). Climate field reconstruction must evidently start from the assumption that relationships between a proxy indicator and synoptic climate patterns have been stationary through time (Jones and Mann 2004). Time-series calibration of nonannually resolved climate-proxy records (e.g. Laird et al. 1996) often optimize the correlation by shifting or tuning the time axis of the record, thus making it difficult to identify possible lags in system response to climate change. This tuning can either be avoided or justified through system modeling of the relationship between a particular proxy indicator, the archive concerned, and climate. Although time-series calibration on nonannual proxies is imperfect, such methodologic compromises made for the purpose of climate-proxy calibration are evidently preferable to having no direct test of the exact meaning of climate-proxy signals at that particular site. Unfortunately, in many available high-resolution records of hydrologic change, proxy signals have not been directly calibrated, and the stationarity of signal response is unknown. Reconstruction must be based instead on a plausible mechanistic scenario of system response to climate at the time-scale of interest, inspired sometimes by modern relationships at seasonal time-scales that do not necessarily apply over decadal and centennial scales.

Chronologic issues

A further obstacle to integration of paleohydrologic data and to resolving their association with climate-forcing functions is the limited age control on reconstructed climate anomalies in nonannually resolved records dated with radioisotopes (e.g. ^{14}C, ^{210}Pb, U/Th). Although this issue is not confined to paleohydrologic records, it is of particular importance here because large-scale hydrologic change is thought to have occurred rapidly during certain periods. It also hampers attempts to link presumed low-frequency change (e.g. regional water-table fluctuation) to high-frequency change (e.g. flooding) recorded in different archives. Evidently, more effort should be directed to optimize age control on any record containing significant climate anomalies. Particularly for recent millennia, chronologic precision can sometimes gain significant improvement through wiggle-matching of multiple closely spaced radiocarbon ages to the dendrochronologic radiocarbon calibration curve (Blaauw et al. 2003), an approach used with success for hydrologic reconstructions extracted from European peat deposits (e.g. Mauquoy et al. 2004). When age control is inadequate, researchers are tempted to draw separate climate events at different sites into one illusory regional-scale event (Baillie 1991, Oldfield 2001). Explicit statistical analysis of the age uncertainties in proxy records can stimulate progress towards more rigorous integration of paleoclimate data (Blaauw *et al.* 2006).

Regional integration of proxy records

North-west Europe

Some of the earliest records of past moisture-balance change come from north-west Europe, where pioneering work on Holocene climate reconstruction was carried out. Best known is undoubtedly the Blytt–Sernander division of post-glacial time based on peat stratigraphic changes (Blytt, 1876; Sernander 1908; Birks, this volume), which introduced terms such as "Boreal" to represent a relatively cool, dry period in the early Holocene, and "Atlantic" to represent the subsequent warmer, wetter mid-Holocene. These terms are still used today but can be misleading when taken too literally, even within north-west Europe. Peat stratigraphy remains one of the key sources of Holocene moisture-balance data within this region, together with lake-level records and an increasingly diverse range of other climate archives such as alpine glaciers, speleothems, and tree rings. Establishing a broad regional picture of the main periods of hydrologic variability by integrating paleodata produced by these various archives faces two key problems, already hinted at above, related to a process-based understanding of interpreting natural climate archives and the climate-proxy indicators they contain. Despite broad conceptual understanding that the diverse proxy indicators all relate to moisture balance, their records may represent variation in distinctly different hydrologic variables and thus may not be directly comparable. In the case of peat records of past surface wetness, summer moisture deficit is the key determinant of proxy variations, but depending on location this may be forced primarily by precipitation in oceanic regions, such as the British Isles, or by an increasing influence of summer temperature in continental Europe (Charman *et al.* 2004; Charman 2007). Lake levels are usually interpreted as being a function of net annual moisture balance (precipitation minus evaporation, $P - E$) but with substantial variation in the relative importance of precipitation and seasonal effects of temperature on evapotranspiration (Digerfeldt *et al.* 1997). The mass balance of mountain glaciers (and thus their records of advance and retreat) tends to respond to climate change in particularly individualistic fashion depending on location, size, and direction of exposure. Significant inertia and delay in the response of large glaciers to climate change may generate different histories of high-frequency advance and retreat among glaciers within a single mountain range. On a regional scale, glaciers in the Alps appear to respond mostly to variations in winter temperature and summer precipitation (Holzhauser *et al.* 2005), whereas maritime glaciers in western Norway are regarded as primarily responding to variation in winter precipitation

(Matthews *et al.* 2005). Vincent *et al.* (2005) point out possible nonstationarity in the parameters controlling glacier mass balance in the Alps, with influence of summer temperature apparently having grown during the 20th century, whereas in the past, winter precipitation had greater influence.

Lake-level changes

Comprehensive datasets of Holocene lake-level change exist from two main areas in north-west Europe: Sweden (Digerfeldt 1988, 1997) and the French pre-Alps (Magny 2004). The Swedish records were combined with other geographically more dispersed records into the European Lake Level database (Yu and Harrison 1995), which now includes a total of 96 north-temperate to boreal sites (46–75°N) and 22 south-temperate sites (37–46°N). These data provide an excellent overview of millennial-scale patterns in lake-level status (categorized as high, intermediate or low: Figure 8.2a) across the European continent and along broad latitudinal and longitudinal gradients. Limits to chronologic control on many records necessitated temporal integration over 500 radiocarbon year intervals, precluding analysis of century-scale events. Taking Europe as a whole, the data show generally wet conditions from ca. 9520 to 6295 years BP, followed by declining lake levels culminating in drought ca. 5070–4485 years BP, and gradually increasing wetness during recent millennia to reach a present-day mean lake status comparable to that of the early Holocene. All ages given as calendar years BP or AD, or calibrated ^{14}C years for radiocarbon chronologies (denoted cal. years BP) unless stated otherwise. A broad latitudinal analysis (Figure 8.2b and c) indicates that the major patterns were continental in scale, except that a greater fraction of south-temperate lakes (roughly, south of the Alps) already stood high at the onset of the Holocene. Dryness at the start of the Holocene was mostly concentrated in a broad band across southern Britain, southern Scandinavia, and into the eastern Baltic, consistent with the effects of a glacial anticyclone over the lingering Scandinavian Ice Sheet (Yu and Harrison 1995). Inference of increased wetness during recent millennia, and particularly in the past 1500 years, mostly derives from Scandinavian and boreal lakes (north of 55°N). South-temperate lakes record an increasing proportion of lakes with low status between 2565 and 930 years BP, and hint at greater wetness only in the past 1000 years. Major variability in the fraction of lakes with intermediate rather than low or high status complicates distillation of broad temporal trends in effective moisture from these data.

   Using time series of similar lake-level scores based on detailed stratigraphic studies and drowned tree-stump dating at 26 lakes in the Jura Mountains and French pre-Alps (45–47°N, 5–9°E), Magny (2004) produced the most comprehensive and rigorous dataset of lake-level change for any area in Europe with century-scale resolution. His regional compilation for the Holocene (Figure 8.3) broadly agrees with the European lake-level database compilation for lakes south of 46°N (Figure 8.2c), in particular their relative highstand from the onset of the Holocene, and evidence for major episodes of increased wetness at 9550–9150, 8300–8050, and 7550–7250 years BP. The transition to generally lower mid- and late

(a) All European lakes

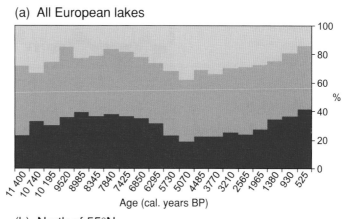

(b) North of 55°N

(c) South of 46°N

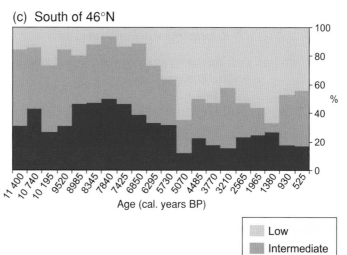

Figure 8.2 Lake-level status of European lakes at 500-^{14}C-year intervals through the Holocene, for (a) the complete European Lake Level Database (ELLDB) of 118 lakes, and for sub-sets of (b) lakes north of 55°N and (c) lakes south of 46°N. Original radiocarbon age categories have been converted to calibrated (cal.) years using the nearest age equivalent in the INCAL04 dataset (Reimer *et al.* 2004). Labels reflect the start of each period, i.e. "525" is the period from 525 calibrated years BP to present. (Based on Yu and Harrison 1995.)

Low

Intermediate

High

Figure 8.3 Summed lake-level scores for 26 lakes in the Jura, French pre-Alps, and on the Swiss Plateau, at 50-year intervals through the Holocene compared with the record of atmospheric Δ^{14}C as proxy for solar activity. (Magny 2004. Copyright Elsevier 2004.)

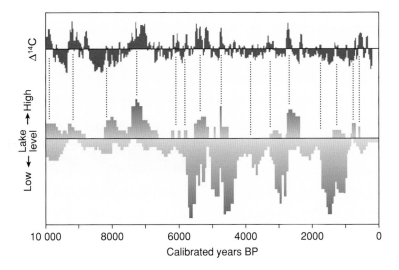

Holocene levels appears to have occurred at the same time in the higher-resolution record from west-central Europe as that evident in the European-wide compilation, ca. 5800 years BP and 5700 years BP respectively. The higher-resolution record of Magny (2004), however, is punctuated by marked reversions to high-stands at 5650–5200 and 4850–4800 years BP almost as significant as those recorded in the early Holocene. Of particular note in the late Holocene is the high lake-level phase at 2700–2400 years BP, and episodes of relative wetness (less severe drought) at 4300–3900, 3500–3200, 2000–1800, 1400–1200, 900–700, and after 400 years BP.

Glacier mass balance variations

There are a sizable number of individual records of glacier fluctuation in the central European Alps, mostly based on dating of moraines and trees buried during ice advance. Here we highlight the late Holocene record from Great Aletsch Glacier in Switzerland (Holzhauser 1997; Holzhauser *et al.* 2005; Figure 8.4d). Periods of pronounced glacier advance occurred at 2800–2450, 1500–1250, 850–750, and 650–50 years BP; and phases of pronounced retreat at 2450–1900, 1250–850, and 550–350 years BP. In northern Norway, the full Holocene record of Bjørnbreen and other glaciers shows mountain ice to have been absent between 9700 and 5700 years BP, except for short-lived glacier advances at 8300–7900 and 7800–7600 years BP. Mid- to late Holocene glacier expansion (neoglaciation) started at 5700 years BP and culminated in periods of maximum glacier extent during 5700–3900, 3600–1600, 1300–800, and 600–100 years BP, albeit with some variation in timing between sites (Matthews *et al.* 2000). The Bjørnbreen record of glacier mass balance is based on the dated lithostratigraphy of alpine stream-bank mires episodically flooded by glacier meltwater, using a model that describes the associated overbank deposition of suspended glaciofluvial sediments. In the context of this chapter the Bjørnbreen record is notable because modeling of variation in snow accumulation

Figure 8.4 Selected annually resolved to century-scale hydrologic records for temperate Europe covering the past 3400 to 4500 years. Blue shaded bars highlight periods of high inferred moisture availability in each record. (a) Mean annual band thickness and 100-yr running average in three speleothems from Northwest Scotland (from Proctor *et al.* (2002) but with time in years before AD 1950, not years before AD 2000 as published). Orange bars highlight periods of drier conditions. (b) Mean standardized peatland surface wetness and 100-yr running average at 12 sites from four regions in northern Britain (from Charman *et al.* 2006). (c) Summed lake-level scores inferred from sediment stratigraphy and drowned tree stumps at 26 lakes in the Jura Mountains, French pre-Alps, and on the Swiss Plateau. (Magny 2004. Copyright Elsevier 2004.) (d) Mass-balance fluctuation (advance and retreat) of the Great Aletsch glacier, Switzerland, based on dating of moraines and fossil trees exposed from beneath retreating ice (modified from Holzhauser *et al.* (2005) with permission from Sage publications).

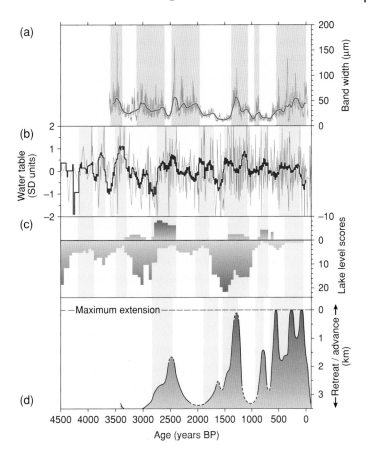

with pollen-inferred summer temperatures allowed Matthews *et al.* (2005) to reconstruct Holocene variations in local equilibrium-line altitude (corrected for post-glacial isostatic uplift) and winter precipitation in maritime northern Europe (Figure 8.5). Winter precipitation appears to have ranged between 40 and 160 percent of the modern value, with late Holocene maxima at 2900–2600 and 1200–800 years BP, and pronounced minima at 4000–3600 and 800–600 years BP.

Speleothem records

Speleothem-based proxy indicators (oxygen and carbon isotopes, fluorescence, annual band thickness) similarly record a variety of climate variables, depending on local cave setting and the above-ground climate regime. Oxygen-isotope signals in some speleothems accurately reflect isotopic variability in past rainfall, but the relationship with temperature and rainfall varies geographically (Fairchild *et al.* 2006). Here we contrast two studies exploiting the potential of speleothems as recorders of hydrologic change. Through direct calibration against instrumental climate data, Proctor *et al.* (2000) found that the ratio of regional mean annual

Figure 8.5 Holocene winter
precipitation record for
coastal northern Norway
reconstructed by combining
the record of glacier advance
and retreat at Bjørnbreen
together with a regional
pollen-based summer-
temperature reconstruction
in a snow-accumulation
model. (Mattews *et al.* 2005.
Copyright Elsevier 2005.)

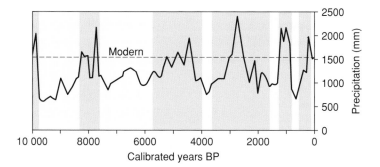

temperature to mean annual rainfall explained > 60 percent of the variation in annual band thickness of a stalagmite from northern Scotland. In a 100-year running average record, drier phases are indicated around 3500, *c.* 3100–2600, *c.* 2500–2000, *c.* 1350–1100, *c.* 950, and *c.* 555–0 years BP. (Proctor *et al.* 2002; Figure 8.4a). Niggemann *et al.* (2003) interpreted speleothem δ^{18}O in west-central Germany as primarily reflecting winter precipitation. From this they inferred periods of enhanced precipitation ca. 4800, 3500, 2700, and 2400 years BP (after 1500 years BP the record is less robust).

Peatland surface wetness changes

Changes in the surface wetness of peatlands are recorded by a number of biologic proxy indicators (plant macrofossils, testate amoebae) and by the degree of peat decay (humification). Testate amoebae permit quantitative inference of water-table depth, when supported by a calibration dataset that describes (typically using weighted-averaging techniques) species distribution in relation to a hydrologic gradient at the local to regional scale. This approach has the advantage that local water balance is reconstructed directly. Inference of climate variables (temperature, precipitation) from these reconstructions, however, depends on calibrating local water balance variations in relation to climate (e.g. Charman *et al.* 2004). This approach contrasts with some other biologic paleoclimate proxies (e.g. pollen, chironomids) where proxies are calibrated directly with the climatic variables, and often over larger geographic scales.

Peat records from those regions across continental Europe where peatlands are present are growing in number, but most records currently available are from the UK. Here we focus on the record from northern Britain, where there is a geographically dense set of records that can be summarized as an average regional series (Charman *et al.* 2006) or by the clustering of "wet shifts" (Barber and Charman 2003). Figure 8.4b shows the averaged series created by identifying correlative events in sets of detrended, normalized local peat surface-wetness records from each of four latitudinal regions (northern Scotland, central Scotland, the England–Scotland border region, and northern England). Individual site radiometric chronologies were tuned within chronologic errors and using independent age markers (tephra horizons, pollen markers, spheroidal carbonaceous particles), so that a "stacked" or averaged record could be derived. This method assumes that significant differences in century-scale climate change are unlikely to have occurred within a region, but are allowed between regions. Although such apparent interregional differences in the late Holocene sequence of wet and dry events may be real and significant (Langdon and Barber 2005), the integrated surface-wetness record for the whole of northern Britain (54–58°N, 5–0°W; Figure 8.4b) appears to provide a useful basis for comparisons with late Holocene hydrologic records with similar time resolution from continental Europe. According to this record, pronounced shifts to wetter conditions occurred at 3600, 2760, and 1600 years BP, and somewhat less pronounced events at 4200, 3800, 2000, 1260, 860, 550, and 260 years BP.

Is there a coherent pattern of moisture-balance variability between archives?

British peat surface-wetness records and central European lake-level data show coherent hydrologic changes at the sub-millennial time-scale during the late Holocene (Figure 8.4b and c). Coincidence is evident between markedly wet episodes in both regions at 2750–2350 and 750–650 years BP, and the pronounced dry episode at 3300–2750 years BP. Other wet events recorded in British peat are reflected in the central European lake-level record as episodes of less pronounced drought: 4150–3900, 3500–3300, 2000–1700, and 500–100 years BP. The most significant discrepancy involves the period 1500–1100 years BP, when inferred drought during 1700–1500 years BP switched to increased surface wetness in Britain but persisted with mostly (but not exclusively: Figure 8.4c) low lake levels in central Europe and only a modest return to higher lake levels. Some of the cited changes also appear in the European Lake Level database (Figure 8.2). For example, in the continent-wide lake-level compilation, but particularly in lakes south of 46°N (Figure 8.2c), greater moisture is evident from 3770 years BP, with a further increase by 2565 years BP. Central European drought from 1750 to 1000 years BP, ending in a rapid return to wetter conditions, is also evident in the low-resolution dataset as a peak and then reduction of the fraction of lakes south of 46°N that experienced low status (1380–930 years BP). Strong overall correspondence between the British peat and the west-central European lake-level records of water balance suggests a predominance of regionally coherent trends in effective moisture. A possible explanation is that both peat surface wetness and lake level are controlled mostly not by the total annual moisture balance but by the duration and intensity of a seasonal $P - E$ deficit (i.e., the extended summer period): much of the excess precipitation during periods of positive $P - E$ balance is lost as run-off from peatlands or increased (sub-) surface outflow from lakes.

The main periods of glacier mass balance in the Alps also show broad correspondence with the lake-level and peat surface-wetness data. In particular the advances at 2750–2450 and 1550–1250 years BP and during the LIA at 650–50 years BP are synchronous with wet periods inferred from the British peat record, or slightly precede them in the case of the 900–750 years BP advance. Most of these glacier fluctuations also show strong correspondence with lake-level variation in the adjacent Jura, the French pre-Alps, and on the Swiss plateau (Holzhauser *et al.* 2005). One possible link between these different archives is summer precipitation and temperature; high precipitation and low temperatures during summer months would increase glacier mass balance, similar to the way they increase available moisture for lakes and peatlands. One notable discrepancy between the regional glacier and lake-level records is the apparent coincidence of major glacier advance at ca. 1450–1250 years BP, with summer drought inferred from a predominance of low lake-level indicators at that time and only minor increases in higher lake-level indicators during the latter part of this phase (Figure 8.4c and d). This is the same period when a discrepancy exists between lake and peat records (cf. above), suggesting a more muted signal in the lake-level records than the other proxies.

It becomes more difficult to argue convincingly about potential climate links between records that are understood to reflect mostly summer (or extended summer) conditions and those reflecting winter climate or an annual mean. Niggeman *et al.* (2003) suggest that $\delta^{18}O$ variation in their German speleothem is mostly a function of winter conditions; cold, dry winters result in low summer water supply to the cave and more enriched $\delta^{18}O$ in drip waters. Proctor *et al.* (2000) find annual band thickness in their Scottish speleothem to correlate strongly with the ratio of annual T over P. In this region, however, where winter rainfall exceeds summer rainfall, the annual signal may be determined more by winter than summer conditions. We should therefore not necessarily expect to see strong correspondence between "annual" proxy indicators in the European speleothem records and "summer" proxy indicators in the peat and lake records. The most obvious comparison to make is between a 100-year running mean of the Scottish speleothem band-thickness record and the peat record from northern Britain. The relationship between these records is not consistent through time (Figure 8.4a and b). In general, the records are in phase between 3500 and 1700 years BP but antiphase after this period. Also instructive is a comparison between the Scottish speleothem record and the late Holocene record of glacier mass balance in maritime Norway (Figure 8.5), which is similarly understood to mostly reflect winter precipitation. Correspondence between episodes of greater inferred wetness in both regions is not immediately evident, however. The only two periods of significant Norwegian glacier retreat during the past 4000 years (1600–1300 and 800–600 years BP) correspond with broad minima (i.e. wet condition in speleothem band thickness (1700–1400 and 800–600 years BP).

In summary, periods of greater moisture excess in north-west Europe over the past 4000 years (Figure 8.4) are consistently recorded in three types of proxy record (peat surface wetness, lake levels, Alpine glaciers) that are principally a function of warm season water balance. Strong correspondence between them suggests that changes in summer $P - E$ were coherent throughout temperate Europe at least from the Alps to the north of Scotland (46–58°N). Judging from the available records with proxies responding more probably to annual and/or winter precipitation, it seems likely that warm-season wetness was associated with decreased winter precipitation since *c.* 1700 years BP (correspondence between Scottish speleo-them and north Britain peat records). This is supported by the centenial-scale correlation between wet summers and drier or unchanged winters during the Little Ice Age in north-west Europe (e.g. Figure 8.6; Lamb, 1985).

Considering that different climatic variables control glacier mass balance in the Alps and Norway it is not surprising that glacier mass balance records from these two regions show only limited convergence. Glacier advance appears to have taken place in both regions around 2800–2600 years ago and during the Little Ice Age (LIA), although the timing of maximum glacier extent is variable and the Alpine records reflect a sequence of higher-frequency changes. Glacier expansion in Norway may result from enhanced winter zonal circulation, which in recent centuries is related to a strongly positive North Atlantic Oscillation (NAO) (Wanner *et al.* 2001). Summer precipitation variability, however, shows no or only very weak association with the NAO, and the warmer winters in central Europe associated with a positive NAO would be expected to have caused glacier recession in the Alps. This implies that synoptic weather patterns during recent centuries are not necessarily a good guide to longer-term climate variability: wetter LIA winters in Norway must have been accompanied by cooler and wetter summers in the Alps. To generate this combination of wetter summers and wetter winters, zonal circulation must not only have been intensified in winter, but also must have shifted southward with persistence of the moisture-bearing Westerlies into the late spring and early autumn. Wet summers can also be generated by higher convective rainfall, but this is associated with higher rather than lower temperatures.

Precipitation changes over the past 500 years from annually resolved proxy records

Until recently, annually resolved records of European precipitation were restricted to instrumental records (very sparse before AD 1800) and a few long documentary time series (Brázdil *et al.* 2005), both of which are exceptionally rich data sources even in a global context. Now a number of studies have extended the time frame of annually resolved precipitation reconstructions to the past 400–500 years. Pauling *et al.* (2006) combine instrumental time series with precipitation indices based on documentary evidence and natural proxy data (tree rings, corals, ice cores, and a speleothem) to provide Europe-wide (30°W to 40°E/30° to 71°N) reconstructions of season-specific precipitation from AD 1500. Although estimated error is evidently higher in the earlier portions of the reconstruction, correlation of this Europe-wide dataset with independent tree-ring datasets (Brázdil *et al.* 2002; Wilson *et al.* 2005) for two regions in central Europe is good. Reminiscent of the conclusion reached by Vincent *et al.* (2005), one notable finding of this study is the occurrence of major nonstationarities in long-term regional precipitation patterns across Europe. Another notable finding is the considerable difference in long-term (multi-decadal) precipitation trends among seasons. Winter (DJF) precipitation (Figure 8.6a) displays pronounced decade-scale fluctuations in the 18th century

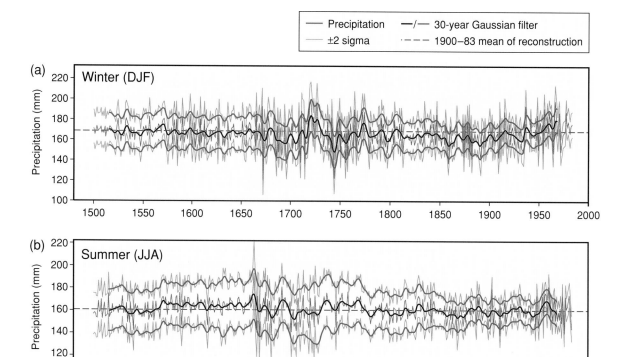

Figure 8.6 Spatially averaged time series and uncertainty range of Europe-wide (30°W–40°E/30–71°N) seasonal precipitation, based on instrumental time series, precipitation indices from documentary evidence, and annually resolved proxy data (tree rings, corals, ice cores, and a speleothem): (a) winter (DJF) and (b) summer (JJA). (From Pauling *et al.* (2006) with kind permission from Springer Science and Business Media.)

before assuming a declining trend in the 19th century and a mostly rising trend since ca. AD 1895. Fluctuation in spring, summer (Figure 8.6b), and autumn precipitation is greatest in either the second half of the 17th century or the 18th century, followed by muted variation and no pronounced trends in the 19th and 20th centuries. These distinct long-term trends in seasonal precipitation patterns support our contention that a thorough understanding of any seasonal bias in proxy data is essential for a meaningful comparison of multi-proxy and multi-archive records at millennial and century time-scales.

Comparison of the central-European lake-level compilation (Magny 2004) with the annually resolved precipitation record of Pauling *et al.* (2006) is not possible because of the large difference in time resolution, but peat surface-wetness records can potentially provide decade-scale resolution (Hendon and Charman 2004). A superficial visual comparison between the annually resolved European summer precipitation time series and the peat-based water-table compilation from northern Britain (Figure 8.4b) suggests matching dry periods in the late 17th and early 18th centuries. Increasing summer precipitation in the course of the 16th century appears to post-date a late 15th century (ca. AD 1480) rise in peatland water tables, but in fact can be synchronous within dating errors on the peat record. Extending annually resolved precipitation records further back in time, combined with a move towards peat-based reconstructions with decadal or better resolution, provides an opportunity to establish multi-proxy reconstructions for region-specific European precipitation over longer time periods with greater confidence.

Relations between flood records and long-term hydrologic change

A topic of growing concern in the context of future climate change is the extent to which short-lived extreme events are related to longer-term, low-frequency climate variability. Multi-archive integration of paleohydrologic proxy data provides an opportunity to improve our understanding of the relationship between extreme hydrologic events (extreme drought, flooding) and long-term trends in moisture balance. As discussed above, our ability to identify coherent patterns of century-scale hydrologic variability with any confidence is still plagued by the problems of incomplete spatial coverage, proxy- and archive-specific responses to climate variables, and chronologic limitations. Likewise, available long-term records of extreme hydrologic events are not always good quality and may be lacking from those regions where the low-frequency changes are best constrained. The large, quality-checked databases of past fluvial activity now being assembled are providing the first regionally integrated records of changing flood frequency over centennial to millennial time-scales. Methods are now also being developed to infer episodes of extreme drought from tree-ring records. Such developments allow tentative speculation on the links between these inventories of extreme events and variation in "average" moisture-balance conditions at least over the late Holocene (Macklin *et al.* 2006).

The principal integrated datasets on past fluvial activity come from Britain, Spain, and Poland (Macklin and Lewin 2003; Macklin *et al.* 2005, 2006; Johnstone *et al.* 2006; Thorndycraft and Benito 2006). Attempts to relate phases of increased flood frequency observed in these records to longer-term hydrologic change have met with mixed success, although links have been made to the integrated Holocene lake-level record for west-central Europe and millennial-scale North Atlantic climate variability as reflected in the record of ice-rafted debris (Macklin *et al.* 2005, 2006; Thorndycraft and Benito 2006). Problems of interpretation relate to the type of event recorded in various fluvial depositional environments (alluvial overbank and channel gravels, flood-basin facies, slackwater flood deposits), the possibility that enhanced fluvial activity is due to (continuous or interrupted) human impact on the landscape, and more generally the problematic dating of individual fluvial events. Focus on the late Holocene portions of the records from Spain (Figure 8.7), Poland, and Britain suggests that Europe-wide flooding episodes centered on 3500, 2750, 2550, 1900, 1300, 870, 660, and 570 years BP (approximate ages; Macklin *et al.* 2006) tend to correspond to phases of increased moisture balance revealed in

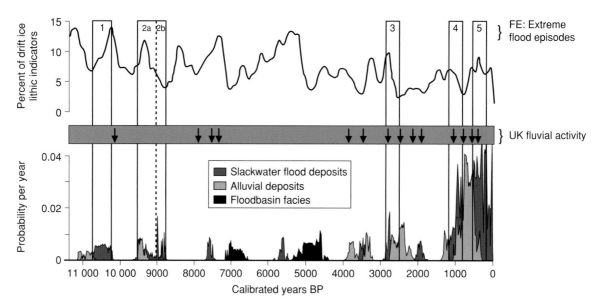

Figure 8.7 Holocene fluvial chronology of Spain in relation to the North Atlantic drift-ice record (Bond *et al.* 2001). (From Thorndycraft and Benito (2006). Copyright Elsevier (2006).)

low-frequency proxy records (Figure 8.4b). This supports the idea that flooding is linked to (multi-) decadal shifts in summer water balance. In the case of Britain these changes may have been caused by semi-permanent southward shifts in westerly circulation, which increased summer precipitation, groundwater levels, and soil moisture. This would increase catchment susceptibility to flooding from extreme rainfall events, even if these events were no more frequent during these periods with cooler, wetter summers.

In an ingenious application of dendroclimatology, Masson-Delmotte *et al.* (2005) used the stable-isotope composition of tree-ring cellulose to reconstruct drought frequency and precipitation seasonality over the past 400 years in western France (Brittany: 48°N, 2°W). Using multiple linear regression the authors derived a relationship of oak latewood $\delta^{18}O$, $\delta^{13}C$ (corrected for progressive atmospheric ^{13}C depletion due to fossil-fuel emissions) and ring width to water stress, i.e. the ratio of actual evapotranspiration to potential evapotranspiration calibrated over the reference period 1951–1996 (Raffalli-Delerce *et al.* 2003). At the decadal time-scale, the resulting 400-year water-stress reconstruction (Figure 8.8a) shows significant (inverse) correlation with historical wine-harvest dates, a regional proxy indicator for dry and warm summer conditions. Combined with instrumental rainfall records, the data suggest a widespread change in the seasonality of western European precipitation during the early 19th century towards drier winters and wetter summers. The authors also produced a drought recurrence record based on the frequency of years with water stress values exceeding one

Figure 8.8 (a) Summer water stress (ratio of actual to potential evapotranspiration) in western France since AD 1600, reconstructed using oak tree-ring width and isotopic composition calibrated against the reference period 1951–1996. (b) Recurrence of extreme summer drought (> 1 SD above the mean of 1951–1996) calculated from the water stress record. The long-term mean drought recurrence is once every four years (index value 0.25); shaded bars indicate gaps in isotopic data. (From Masson-Delmotte *et al.* (2005) with kind permission from Springer Science and Business Media.)

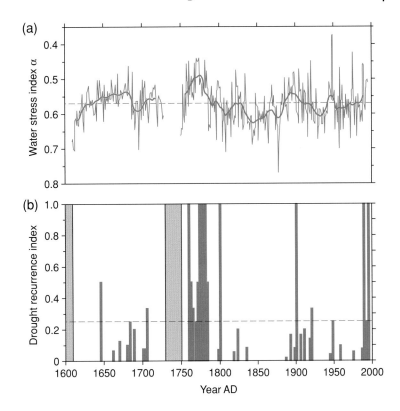

standard deviation above the 1951–1996 mean (Figure 8.8b). The most notable climate anomalies in this record are the high drought recurrence in the late 20th century, exceeded only by a period of persistent extreme summer drought ca. 1765–1780.

Tropical Africa, extending into monsoonal Asia

Millennial-scale moisture-balance variation through the Holocene

Street and Grove (1976; Figure 8.9) compiled the first millennial-scale lake-status dataset for tropical Africa comparable to the European Lake Level database; Kutzbach and Street-Perrott (1985) expanded this dataset to tropical lakes world-wide. Many contributing lake-level chronologies have since been modified to correct for lake carbon reservoir ages and the reworking of plant macrofossils in the beach and lowstand deposits where they were found. This aggregate data-set, however, still succeeds in illustrating the two drastic dry-to-humid transitions in northern and equatorial Africa around 15 ± 0.5 and 11.5–10.8 ka (ca. 12 700 and 10 000–9500 ^{14}C years BP, respectively on Figure 8.9). The dry-to-humid transition at ca. 15 ka represents the start of the so-called African Humid Period

Figure 8.9 Frequency of stratigraphic evidence for intermediate or high lake status in seven regions of (sub-) tropical north and equatorial Africa since 30 000 ^{14}C years BP, highlighting the early Holocene Humid Period. (From Street and Grove (1976). Reprinted by permission from Macmillan Publishers Ltd: *Nature*.)

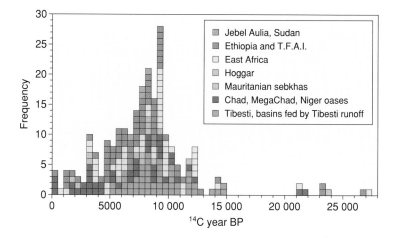

(deMenocal *et al.* 2000), when fairly abrupt intensification of the northern Summer Monsoon (e.g. Williams *et al.* 2006) ended a continent-wide drought linked to Heinrich event 1 (Stager *et al.* 2002; Lamb *et al.* 2007a). Most proxy data agree that tropical Africa was also dry during the Last Glacial Maximum (LGM; ca. 21–18 ka) and the Younger Dryas episode (YD; ca. 12.8–11.5 ka) (Gasse 2000; Barker and Gasse 2003; but see Garcin *et al.* 2006). Paleoclimate research over the past two decades, mostly based on lake-sediment records scattered across northern and eastern Africa, has elucidated the broad latitudinal patterns of Holocene moisture-balance variation across the African continent (Gasse 2000). Throughout northern (sub-) tropical and equatorial Africa these provide strong support for precessional insolation forcing to have been the dominant driver of Atlantic and Indian Ocean monsoon dynamics and moisture advection at this timescale, through its control on the mean latitudinal position of the Intertropical Convergence Zone (ITCZ; Haug *et al.* 2001). To what extent (sub-) tropical southern Africa experienced the predicted anti-phased patterns of Holocene hydrologic change has only partly been resolved (Barker *et al.* 2004; Scott and Lee-Thorp 2004). Many southern lakes recorded a dry LGM followed by wet conditions between ca. 15 (17) and ca. 13 ka, indicating that immediately prior to the Holocene glacial boundary conditions dominated over low-latitude insolation forcing (Barker and Gasse 2003). These same lakes (e.g. Lake Rukwa at 8°S, Lake Malawi at 12°S) show decreasing water levels during or immediately after the YD and into the early Holocene (Gasse *et al.* 2002; Thevenon *et al.* 2002), reflecting a less favorable regional moisture balance, due at least partly to northward displacement of the meteorologic Equator. They also record a return of modestly wetter conditions over the past 3000–5000 years, opposite to the pronounced drying trend experienced by the Northern Hemisphere tropics during the middle and late Holocene (Gasse 2000). Early Holocene establishment and persistence of woodland vegetation tolerant to a prolonged dry season in south-eastern tropical Africa suggested to Garcin *et al.* (2007a) that resumption of the African monsoon at the YD–Holocene transition must have been accompanied by greater latitudinal amplitude of seasonal ITCZ migration over eastern Africa.

Century-scale moisture-balance variation during the early to middle Holocene

A current major issue in African paleoclimate research is the regional expression of sub-Milankovitch climate variability. Speleothem records of Indian and Asian monsoon intensity (Neff *et al.* 2001; Fleitmann *et al.* 2003; Wang *et al.* 2005) suggest Holocene hydrologic change in the Northern Hemisphere sub-tropics to have been a rather smooth adjustment to the gradual southward migration of mean annual ITCZ position in response to the orbitally induced latitudinal shift in summer solar insolation (Figure 8.10a–c). For example the Dongge Cave record from sub-tropical China (Wang *et al.* 2005; Figure 8.10a) closely tracks summer insolation at 30°N, except for eight relatively modest century-scale events inferred to reflect episodes of weakened monsoon activity, and minor additional modulation by multi-decadal solar-activity changes. Six of these weak monsoon events coincide, within dating error, with the Northern Hemisphere cold episodes first recorded as episodes of peak ice-rafted debris abundance in marine sediments (Bond *et al.* 2001). A high-resolution marine sediment record from the Arabian Sea (Gupta *et al.* 2003) also mainly recorded the long-term trend of decreasing Indian monsoon intensity, however, with somewhat stronger signatures of inferred weak monsoon episodes superimposed on it (Figure 8.10d).

The above-mentioned proxy records of predominantly gradual monsoon weakening during the Holocene contrast with many lake-level records of hydrologic variability in eastern Africa and south Asia, which are punctuated by abrupt century-scale dry spells with a magnitude rivaling that of the main Holocene drying trend (Gasse and van Campo 1994; Figure 8.10f and g). The two best-documented drought episodes, at ca. 8400–8000 and 4200–4000 years BP (Gasse 2000), broadly match the timing of the two most pronounced weak monsoon episodes in the Dongge Cave speleothem record, and thus can be considered coincident with Bond events 5 and 3 (Mayewski *et al.* 2004). The 4200–4000 years BP drought similarly had widespread impact, having been linked to the demise of ancient cultures from Egypt (the Old Kingdom: deMenocal 2001) over Mesopotamia (the Akkadian Empire: Weiss 2000) and India (Indus valley civilization: Staubwasser *et al.* 2003) to China (Wu and Liu 2004).

Lake-status data from the Sahara, the Arabian Desert, and adjacent sub-arid regions (Hoelzmann *et al.* 2004) suggest that across these regions the early Holocene wet period ended in a succession of two pronounced dry spells at 6700–5500 and 4000–3600 years BP. Even more extreme, the marine sediment record of eolian deposition in the sub-tropical North Atlantic Ocean adjacent to the West African Sahara suggests that it ended abruptly in a single episode of rapid

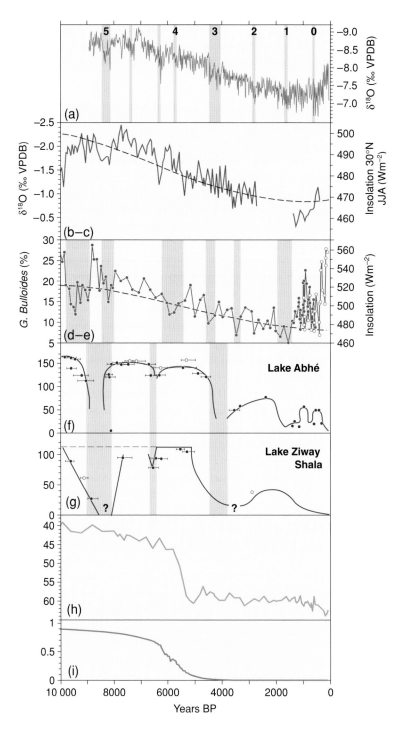

aridification ca. 5500 BP (deMenocal *et al.* 2000; Figure 8.10h). An important question is whether these disparate records of Holocene moisture-balance change truly reflect large-scale regional differences in climate history, or are a function of proxy- and archive-specific factors. For example, the otherwise widespread hyper-arid spell at 8400–8000 years BP has so far not been recorded at lake sites in the eastern Sahara desert, presumably because of the buffering effect of fully recharged regional aquifers at that time (Hoelzmann *et al.* 2004). Model simulations of mid-Holocene aridification of the Sahara (Claussen *et al.* 1999) suggest that the nonlinear climate response to gradual insolation forcing observed in the marine dust record results from vegetation–atmosphere feedbacks, namely the albedo effect of changing vegetation on monsoonal rainfall penetrating the continent (Figure 8.10i; see also Renssen *et al.* (2006) and Claussen, this volume). Notably in this study the exact timing of Sahara aridification (5800–5300 years BP in various model runs) depended on prescribed values for North Atlantic sea-surface temperature (SST) and sea-ice cover, which through a vegetation–snow–albedo feedback on boreal vegetation affected the meridional temperature gradient and hence Northern Hemisphere summer monsoon intensity.

Exactly when aridification culminated in the (near-) complete loss of vegetation cover at the local scale would have depended both on the absolute rainfall amount and on the rate of lowering of the regional groundwater table, possibly generating a simultaneous threshold response over large areas. Conversely, the hydrologic

Figure 8.10 (*Opposite*) Selected high-resolution moisture-balance records for subtropical monsoon regions in Africa and Asia covering the Holocene. Shaded bars are weak monsoon events in (a) and (d), and arid episodes in (f) and (g); insolation is the hatched line in (c) and (e). (a) Asian monsoon strength inferred from stalagmite $\delta^{18}O$ in Dongge Cave (sub-tropical China), with indication of weak monsoon events coeval with North Atlantic cold events (numbered). (From Wang *et al.* (2005). Reprinted with permission from AAAS.). (b) Indian monsoon strength inferred from stalagmite $\delta^{18}O$ in southern Oman compared with (c) summer (JJA) insolation at 30°N (from Fleitmann *et al.* (2003). Reprinted with permission from AAAS), with insolation curve based on Berger and Loutre (1991). (d) Indian monsoon strength inferred from *Globigerina bulloides* abundance in sediments from an upwelling zone in the Arabian Sea compared with (e) July insolation at 65°N (from Gupta *et al.* 2003 – reprinted by permission from Macmillan Publishers Ltd: *Nature*), with insolation curve based on Berger (1978). (f and g) Rainfall in north-east Africa inferred from the surface elevations of Lakes Abhé and Ziway Shala, Ethiopia. (From Gasse (2000). Copyright Elsevier (2000).) (h) Terrigenous input in marine sediments from the eastern tropical Atlantic Ocean. (From deMenocal *et al.* (2000). Copyright Elsevier (2000).) (i) Fractional vegetation cover in the Sahara desert simulated using a CLIMBER-2 coupled atmosphere–ocean–vegetation model of intermediate complexity (From Claussen *et al.* (1999). Copyright [1999] American Geophysical Union. Reproduced by permission of American Geophysical Union.)

response of individual Saharan lakes to an abrupt decrease in rainfall may have been either rapid or gradual (Salzmann and Waller 1998), depending on whether there was continued groundwater input from surrounding aquifers long after recharge was reduced. In the driest continental areas, an abrupt humid-to-arid transition recorded sometime between 7000 and 4000 years BP in a hydrologically favored lake basin can simply mean that summer monsoon rainfall no longer reached the area (Fleitmann *et al.* 2007a). A synoptic compilation of geologic and archaeologic data from the eastern Sahara over the past 12 000 years shows that Holocene aridification progressed gradually from ca. 7300 years BP to the present day (Küper and Kröpelin 2006). Similarly, palynologic and paleolimnologic data from the western Sahel region of North Africa (10°N, 12°E) indicate that the early Holocene humid period there ended in a gradual decline of the precipitation/evaporation ratio, not obviously interrupted by abrupt climatic events (Salzmann *et al.* 2002).

Century-scale moisture-balance variation during the late Holocene

The three most recent weak monsoon events in the Asian monsoon record at Dongge Cave (Wang *et al.* 2005) occur at 2800–2700, 1500–1600, and 600–400 years BP. The latter corresponds with the most recent Bond cold event in the North Atlantic, but is rather tenuously expressed, short-lived, and is immediately followed by a pronounced intensification of the Asian Monsoon over the past 400 years, also recorded in Arabian Sea sediments (Anderson *et al.* 2002). Similar century-scale hydrologic events affecting the African continent during the past 3000–4000 years are difficult to correlate spatially because of poor geographic coverage and limited age control on available climate-proxy records, and frequent hiatuses in lake-sediment records due to intermittent complete desiccation. The speleothem record of Indian monsoon activity in Oman (Fleitmann *et al.* 2003) is also interrupted between 3000 and 1500 years BP, suggesting that this extended period may have been the driest of the Holocene, excluding short-lived hyper-arid spells at other times. In the central Sahara, arid to hyper-arid conditions have prevailed since 3200 years BP (Hoelzmann *et al.* 2004).

Pollen data of Sahel vegetation in Nigeria (Salzmann and Waller 1998) and a reappraisal of the principal late Holocene regression of Lake Bosumtwi, Ghana (Russell *et al.* 2003) suggest that pronounced drying in tropical West Africa started at or shortly before 3200–3300 years BP. Lake-status and pollen records from the Atlantic side of intertropical Africa typically show dry conditions between 3000–2800 and 2000 years BP (Stager and Anfang-Sutter 1999; Vincens *et al.* 1999; Wirrmann *et al.* 2001; Giresse *et al.* 2004), often with a clear aridity maximum around 2800–2500 years BP (Reynaud-Farrera *et al.* 1996; Maley and Brenac 1998). Peak late Holocene aridity in West Africa is thus coeval with a weakened Asian monsoon and North Atlantic cold event 2.

Using percent Mg in authigenic calcite as a proxy indicator for the water balance of Lake Edward, Russell and Johnson (2005b) constructed a chronologically

Figure 8.11 Percent Mg in authigenic calcite from Lake Edward (Uganda) as proxy indicator for rainfall and drought in central equatorial Africa over the past 5500 years. (a) Composite record based on integration of data from successively deeper cores; (b) same record after removal of the long-term trend. (Modified from Russell and Johnson (2005b). Copyright Elsevier (2005).)

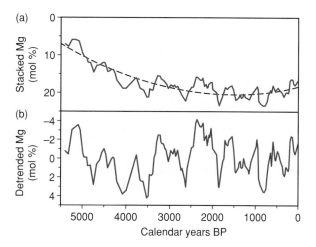

well-constrained record of rainfall and drought for central equatorial Africa (the western shoulder of the East African Plateau) for the past 5400 years. It reveals a chronology of century-scale drought events at 4100–3950, 3600–3400, 2750–2600, 2050–1850, 1500, and 900–700 years BP, superimposed on a mid-Holocene drying trend that may have reversed to slightly wetter average conditions in the past two millennia (Figure 8.11). The driest conditions in this region thus appear to have occurred during the 2050–1850 years BP event, not the ca. 2650 years BP event. Integration with proxy water-balance records from lakes Tanganyika (Alin and Cohen 2003), Turkana (Halfman *et al.* 1994), Naivasha (Verschuren 2001) and small crater lakes in western Uganda (Russell *et al.* 2007) indicates that this general pattern may hold true for both western and eastern portions of the East African Plateau, and that drought at ca. 2000 years BP affected much, if not all, of equatorial Africa. The shift to a slightly wetter mean climate regime at the Equator in recent millennia may reflect an intensification of the north-eastern monsoon, due to enhanced Southern Hemisphere summer insolation (Russell and Johnson 2005a).

A review of diverse paleodata covering the past two millennia of tropical African climate history (Verschuren 2004) suggested that drought was also widespread in eastern equatorial Africa during the period ca. AD 900–1250, i.e. broadly coincident with the period of mostly warm conditions in high northern latitudes defined as the Medieval Warm Period (MWP) or Medieval Climate Anomaly (MCA) (Lamb 1985). Distinct regional climatic anomalies within this period together with inadequate dating control on many relevant paleodata continue to hamper a world-wide synthesis of this period (Hughes and Diaz 1994; Bradley *et al.* 2003), but relatively dry conditions also prevailed further away from the Equator in Ethiopia (Lamb *et al.* 2007b) and in south-eastern tropical Africa (Owen *et al.* 1990). In all these locations, Medieval drought ended in the mid- to late-13th century with a pronounced shift to more positive water balance. Although widespread on the African continent, the Medieval dry period post-dates the episode of inter-mittent severe aridity in sub-tropical Central America that has been linked to the collapse of Maya civilization (1200–1050 years BP, or AD 700–900: Hodell *et al.* 2001; Haug *et al.* 2003). Drought in Central America is typically thought to imme-diately precede the local equivalent to the MWP, which started ca. 900 years BP with a shift to a variable but generally more positive water balance, then evolved into a drier climate regime by AD 1100–1200 (Haug *et al.* 2003), i.e. coeval with "High Medieval" warming in north-temperate Europe (Bradley *et al.* 2003). A pronounced drought episode affecting the western Sahel region of North Africa (Holmes *et al.* 1998) is dated to 1200–1000 years BP, however, and thus coeval with Central American drought (Street-Perrott *et al.* 2000). Less well constrained chronologically, Lake Chad in the central Sahel also appears to have registered a transgression, rather than lake-level decline, at about 1000 years BP (Maley 1993). The Central and East African MWP-equivalent drought is therefore clearly distinct from pre-MWP drought in Central America (Russell and Johnson 2005b) and the presumed pre-MWP drought in the North African Sahel. Low-resolution vege-tation reconstructions from Gabon (Ngomanda *et al.* 2005) and Namibia (Scott 1996) reveal more or less continuous drought/warmth for a 500-year period encompassing the 8th through 13th centuries, suggesting that pre-MWP (Atlantic) and MWP-equivalent (Central and East African) drought had an impact on areas along the Atlantic Ocean coast in immediate succession.

The African Little Ice Age

During the time period equivalent with the Little Ice Age (LIA) in high-latitude regions, hydrologic variations in tropical Africa are divisible into an "early" LIA phase ca. AD 1250–1550 and a "main" LIA phase ca. AD 1550–1850. The start of the "early" LIA follows the glaciologic definition of the Little Ice Age, starting with pronounced glacier advance in the European Alps around AD 1250 (Holzhauser 1997) and expansion of sea ice around Iceland. In most of tropical Africa around this time, a marked shift to improved moisture balance ended the African equival-ent of the MWP. The "main" LIA is temporally broadly equivalent to the climati-cally defined Little Ice Age in Europe, i.e. ca. AD 1570–1900, when Northern Hemisphere summer temperatures fell significantly below the modern mean (Matthews and Briffa 2005).

Although there is considerable uncertainty in the chronology of available proxy records, Central Africa from the Equator to 11°S and further south into eastern South Africa all appear to have enjoyed relatively moist conditions only during the early LIA from ca. AD 1250–1300 to 1500–1550, then switched to a sustained drying trend that eventually culminated in severe 18th century drought. This regional picture now includes evidence from Lake Edward (Figure 8.12c; Russell and Johnson 2007), western Ugandan pollen records (Marchant and Taylor 1998) and crater lakes (Russell *et al.* 2007), Lake Tanganyika (Alin and Cohen 2003), Lake Malawi (Brown and Johnson 2005), Lake Masoko in southern Tanzania (Garcin *et al.* 2007b), and the speleothem record from Cold Air Cave, South Africa

Figure 8.12 West-to-east transect of lake-based moisture-balance reconstructions across the East African Plateau for the past 1000 years, compared with shorter reconstructions of the North Atlantic Oscillation and drought along the Atlantic coast of equatorial Africa. (a) 30-year running mean of NAO/AO index. (From Luterbacher *et al.* (2002) with kind permission from Springer Science and Business Media.) (b) Luanda (Angola) drought index based on Portuguese trader data, summed per decade (from Miller 1982). (c) Percent Mg in authigenic calcite from Lake Edward, western Uganda. (Modified from Russell and Johnson (2007). Copyright Elsevier (2007).) (d) Percent shallow-water diatoms in Pilkington Bay, Lake Victoria. (From Stager *et al.* (2005) with kind permission from Springer Science and Business Media. (e) Lithology-inferred water depth of Lake Naivasha, central Kenya (from Verschuren *et al.* 2000). Blue and orange bars highlight the geographic distribution of wet and dry episodes respectively.

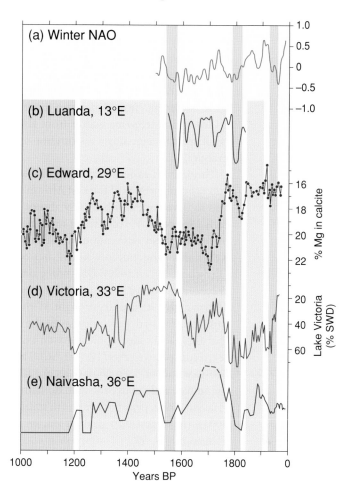

(Holmgren *et al.* 1999, 2003; Lee-Thorp *et al.* 2001). In eastern equatorial Africa, however, these moist conditions persisted until the mid-18th century, i.e. during much of the main LIA, before ending in late 18th century drought. This includes evidence from Lake Naivasha (Figure 8.12e; Verschuren *et al.* 2000) and Loboi Swamp (Ashley *et al.* 2004; Driese *et al.* 2004) in central Kenya, Lake Victoria (Figure 8.12d; Stager *et al.* 2005), and the Mount Kilimanjaro ice-core record (Gasse 2002; Thompson *et al.* 2002). Thus, following widespread drought in Medieval time (ca. AD 1000 to 1250) and widespread wet conditions during the early LIA (ca. AD 1250 to 1500–1550), a pronounced regional gradient developed during the main LIA, with above-average wet conditions in eastern equatorial Africa (Figure 8.12e) coinciding with relative drought in westernmost East Africa and southern tropical Africa (Figure 8.12c).

As recorded at Lake Naivasha in central Kenya, the relative wetness of eastern equatorial Africa between ca. AD 1250 and the late 18th century was interrupted by two decade-scale episodes of severe drought dated to ca. AD 1380–1420 and

1560–1620 (Verschuren *et al.* 2000; Figure 8.12e). Their geographic extent cannot be properly assessed with the currently available proxy records, due either to lack of resolution or because the signature of these decade-scale events is masked by the sustained drying trend of the main LIA. Late 16th century drought in eastern equatorial Africa, however, is broadly coeval with drought reported in Portuguese trade documents from Angola on the tropical Atlantic coast (Miller 1982; Figure 8.12b) and in tree-ring widths from Kwazulu-Natal, south-eastern South Africa (Hall, 1976), suggesting a distribution of impact that was at least sub-continental in scale.

Throughout eastern Africa from Ethiopia to Lake Malawi and in both western and eastern portions of the East African Plateau, the LIA-equivalent period can be said to have ended with a fairly abrupt switch to a positive water balance in the early 19th century, recorded at multiple locations as a marked lake transgression or refilling after desiccation (Verschuren 2004; Bessems *et al.* 2007). In central Kenya this pronounced wet-shift has been dated by ^{210}Pb assay to AD 1818 ± 8 (Verschuren *et al.* 1999). Depending on the region (see above) this wet-shift was preceded by almost 300 years of dry conditions or just a few decades of severe late 18th to early 19th century drought. With aridity perhaps peaking during the 1790s, this East African dry episode may be the regional manifestation of a mega-drought that also had a catastrophic impact on India, when the Indian monsoon failed for seven consecutive years (1790–1796). This is the most extreme climate event in the 560-year Indian documentary record (Grove 1998) and is expressed throughout the Indian and Asian Monsoon domains as a minimum in Nile River flood level (Hassan 1998), peak $\delta^{18}O$ values in stalagmite calcite from Oman (Fleitmann *et al.* 2004), the largest dust peak in glacier ice at Dasuopu on the southern Tibetan Plateau (28°N, 86°E; Thompson *et al.* 2000), and a short-lived lowstand of Lake Huguang in sub-tropical China (22°N, 110°E; Chu *et al.* 2002). Surprisingly it is not immediately evident in the Wang *et al.* (2005) speleothem record from Dongge Cave, 5° north of Lake Huguang. Based on speleothem data from Cold Air Cave (Holmgren *et al.* 2003), sub-tropical southern Africa may have recovered from 18th century drought earlier than tropical Africa, showing peak aridity in the early 18th century and a return to wetter conditions underway well before AD 1800.

High-resolution analysis of the Ba/Ca ratio in a ca. 300-year *Porites* coral from the Indian Ocean coast of Kenya produced a uniquely long-term record of river suspended sediment flux integrated over the Athi–Kabati river system, which drains a sizable portion of southern Kenya (Fleitmann *et al.* 2007b). A continuous rise since ca. 1920 reflects anthropogenic soil erosion due first to British colonial agriculture and since the 1960s due to steadily increasing demographic pressure on suitable agricultural land. Prior to the 20th century, however, long-term Ba/Ca levels are remarkably constant. Brief episodes of increased sediment flux during the 18th and 19th centuries can be inferred to reflect events of natural high soil erosion when torrential rain follows upon a period of severe drought. Such Ba/Ca peaks occur in 1737, intermittently between 1794 and 1826, and in 1884, which are known episodes of severe drought in East Africa also affecting the moisture balance of regional lakes (Figure 8.12e).

Latitudinal linkages

Causes and time-scales of hydrologic links between Europe and Africa

Precessional forcing of Holocene hydrologic variability

The principal long-term trend of Holocene climate change in both north-temperate and tropical regions reflects the global climate forcing exerted by variation in the latitudinal position of peak seasonal insolation, due to precession of the Earth's orbit around the Sun. Northern Hemisphere summer insolation has been above-average between ca. 15 and 5 ka, peaked at 10 ka, and is currently near its minimum; Southern Hemisphere summer insolation has been above average for the past five millennia, and is currently near its peak (Berger 1978). The Holocene history of temperature and precipitation on regional scales, however, has often deviated significantly from the perfect hemispheric anti-phasing which this precessional forcing would predict, due to various amplifying, damping, and feedback processes involving some or all components of the climate system.

North Atlantic Ocean summer sea-surface temperatures (SSTs) reached their Holocene maximum of 15°C by 10.8 ka, which in combination with cold air masses adjacent to the lingering Scandinavian ice sheet enhanced cyclonic activity, and brought mild but moist weather to northern Europe (Snowball *et al.* 2004; Jansen *et al.* this volume). When the ice sheet disappeared and the European continent itself started heating up, cyclonic incursions decreased and climate became drier, starving the Norwegian mountain glaciers as early as 9.7 ka (excepting the advance at 8.2 ka; Figure 8.5) and lowering lake levels across northern European after 7 ka (Figure 8.2b). The Holocene thermal maximum on the continent was reached between 8 and 6 ka, with especially dry summers 7000–6000 years ago. This early Holocene warming allowed the northward expansion of vegetation zones, with boreal forest reaching its furthest position north of the present-day tree line between 10 and 9 ka. Decreasing Northern Hemisphere summer insolation from 10 ka allowed greater seasonal sea-ice cover in the western Nordic Sea as early as 7 ka, but at the same time the continued advection of warm water to the eastern North Atlantic maintained high SSTs there until 5 ka (Snowball *et al.* 2004). Cooling on the adjacent continent set in around 5.8 ka, stimulating rapid expansion of the maritime glaciers in Norway, which were then in receipt of sufficient moisture (Figure 8.5). Glaciers in the Alps initially did not benefit because continental European climate remained dry, as evidenced by widespread low lake levels (Figure 8.2). Reduced summer evaporation due to cooling gradually improved the continental water balance, causing lakes to rise and peat-bog surfaces to become wetter from 4.5 ka, and both the Scandinavian and Alpine glaciers to experience significant advances from 3 to 2.5 ka (Figure 8.4c). The inference of increased wetness during recent millennia mostly derives from northern European lakes (Figure 8.2b). In the Mediterranean region from Spain in the west to Turkey in the east (Europe south of 46°N: Figure 8.2c), and even in west-central Europe (Figure 8.3), lowland continental proxy climate indicators suggest that strongly positive water-balance conditions started at ca. 10 ka but ended at ca. 4 ka with no recovery towards increased moisture availability. There were local exceptions to this pattern; for example, Roberts *et al.* (2004) invoked a reduced evaporation effect due to cooler winters in the Atlas Mountains of north-west Africa during the past 3000 years to explain the regional persistence of high lake levels there.

In the Northern Hemisphere tropics and sub-tropics, orbitally induced insolation forcing of Holocene rainfall was mediated by the accompanying shift in the mean annual latitudinal position of the ITCZ. Greatly increased monsoonal precipitation in the Northern Hemisphere sub-tropics during the early Holocene (the African Humid Period) resulted from the synergistic effect of the ITCZ being pulled further away from the Equator during Northern Hemisphere summer (Haug *et al.* 2001) and an enhanced temperature contrast between the continents and adjacent oceanic moisture sources (COHMAP Members 1988; Kutzbach and Webb 1993). Because the latitude of highest SST is modified in areas of cool-water upwelling (Berger and Loutre 1991), the ITCZ's mean annual latitude also varies around the globe. Consequently changes in the meridional temperature gradient within each ocean basin must be considered separately in order to determine the magnitude and timing of ITCZ migration over the course of the Holocene.

The Mediterranean margin of the Sahara receives winter rainfall from mid-latitude Westerlies displaced southward, whereas its southern margin receives summer rainfall of monsoonal origin. Today there is almost no overlap between these summer and winter rainfall regimes. Consequently, a true arid zone exists between 20° and 30°N. In the early Holocene the boundary between winter and summer rainfall regions is thought to have been located at 21–23°N, and at times the spatial domains of the two rainfall regimes may have overlapped (Hoelzmann *et al.* 2004). Afro-Asian monsoonal rains, however, probably did not fully cross the sub-tropical arid zone. The short Holocene humid interval 9250–7250 years BP in the northern Red Sea region (27°N) is thought to have involved southward extension of winter rainfall from the Mediterranean and local monsoon-type circulation during summer (Arz *et al.* 2003).

Century-scale hydrologic variability

Strong climatic links between the tropical Atlantic and Indian Oceans and Northern Hemisphere high-latitude regions are already evident in the run-up to the Holocene. Insolation-driven intensification of the African and Indian monsoons established wet conditions in tropical and northern sub-tropical Africa in two steps dated to 15 ± 0.5 ka and 11.5–10.8 ka, i.e. matching the two major warming events that introduced post-glacial conditions to northern temperate regions. The first dry-to-humid transition ended widespread tropical African drought associated with Heinrich event 1, and the second dry-to-humid transition ended widespread tropical African drought corresponding to the Younger Dryas chronozone. The first major dry spell punctuating the Holocene Humid

Period in northern intertropical Africa and throughout the African–Indian–Asian monsoon domain at ca. 8.4–8.0 ka has naturally been linked to the 8.2 ka event in Greenland and the circum-North Atlantic region (Alley *et al.* 1997; von Grafenstein *et al.* 1998; Mathews *et al.* 2005), attributed to abrupt freshwater release from the margin of the disintegrating North American ice sheet (Barber *et al.* 1999; Clarke *et al.* 2003; Alley and Agustsdottir 2005). Interestingly, high-resolution speleothem-based records of the Asian and Indian monsoon (Fleitmann *et al.* 2003; Dykoski *et al.* 2005; Wang *et al.* 2005) consistently show structure within the 8.4–8.0 ka event, with two distinct negative anomalies peaking at 8.26 ± 0.06 and 8.08 ± 0.07 ka (Fleitmann *et al.* 2007a). Which one of these two weak monsoon episodes is dynamically related to the North Atlantic cold event has not been elucidated, but at this century time-scale Indian and Asian monsoon dynamics clearly bear the signature of climatic variability originating in the North Atlantic Ocean.

Latitudinal linkages of millennial-scale hydrologic events during the middle to late Holocene are less evident because of poor chronologic control on North Atlantic cold events 0 to 4 as originally defined (Figures 8.10 and 8.13b; Bond *et al.* 2001) and their uncertain relationship to millennial-scale variability in the Greenland ice-core record (Figure 8.13a; O'Brien *et al.* 1995). By comparison, the events of weak Indian and Asian monsoon activity that supposedly are dynamically related to the North Atlantic cold events are much better constrained chronologically (Wang *et al.* 2005; Figures 8.10a and 8.13c), except that the Dongge Cave speleothem record does not give strong evidence of weak Asian monsoon events coincident with the LIA (event 0) and the Dark Ages cold episode (1800–1400 years BP; event 1). Wang *et al.* (2005) also confirmed the significant influence of small (< 1 percent) variations in century-scale solar output on Asian monsoon intensity, a sun–climate connection that was previously hinted at by many less well-dated climate proxy records tuned to the atmospheric $\Delta^{14}C$ record of solar forcing. These included the original record of a 1500-year quasi-cycle in ice rafting on the North Atlantic Ocean (Bond *et al.* 2001), mass balance of Scandinavian glaciers (Denton and Karlén 1973), temperature (Heiri *et al.* 2004) and lake-level variations (Magny 2004; Figures 8.3 and 8.13e) in central Europe, and Indian monsoon intensity in the Arabian Sea (Agnihotri *et al.* 2002; Gupta *et al.* 2003, 2005) and on adjacent land areas (Neff *et al.* 2001; Fleitmann *et al.* 2003). Modeling studies indicate that the reported millennial-scale climate anomalies are caused by beat modulation of the century-scale periodicity in solar output (Clemens 2005; see also Braun *et al.* 2005).

In equatorial Africa, Russell and Johnson (2005a) found temporal correlation between drought in central equatorial Africa and both cold and warm extremes in the North Atlantic 1500-year cycle, when employing the derivative of their 5400-year Lake Edward percent Mg record (see above) as principal indicator of African moisture balance. These authors proposed that the reconstructed average drought recurrence time of 750 years results because either southward or northward displacement of the ITCZ from its mean position during cold or warm events in the North Atlantic would have starved the equatorial region of its normal share of

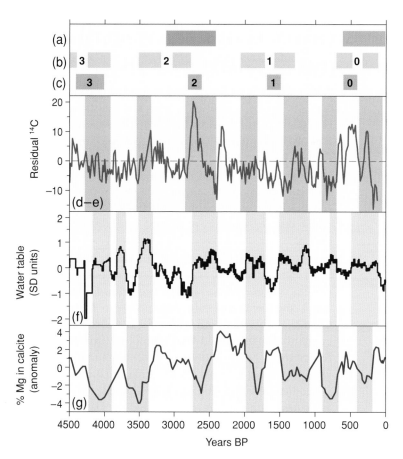

Figure 8.13 Timing of century-scale moisture-balance reconstructions in northwest Europe and tropical Africa in relation to atmospheric $\Delta^{14}C$ as proxy of solar activity variation and proxy records from elsewhere. (a) Holocene cold events in Greenland (from O'Brien *et al.* 1995). (b) Ice-rafting events in the North Atlantic Ocean (from Bond *et al.* 2001). (c) Episodes of weakened Asian monsoon (from Wang *et al.* (2005)). (d) Lake highstand episodes in west-central Europe (blue shaded bars) in relation to (e) atmospheric $\Delta^{14}C$ variation (from Magny (2004)). (f) Northern British peatland surface wetness, with inferred wet episodes highlighted (from Charman *et al.* 2006). (g) Percent Mg in authigenic calcite from Lake Edward (western Uganda) with inferred dry episodes highlighted (from Russell and Johnson (2005)).

rainfall. Holocene climate variability at similar ca. 700-year intervals has also been reported from the Indian Ocean (von Rad *et al.* 1999; Staubwasser *et al.* 2003), and attributed to solar forcing. Even so, the direct effect of solar forcing on tropical temperature and rainfall is probably much smaller than its indirect effect through amplifying processes in high-latitude regions, for example, forced shifts of the North Atlantic Ocean to a predominantly high or low AO/NAO index state (Shindell *et al.* 2001). In an earlier analysis of the high-resolution Lake Edward record, Russell and Johnson (2005b) had found no direct correlation with the atmospheric $\Delta^{14}C$ record of solar forcing. The most prominent $\Delta^{14}C$ anomaly of the late Holocene, at 2800–2600 years BP, corresponds with prolonged drought both in central Africa and tropical West Africa, but the most prominent central African drought of the late Holocene, at 2050–1850 years BP, occurred in a time of rather stable solar output (Figure 8.13g). Van Geel *et al.* (1998) pointed to the synchrony of the 2800–2600 years BP drought in Cameroon with the prominent shift to a cooler, wetter climate in north-temperate Europe (Figure 8.4) to associate both with reduced solar activity at that time (cf. Beer and van Geel, this volume). Looking at all proxy records of hydrologic change, a solar influence on central European lake levels (Magny 2004) and British peat surface wetness (Charman *et al.* 2006) during the late Holocene is evident but not strong. Coincidence of central African drought with phases of high surface wetness in British peatlands is high, however, throughout the past 4500 years (Figure 8.13f and g), suggesting a rather direct connection between these two regions in the climate processes modulating century-scale climate variability. One possibility is that variability in North Atlantic sea-ice extent, through its influence on ocean thermohaline circulation and cross-equatorial heat transport, creates sustained changes in mean annual ITCZ position (Vellinga and Wood 2002; Lohmann 2003; Li *et al.* 2005). When during cold events in the North Atlantic moisture advection onto the European continent is enhanced and the ITCZ assumes a more southerly position, westerly winds may not penetrate as far eastward into Africa (McHugh 2004) and potentially cause drought in the westernmost portions of East Africa that partly depend on this Atlantic moisture (Russell *et al.* 2007).

Conclusions

Marked regional correspondence exists both within and between high and low latitude regions in Holocene moisture-balance changes at millennial and century time-scales. Apparent regional variation in the principal Holocene trends of monsoon intensity across the Northern Hemisphere sub-tropics, from a gradual response to precessional insolation forcing to a succession of millennial-scale climate swings, must have both a real regional climate component and a component dependent on the archive or proxy indicator recording the climate change. Consistent century-scale water-table changes in north-west Europe reconstructed from peatlands have occurred in synchrony with lake-level changes in west-central Europe, and somewhat less clearly with mass balance changes of Alpine glaciers. The relative magnitude of individual events is archive-dependent and regionally diverse. Some episodes of high lake level and aquifers (at ca. 3400, 2750, 1250, and 700 years BP) and inferred cool weather correspond to periods of reduced solar activity, suggesting a causative link. The anomalies are not proportional to the proposed forcing, however, and some hydrologic events have no equivalent in the record of solar activity. Records from Iceland to tropical Africa reveal a pronounced shift to wetter conditions at ca. AD 1250, terminating a warm and/or dry Medieval Optimum with the onset of the most recent North Atlantic cool event. Strong inverse correlations between British peat surface wetness and moisture balance in central equatorial Africa at century time-scales over the past 5000 years suggest that North Atlantic cool episodes (not all of which are recorded as drift-ice events) affected cross-equatorial heat transport and the mean position of the Intertropical Convergence Zone. During the main phase of the Little Ice Age (ca. AD 1550 to 1800) the current west-to-east gradient across equatorial Africa from a humid climate in the west to a semi-arid climate in the east was reduced or possibly reversed. This modification of geographic climate patterns is a reflection of the contrasting effects of external climate forcing, such as reduced solar output, on rainfall advection onto the African continent from the equatorial Atlantic and Indian Oceans, mediated respectively by events in the North Atlantic and Pacific Oceans.

Acknowledgments

The authors wish to thank the HOLIVAR Steering Committee for the opportunity to produce this overview of Holocene hydrologic variability across continents and time-scales, two anonymous referees for constructive criticism of the draft manuscript, and the funding agencies which support our research on this topic.

References

Agnihotri R., Dutta K., Bhushan R. & Somayajulu B.L.K. (2002) Evidence for solar forcing on the Indian monsoon during the last millennium. *Earth and Planetary Science Letters*, **198**, 521–527.

Alin S.R. & Cohen A.S. (2003) Lake-level history of Lake Tanganyika, East Africa, for the past 2500 years based on ostracode-inferred water-depth reconstruction. *Palaeogeography, Palaeoclimatology, Palaeoecology*, **199**, 31–49.

Alley R.B. & Agustsdottir A.M. (2005) The 8k event: cause and consequences of a major Holocene abrupt climate change. *Quaternary Science Reviews*, **24**, 1123–1149.

Alley R.B., Mayewski P.A., Sowers T., Stuiver M., Taylor K.C. & Clark P.U. (1997) Holocene climatic instability: a prominent, widespread event 8200 yr ago. *Geology*, **25**, 483–486.

Anderson D.M., Overpeck J.T. & Gupta A.K. (2002) Increase in the Asian southwest monsoon during the past four centuries. *Science*, **297**, 596–599.

Arz H.W., Lamy F., Patzold J., Muller P.J. & Prins M. (2003) Mediterranean moisture source for an early-Holocene humid period in the northern Red Sea. *Science*, **300**, 118–121.

Ashley G.M., Mworia J.M., Muasya A.M., *et al.* (2004) Sedimentation and recent history of a freshwater wetland in a semi-arid environment: Loboi Swamp, Kenya, East Africa. *Sedimentology*, **51**, 1301–1321.

Baillie M.G.L. (1991) Suck in and smear: two related chronological problems for the 90s. *Journal of Theoretical Archaeology*, **2**, 12–16.

Barber D.C., Dyke A., Hillaire-Marcel C., *et al.* (1999) Forcing of the cold event of 8200 years ago by catastrophic drainage of Laurentide lakes. *Nature*, **400**, 344–348.

Barber K.E. & Charman D.J. (2003) Holocene palaeoclimate records from peatlands. In: *Global Change in the Holocene* (Eds A.W. Mackay, R.W. Battarbee, H.J.B. Birks & F. Oldfield), pp. 210–226. Edward Arnold, London.

Barker P. & Gasse F. (2003) New evidence for a reduced water balance in East Africa during the Last Glacial Maximum: implications for model-data comparison. *Quaternary Science Reviews*, **22**, 823–837.

Barker P., Talbot M.R., Street-Perrott F.A., Marret J., Scourse J. & Odada E. (2004) Late Quaternary climatic variability in intertropical Africa. In: *Past Climate Variability through Europe and Africa* (Eds R.W. Battarbee, F. Gasse & C.E. Stickley), pp. 117–138. Springer-Verlag, Berlin.

Beer J. & van Geel B. (this volume) Holocene climate change and the evidence for solar and other forcings. In: *Global Warming and Natural Climate Variability: a Holocene Perspective* (Eds R.W. Battarbee & H.A. Binney), pp. 138–162. Blackwell, Oxford.

Berger A.L. (1978) Long-term variations of caloric insolation resulting from earths orbital elements. *Quaternary Research*, **9**, 139–167.

Berger A. & Loutre M.F. (1991) Insolation values for the climate of the last 10 000 000 years. *Quaternary Science Reviews*, **10**, 297–317.

Bessems I., Verschuren D., Russell J.M., Hus J. & Cumming B.F. (2007). Paleolimnological evidence for widespread late-18th century drought across equatorial East Africa. *Palaeogeography, Palaeoclimatology, Palaeoecology*, doi:10.1016/j.palaeo.2007.10.002.

Birks J. (this volume) Holocene climate research – progress, paradigms, and problems. In: *Global Warming and Natural Climate Variability: a Holocene Perspective* (Eds R.W. Battarbee & H.A. Binney), pp. 7–57. Blackwell, Oxford.

Blaauw M., Heuvelink G.B.M., Mauquoy D., van der Plicht J. & van Geel B. (2003) A numerical approach to C-14 wiggle-match dating of organic deposits: best fits and confidence intervals. *Quaternary Science Reviews*, **22**, 1485–1500.

Blaauw M., Christen J.A., Mauquoy D., van Geel B. & van der Plicht J. (2006) Simultaneous or sucked-in? Testing the timing of radiocarbon-dated climate events between sites. *HOLIVAR Final Open Science Conference*, London, June.

Blytt A. (1876) *Essay on the Immigration of the Norwegian Flora during Alternating Rainy and Dry Periods*. Cammermeyer, Kristiania, Oslo.

Bond G., Kromer B., Beer J., *et al.* (2001) Persistent solar influence on North Atlantic climate during the Holocene. *Science*, **294**, 2130–2136.

Bradley R.S., Dodson J., Duplessy J.-C., Gasse F., Liu T.-S. & Markgraf V. (1995) PANASH–PEP science and implementation. In: *PAGES Series 95-1:*

Paleoclimates of the Northern and Southern Hemispheres, pp. 1–22. PAGES IPO, Bern.

Bradley R.S., Hughes M.K. & Diaz H.F. (2003) Climate in medieval time. *Science*, **302**, 404–405.

Braun H., Christl M., Rahmstorf S., *et al.* (2005) Possible solar origin of the 1470-year glacial climate cycle demonstrated in a coupled model. *Nature*, **438**, 208–211.

Brázdil R., Stepankova P., Kyncl T. & Kyncl J. (2002) Fir tree-ring reconstruction of March–July precipitation in southern Moravia (Czech Republic), 1376–1996. *Climate Research*, **20**, 223–239.

Brown E.T. & Johnson T.C. (2005) Coherence between tropical East African and South American records of the Little Ice Age. *Geochemistry, Geophysics, Geosystems*, **6**, doi: 10.1029/2005GC000959.

Charman D.J. (2007) Summer water deficit variability controls on peatland water-table changes: implications for Holocene palaeoclimate reconstructions. *The Holocene*, **17**, 217–227.

Charman D.J., Brown A.D., Hendon D. & Karofeld E. (2004) Testing the relationship between Holocene peatland palaeoclimate reconstructions and instrumental data at two European sites. *Quaternary Science Reviews*, **23**, 137–143.

Charman D.J., Blundell A., Chiverrell R.C., Hendon D. & Langdon P.G. (2006) Compilation of non-annually resolved Holocene proxy climate records: stacked Holocene peatland palaeo-water table reconstructions from northern Britain. *Quaternary Science Reviews*, **25**, 336–350.

Chu G.Q., Liu J.Q., Sun Q., *et al.* (2002) The "Mediaeval Warm Period" drought recorded in Lake Huguangyan, tropical South China. *The Holocene*, **12**, 511–516.

Clarke G., Leverington D., Teller J. & Dyke A. (2003) Superlakes, megafloods, and abrupt climate change. *Science*, **301**, 922–923.

Claussen M. (this volume) Holocene rapid land-cover changes – evidence and theory. In: *Global Warming and Natural Climate Variability: a Holocene Perspective* (Eds R.W. Battarbee & H.A. Binney), pp. 232–253. Blackwell, Oxford.

Claussen M., Kubatzki C., Brovkin V., Ganopolsjki A., Hoelzmann P. & Pachur H.-J. (1999) Simulation of an abrupt change in Saharan vegetation in the mid-Holocene. *Geophysical Research Letters*, **26**, 2037–2040.

Clemens S.C. (2005) Millennial-band climate spectrum resolved and linked to centennial-scale solar cycles. *Quaternary Science Reviews*, **24**, 521–531.

COHMAP Members (1988) Climatic changes of the last 18 000 years: observations and model simulations. *Science*, **241**, 1043–1052.

DeMenocal P.B. (2001) Cultural responses to climate change during the late Holocene. *Science*, **292**, 667–673.

DeMenocal P., Ortiz J., Guilderson T., *et al.* (2000) Abrupt onset and termination of the African Humid Period: rapid climate responses to gradual insolation forcing. *Quaternary Science Reviews*, **19**, 347–361.

Denton G.H. & Karlén W. (1973) Holocene climatic variations – their pattern and possible cause. *Quaternary Research*, **3**, 155–205.

Digerfeldt G. (1988) Reconstruction and regional correlation of Holocene lake-level fluctuations in Lake Bysjon, south Sweden. *Boreas*, **17**, 165–182.

Digerfeldt G. (1997) Reconstruction of Holocene lake-level changes in Lake Kalsjön, southern Sweden, with a contribution to the local palaeohydrology at the Elm Decline. *Vegetation History and Archaeobotany*, **6**, 9–14.

Digerfeldt G., de Beaulieu J. L., Guiot J. & Mouthon J. (1997) Reconstruction and paleoclimatic interpretation of Holocene lake-level changes in Lac de Saint-Leger, Haute-Provence, Southeast France. *Palaeogeography, Palaeoclimatology, Palaeoecology*, **136**, 231–258.

Driese S.G., Ashley G.M., Li Z.H., Hover V.C. & Owen R.B. (2004) Possible Late-Holocene equatorial palaeoclimate record based upon soils spanning the Medieval Warm Period and Little Ice Age, Loboi Plain, Kenya. *Palaeogeography, Palaeoclimatology, Palaeoecology*, **213**, 231–250.

Dykoski C.A., Edwards R.L., Cheng H., *et al.* (2005) A high-resolution, absolute-dated Holocene and deglacial Asian monsoon record from Dongge Cave, China. *Earth and Planetary Science Letters*, **233**, 71–86.

Fairchild I.J., Smith C.L., Baker A., *et al.* (2006) Modification and preservation of environmental signals in speleothems. *Earth-Science Reviews*, **75**, 105–153.

Fleitmann D., Burns S.J., Mudelsee M., *et al.* (2003) Holocene forcing of the Indian monsoon recorded in a stalagmite from southern Oman. *Science*, **300**, 1737–1739.

Fleitmann D., Burns S.J., Neff U., Mudelsee M., Mangini A. & Matter A. (2004) Palaeoclimatic interpretation of high-resolution oxygen isotope profiles derived from annually laminated speleothems from southern Oman. *Quaternary Science Reviews*, **23**, 935–945.

Fleitmann D., Burns S.J., Mangini A., *et al.* (2007a) Holocene ITCZ and Indian monsoon dynamics recorded in stalagmites from Oman and Yemen (Socotra). *Quaternary Science Reviews*, **26**, 170–188.

Fleitmann D., Dunbar R.B., McCulloch M., *et al.* (2007b) East African soil erosion recorded in a 300-year old coral colony from Kenya. *Geophysical Research Letters*, **34**, L04401.

Garcin Y., Vincens A., Williamson D., Guiot J. & Buchet G. (2006) Wet phases in tropical southern Africa during the last glacial period. *Geophysical Research Letters*, **33**, L07703.

Garcin Y., Vincens A., Williamson D., Buchet G. & Guiot J. (2007a) Abrupt resumption of the African Monsoon at the Younger Dryas-Holocene climatic transition. *Quaternary Science Reviews*, **26**, 690–704.

Garcin Y., Williamson D., Bergonzini L., *et al.* (2007b) Solar and anthropogenic imprints on Lake Masoko (southern Tanzania) during the last 500 years. *Journal of Paleolimnology*, **37**, 475–490.

Gasse F. (2000) Hydrological changes in the African tropics since the Last Glacial Maximum. *Quaternary Science Reviews*, **19**, 189–211.

Gasse F. (2002) Kilimanjaro's secrets revealed. *Science*, **298**, 548–549.

Gasse F. & Van Campo E. (1994) Abrupt post-glacial climate events in West Asia and North Africa monsoon domains. *Earth and Planetary Science Letters*, **126**, 435–456.

Gasse F., Barker P.A. & Johnson T.C. (2002) A 24 000 yr diatom record from the northern basin of Lake Malawi. In: *East African Lakes: Limnology, Paleolimnology and Biodiversity* (Eds E.O. Odada & D.O. Olago), pp. 393–414. Kluwer Academic Publishers, Dordrecht.

Giresse P., Maley J. & Kossoni A. (2004) Sedimentary environmental changes and millennial climatic variability in a tropical shallow lake (Lake Ossa, Cameroon) during the Holocene. *Palaeogeography, Palaeoclimatology, Palaeoecology*, **218**, 257–285.

Grove R.H. (1998) Global impact of the 1789–1793 El Niño. *Nature*, **393**, 318–319.

Gupta A.K., Anderson D.M. & Overpeck J.T. (2003) Abrupt changes in the Asian southwest monsoon during the Holocene and their links to the North Atlantic Ocean. *Nature*, **421**, 354–357.

Gupta A.K., Das M. & Anderson D.M. (2005) Solar influence on the Indian summer monsoon during the Holocene. *Geophysical Research Letters*, **32**, L17703.

Halfman J.D., Johnson T.C. & Finney B.P. (1994) New AMS dates, stratigraphic correlations and decadal climatic cycles for the past 4 ka at Lake Turkana, Kenya. *Palaeogeography, Palaeoclimatology, Palaeoecology*, **111**, 83–98.

Hall M. (1976) Dendroclimatology, rainfall and human adaptation in the later Iron Age of Natal and Zululand. *Annals of the Natal Museum*, **22**, 693–703.

Hassan F.A. (1998) Climatic change, Nile floods and civilization. *Nature and Resources*, **34**, 34–40.

Haug G.H., Hughen K.A., Sigman D.M., Peterson L.C. & Rohl U. (2001) Southward migration of the intertropical convergence zone through the Holocene. *Science*, **293**, 1304–1308.

Haug G.H., Gunther D., Peterson L.C., Sigman D.M., Hughen K.A. & Aeschlimann B. (2003) Climate and the collapse of Maya civilization. *Science*, **299**, 1731–1735.

Heiri O., Tinner W. & Lotter A.F. (2004) Evidence for cooler European summers during periods of changing meltwater flux to the North Atlantic. *Proceedings of the National Academy of Sciences USA*, **101**, 15285–15288.

Hendon D. & Charman D.J. (2004) High-resolution peatland water-table changes for the past 200 years: The influence of climate and implications for management. *The Holocene*, **14**, 125–134.

Hodell D.A., Brenner M., Curtis J.H. & Guilderson T. (2001) Solar forcing of drought frequency in the Maya lowlands. *Science*, **292**, 1367–1370.

Hoelzmann P., Gasse F., Dupont L.M., *et al.* (2004) Palaeoenvironmental changes in the arid and subarid belt (Sahara–Sahel–Arabian Peninsula) from 150 kyr to present. In: *Past Climate Variability through Europe and Africa* (Eds R.W. Battarbee, F. Gasse & C.E. Stickley), pp. 219–256. Springer-Verlag, Berlin.

Holmes J.A., Fothergill P.A., Street-Perrott F.A. & Perrott R.A. (1998) A high-resolution Holocene ostracod record from the Sahel zone of Northeastern Nigeria. *Journal of Paleolimnology*, **20**, 369–380.

Holmgren K., Karlén W., Lauritzen S.E., *et al.* (1999) A 3000-year high-resolution stalagmite record of palaeoclimate for North-Eastern South Africa. *The Holocene*, **9**, 271–278.

Holmgren K., Lee-Thorp J.A., Cooper G.R.J., *et al.* (2003) Persistent millennial-scale climatic variability over the past 25 000 years in Southern Africa. *Quaternary Science Reviews*, **22**, 2311–2326.

Holzhauser H. (1997) Fluctuations of the Grosser Aletsch Glacier and the Gorner Glacier during the last 3200 years: new results. *Palaeoclimate Research*, **24**, 35–58.

Holzhauser H., Magny M. & Zumbuhl H.J. (2005) Glacier and lake-level variations in west-central Europe over the last 3500 years. *The Holocene*, **15**, 789–801.

Hughes M.K. & Diaz H.F. (1994) Was there a medieval warm period, and if so, where and when? *Climatic Change*, **26**, 109–142.

IPCC (2007) *Climate Change 2007 – The Physical Science Basis Working Group I.* Contribution to the Fourth Assessment Report of the Intergovernmental Panel on Climate Change, Cambridge University Press.

Jansen E., Andersson C., Moros M., Kerim, H., Nisancioglu K.H., Nyland B.F. & Telford R.J. (this volume) The early to mid-Holocene thermal optimum in the North Atlantic. In: *Global Warming and Natural Climate Variability: a Holocene Perspective* (Eds R.W. Battarbee & H.A. Binney), pp. 123–137. Blackwell, Oxford.

Johnstone E., Macklin M.G. & Lewin J. (2006) The development and application of a database of radiocarbon-dated Holocene fluvial deposits in Great Britain. *Catena*, **66**, 14–23.

Jones P.D. & Mann M.E. (2004) Climate over past millennia. *Reviews of Geophysics*, **42**, No. RG2002.

Jones P.D., Briffa K.R., Barnett T.P. & Tett S.F.B. (1998) High-resolution palaeoclimatic records for the last millennium: interpretation, integration and comparison with General Circulation Model control-run temperatures. *The Holocene*, **8**, 455–471.

Küper R. & Kröpelin S. (2006) Climate-controlled Holocene occupation in the Sahara: motor of Africa's evolution. *Science*, **313**, 803–807.

Kutzbach J.E. & Street-Perrott F.A. (1985) Milankovitch forcing of fluctuations in the level of tropical lakes from 18 to 0 kyr BP. *Nature*, **317**, 130–134.

Kutzbach J.E. & Webb III T. (1993) Conceptual basis for understanding Late Quaternary climates. In: *Global Climates since the Last Glacial Maximum* (Ed. H.E. Wright, Jr.), pp. 5–11. University of Minnesota Press, Minneapolis.

Laird K.R., Fritz S.C., Grimm E.C. & Mueller P.G. (1996) Century-scale paleoclimatic reconstruction from Moon Lake, a closed-basin lake in the northern Great Plains. *Limnology and Oceanography*, **41**, 890–902.

Lamb H.H. (1985) *Climate, History and the Modern World.* Methuen, London, 387 pp.

Lamb H.F., Bates C.R., Coombes P.V., *et al.* (2007a) Late Pleistocene desiccation of Lake Tana, source of the Blue Nile. *Quaternary Science Reviews*, **26**, 287–299.

Lamb H.F., Leng M.J., Telford R.J., Ayenew T. & Umer M. (2007b) Oxygen and carbon isotope composition of authigenic carbonate from an Ethiopian lake: a climate record of the last 2000 years. *The Holocene*, **17**, 517–526.

Langdon P.G. & Barber K.E. (2005) The climate of Scotland over the last 5000 years inferred from multi-proxy peatland records: Inter-site correlations and regional variability. *Journal of Quaternary Science*, **20**, 549–566.

Lee-Thorp J.A., Holmgren K., Lauritzen S.-E., *et al.* (2001) Rapid climate shifts in the southern African interior throughout the mid to late Holocene. *Geophysical Research Letters*, **28**, 4507–4510.

Li C., Battisti D.S., Schrag D.P. & Tziperman E. (2005) Abrupt climate shifts in Greenland due to displacements of the sea ice edge. *Geophysical Research Letters*, **32**, L19702.

Lohmann G. (2003) Atmospheric and oceanic freshwater transport during weak Atlantic overturning circulation. *Tellus Series A*, **55**, 438–449.

Luterbacher J., Xoplaki E., Dietrich D., *et al.* (2002) Reconstruction of sea level pressure fields over the Eastern North Atlantic and Europe back to 1500. *Climate Dynamics*, **18**, 545–561.

Macklin M.G. & Lewin J. (2003) River sediments, great floods and centennial-scale Holocene climate change. *Journal of Quaternary Science*, **18**, 101–105.

Macklin M.G., Johnstone E. & Lewin J. (2005) Pervasive and long-term forcing of Holocene river instability and flooding in Great Britain by centennial-scale climate change. *The Holocene*, **15**, 937–943.

Macklin M.G., Benito G., Gregory K.J., *et al.* (2006) Past hydrological events reflected in the Holocene fluvial record of Europe. *Catena*, **66**, 145–154.

Magny M. (2004) Holocene climate variability as reflected by mid-European lake-level fluctuations and its probable impact on prehistoric human settlements. *Quaternary International*, **113**, 65–79.

Maley J. (1993) Chronologie calendaire des principales fluctuations du lac Tchad au cours du dernier millénaire. In: *Datation et Chronologie dans le Bassin du Lac Tchad* (Eds D. Barreteau D. and C. Von Graffenried), pp. 161–163. ORSTOM, Paris.

Maley J. & Brenac P. (1998) Vegetation dynamics, palaeoenvironments and climatic changes in the forests of western Cameroon during the last 28 000 years BP. *Review of Palaeobotany and Palynology*, **99**, 157–187.

Marchant R. & Taylor D. (1998) Dynamics of montane forest in Central Africa during the late Holocene: a pollen-based record fromWestern Uganda. *The Holocene*, **8**, 375–381.

Masson-Delmotte V., Raffalli-Delerce G., Danis P.A., *et al.* (2005) Changes in European precipitation seasonality and in drought frequencies revealed by a four-century-long tree-ring isotopic record from Brittany, western France. *Climate Dynamics*, **24**, 57–69.

Matthews J.A. & Briffa K.R. (2005) The "Little Ice Age": Re-evaluation of an evolving concept. *Geografiska Annaler Series A – Physical Geography*, **87**, 17–36.

Matthews J.A., Dahl S.O., Nesje A., Berrisford M.S. & Andersson C. (2000) Holocene glacier variations in central Jotunheimen, southern Norway based on distal glaciolacustrine sediment cores. *Quaternary Science Reviews*, **19**, 1625–1647.

Matthews J.A., Berrisford M.S., Dresser Q.P., *et al.* (2005) Holocene glacier history of Bjornbreen and climatic reconstruction in central Jotunheimen, Norway, based on proximal glaciofluvial stream-bank mires. *Quaternary Science Reviews*, **24**, 67–90.

Mauquoy D., van Geel B., Blaauw M., Speranza A.O.M. & van der Plicht J. (2004) Changes in solar activity and Holocene climate shifts derived from C-14 wiggle-match dated peat deposits. *The Holocene*, **14**, 45–52.

Mayewski P.A., Rohling E.E., Stager J.C., *et al.* (2004) Holocene climate variability. *Quaternary Research*, **62**, 243–255.

McHugh M.J. (2004) Near-surface zonal flow and East African precipitation receipt during austral summer. *Journal of Climate*, **17**, 4070–4079.

Miller J. (1982) The significance of drought, disease and famine in the agriculturally marginal zones of West-Central Africa. *Journal of African History*, **23**, 17–61.

Moberg A., Sonechkin D.M., Holmgren K., Datsenko N.M. & Karlén W. (2005) Highly variable Northern Hemisphere temperatures reconstructed from low- and high-resolution proxy data. *Nature*, **433**, 613–617.

Neff U., Burns S.J., Mangini A., Mudelsee M., Fleitmann D. & Matter A. (2001) Strong coherence between solar variability and the monsoon in Oman between 9 and 6 kyr ago. *Nature*, **411**, 290–293.

Ngomanda A., Chepstow-Lusty A., Makaya M., *et al.* (2005) Vegetation changes during the past 1300 years in western equatorial Africa: a high-resolution pollen record from Lake Kamalété, Lope Reserve, Central Gabon. *The Holocene*, **15**, 1021–1031.

Niggemann S., Mangini A., Mudelsee M., Richter D.K. & Wurth G. (2003) Sub-Milankovitch climatic cycles in Holocene stalagmites from Sauerland, Germany. *Earth and Planetary Science Letters*, **216**, 539–547.

O'Brien S.R., Mayewski P.A., Meeker L.D., Meese D.A., Twickler M.S. & Whitlow S.I. (1995) Complexity of Holocene climate as reconstructed from a Greenland ice core. *Science*, **270**, 1962–1964.

Oldfield F. (2001) A question of timing: a comment on Hong, Jiang, Lui, Zhou, Beer, Li, Leng, Hong and Qin. *The Holocene*, **11**, 123–124.

Owen R., Crossley R., Johnson T., *et al.* (1990) Major low levels of Lake Malawi and their implications for speciation rates in cichlid fishes. *Proceedings of the Royal Society of London B*, **240**, 519–553.

Pauling A., Luterbacher J., Casty C. & Wanner H. (2006) Five hundred years of gridded high-resolution precipitation reconstructions over Europe and the connection to large-scale circulation. *Climate Dynamics*, **26**, 387–405.

Proctor C.J., Baker A., Barnes W.L. & Gilmour R.A. (2000) A thousand year speleothem proxy record of North Atlantic climate from Scotland. *Climate Dynamics*, **16**, 815–820.

Proctor C.J., Baker A. & Barnes W.L. (2002) A three thousand year record of North Atlantic climate. *Climate Dynamics*, **19**, 449–454.

Raffalli-Delerce G., Masson-Delmotte V., Dupouey J.L., Stievenard M., Breda N. & Moisselin J.M. (2004) Reconstruction of summer droughts using tree-ring cellulose isotopes: a calibration study with living oaks from Brittany (western France). *Tellus Series B – Chemical and Physical Meteorology*, **56**, 160–174.

Reimer P.J., Baillie M.G.L., Bard E., *et al.* (2004) IntCal04 terrestrial radiocarbon age calibration, 0–26 cal kyr BP. *Radiocarbon*, **46**, 1029–1058.

Renssen H., Brovkin V., Fichefet T. & Goosse H. (2006) Simulation of the Holocene climate evolution in Northern Africa: the termination of the African Humid Period. *Quaternary International*, **150**, 95–102.

Reynaud-Ferrara I., Maley J. & Wirrmann D. (1996) Végétation et climat dans les forêts du Sud-Ouest Cameroun depuis 4770 ans BP: analyse pollinique de sédiments du lac Ossa. *Comptes Rendus de l'Académie des Sciences de Paris*, **332**, 749–755.

Roberts N., Stevenson A.C., Davis B., Cheddadi R., Brewer S. & Rosen A. (2004) Holocene climate, environment and cultural change in the circum-Mediterranean region. In: *Past Climate Variability through Europe and Africa* (Eds R.W. Battarbee, F. Gasse and C.E. Stickley), pp. 343–362. Springer-Verlag, Berlin.

Russell J.M. & Johnson T.C. (2005a) Late Holocene climate change in the North Atlantic and Equatorial Africa: millennial-scale ITCZ migration. *Geophysical Research Letters*, **32**, doi:10.1029/2005GL023295.

Russell J.M. & Johnson T.C. (2005b) A high-resolution geochemical record from Lake Edward, Uganda-Congo, and the timing and causes of tropical African drought during the late Holocene. *Quaternary Science Reviews*, **24**, 1375–1389.

Russell J.M. & Johnson T.C. (2007) Little Ice Age drought in equatorial Africa: ITCZ migrations and ENSO variability. *Geology*, **35**, 21–24.

Russell J.M., Talbot M.R. & Haskell B.J. (2003) Mid-Holocene climate change in Lake Bosumtwi, Ghana. *Quaternary Research*, **60**, 133–141.

Russell J.M., Verschuren D. & Eggermont H. (2007) Spatial complexity of Little Ice Age climate in East Africa: sedimentary records from two crater lake basins in western Uganda. *The Holocene*, **17**, 183–193.

Salzmann U. & Waller M. (1998) The Holocene vegetational history of the Nigerian Sahel based on multiple pollen profiles. *Review of Palaeobotany and Palynology*, **100**, 39–72.

Salzmann U., Hoelzmann P. & Morczinek I. (2002) Late Quaternary climate and vegetation of the Sudanian zone of northeast Nigeria. *Quaternary Research*, **58**, 73–83.

Scott L. (1996) Palynology of hyrax middens: 2000 years of palaeoenvironmental history in Namibia. *Quaternary International*, **33**, 73–79.

Scott L. & Lee-Thorp J.A. (2004) Holocene climatic trends and rhythms in Southern Africa. In: *Past Climate Variability through Europe and Africa* (Eds R.W. Battarbee, F. Gasse & C.E. Stickley), pp. 69–91. Springer-Verlag, Berlin.

Sernander R. (1908) On the evidence of postglacial changes of climate furnished by the peat-mosses of Northern Europe. *Geologiska Föreningens i Stockholm Förhandlingar*, **30**, 465–473.

Shindell D.T., Schmidt G.A., Mann M.E., Rind D. & Waple A. (2001) Solar forcing of regional climate change during the Maunder Minimum. *Science*, **294**, 2149–2152.

Snowball I., Korhola A., Briffa K.R. & Koç N. (2004) Holocene climate dynamics in Fennoscandia and the North Atlantic. In: *Past Climate Variability through Europe and Africa* (Eds R.W. Battarbee, F. Gasse & C.E. Stickley), pp. 465–494. Springer-Verlag, Berlin.

Stager J.C. & Anfang-Sutter R. (1999) Preliminary evidence of environmental changes at Lake Bambili (Cameroon, West Africa) since 24 000 BP. *Journal of Paleolimnology*, **22**, 319–330.

Stager J.C., Mayewski P.A. & Meeker L.D. (2002) Cooling cycles, Heinrich event 1, and the desiccation of Lake Victoria. *Palaeogeography, Palaeoclimatology, Palaeoecology*, **183**, 169–178.

Stager J.C., Ryves D., Cumming B.F., Meeker L.D. & Beer J. (2005) Solar variability and the levels of Lake Victoria, East Africa, during the last millennium. *Journal of Paleolimnology*, **33**, 243–251.

Staubwasser M., Sirocko F., Grootes P.M. & Segl M. (2003) Climate change at the 4.2 ka BP termination of the Indus valley civilization and Holocene south Asian monsoon variability. *Geophysical Research Letters*, **30**, Art. No. 1425.

Street F.A. & Grove A.T. (1976) Environmental and climatic implications of late Quaternary lake-level fluctuations in Africa. *Nature*, **261**, 385–390.

Street-Perrott F.A., Holmes J.A., Waller M.P., *et al.* (2000) Drought and dust deposition in the West African Sahel: a 5500-year record from Kajemarum Oasis, northeastern Nigeria. *The Holocene*, **10**, 293–302.

Thevenon F., Williamson D. & Taieb M. (2002) A 22 kyr BP sedimentological record of Lake Rukwa (8°S, SW Tanzania): environmental, chronostratigraphic and climatic implications. *Palaeogeography, Palaeoclimatology, Palaeoecology*, **187**, 285–294.

Thompson L.G., Yao T., Mosley-Thompson E., Davis M.E., Henderson K.A. & Lin P.N. (2000) A high-resolution millennial record of the South Asian Monsoon from Himalayan ice cores. *Science*, **289**, 1916–1919.

Thompson L.G., Mosley-Thompson E., Davis M.E., *et al.* (2002) Kilimanjaro ice core records: evidence of Holocene climate change in tropical Africa. *Science*, **298**, 589–593.

Thorndycraft V.R. & Benito G. (2006) The Holocene fluvial chronology of Spain: evidence from a newly compiled radiocarbon database. *Quaternary Science Reviews*, **25**, 223–234.

Trenberth K.E. & Shea D.J. (2005) Relationships between precipitation and surface temperature. *Geophysical Research Letters*, **32**, Art. No. L14703.

Van Geel B., van der Plicht J., Kilian M.R., *et al.* (1998) The sharp rise of Delta C-14 ca. 800 cal BC: possible causes, related climatic teleconnections and the impact on human environments. *Radiocarbon*, **40**, 1163–1164.

Vellinga M. & Wood R.A. (2002) Global climatic impacts of a collapse of the Atlantic thermohaline circulation. *Climatic Change*, **54**, 251–267.

Verschuren D. (1999) Influence of depth and mixing regime on sedimentation in a fluctuating tropical soda lake. *Limnology and Oceanography*, **44**, 1103–1113.

Verschuren D. (2001) Reconstructing fluctuations of a shallow East African lake during the past 1800 years, from sediment stratigraphy in a submerged crater basin. *Journal of Paleolimnology*, **25**, 297–311.

Verschuren D. (2004) Decadal to century-scale climate variability in tropical Africa during the past 2000 years. In: *Past Climate Variability through Europe and Africa* (Eds R.W. Battarbee, F. Gasse & C.E. Stickley), pp. 139–158. Springer-Verlag, Berlin.

Verschuren D., Laird K.R. & Cumming B. (2000) Rainfall and drought in equatorial East Africa during the past 1100 years. *Nature*, **403**, 410–414.

Vincens A., Schwartz D., Elenga H., *et al.* (1999) Forest response to climate change in Atlantic Equatorial Africa during the last 4000 years BP and inheritance on the modern landscape. *Journal of Biogeography*, **26**, 879–885.

Vincent C., Le Meur E., Six D. & Funk M. (2005) Solving the paradox of the end of the Little Ice Age in the Alps. *Geophysical Research Letters*, 32, L09706.

Von Grafenstein U., Erlenkeuser H., Muller J., Jouzel J. & Johnsen S. (1998) The cold event 8200 years ago documented in oxygen isotope records of precipitation in Europe and Greenland. *Climate Dynamics*, **14**, 73–81.

Von Rad U., Schulz H., Riech V., den Dulk M., Berner U. & Sirocko F. (1999) Multiple monsoon-controlled breakdown of oxygen-minimum conditions during the past 30 000 years documented in laminated sediments off Pakistan. *Palaeogeography, Palaeoclimatology, Palaeoecology*, **152**, 129–161.

Wang Y., Cheng H., Edwards R.L., *et al.* (2005) The Holocene Asian monsoon: links to solar changes and North Atlantic climate. *Science*, **308**, 854–857.

Wanner H., Brönniman S., Casty C., *et al.* (2001) North Atlantic Oscillation – concepts and studies. *Surveys in Geophysics*, **22**, 321–381.

Weiss H. (2000) Beyond the Younger Dryas: collapse as adaptation to abrupt climate change in ancient West Asia and the Eastern Mediterranean. In: *Confronting Natural Disaster: Engaging the Past to Understand the Future* (Eds G. Bawden and R. Reycraft), pp. 75–98. University of New Mexico Press, Albuquerque.

Weiss H. & Bradley R.S. (2001) What drives societal collapse? *Science*, **291**, 609–610.

Williams M.A.J., Talbot M.R., Aharon P., Salaam Y.A., Williams F. & Brendeland K.I. (2006) Abrupt return of the summer monsoon 15 000 years ago: new evidence from the lower White Nile valley and Lake Albert. *Quaternary Science Reviews*, **25**, 2651–2665.

Wilson R.J.S., Luckman B.H. & Esper J. (2005) A 500 year dendroclimatic reconstruction of spring-summer precipitation from the lower Bavarian Forest region, Germany. *International Journal of Climatology*, **25**, 611–630.

Wirrmann D., Bertaux J. & Kossoni A. (2001) Late Holocene paleoclimatic changes in Western Central Africa inferred from mineral abundance in dated sediments from Lake Ossa (southwest Cameroon). *Quaternary Research*, **56**, 275–287.

Wu W.X. & Liu T.S. (2004) Possible role of the "Holocene Event 3" on the collapse of Neolithic Cultures around the Central Plain of China. *Quaternary International*, **117**, 153–161.

Yu G. & Harrison S.P. (1995) *Lake Status Records from Europe, Database Documentation*. IGBP PAGES/World Data Center-A for Paleoclimatology Data Contribution Series # 95-009. NOAA/NGDC Paleoclimatology Program, Boulder CO.

Zhang Z.H., Mann M.E. & Cook E.R. (2004) Alternative methods of proxy-based climate field reconstruction: application to summer drought over the conterminous United States back to AD 1700 from tree-ring data. *The Holocene*, **14**, 502–516.

9 Holocene rapid land-cover changes – evidence and theory

Martin Claussen

Keywords

Atmosphere–vegetation interaction, Holocene climate change, rapid climate change, Holocene land-cover change, climate modeling, Holocene arctic tree line, Holocene aridification of North Africa

Introduction

It is now commonly accepted that climate can change rapidly in response to a subtle forcing. There are numerous examples in climate archives which show that large, widespread, abrupt climate changes have repeatedly occurred in the past and will occur in the future (Rial *et al.* 2004). Abrupt climate changes happen naturally, but it is conceivable that human interference with the climate system is increasing the probability of large, abrupt events (Alley *et al.* 2003). If such events come unexpectedly, the economic and ecological impacts could be large and potentially serious (deMenocal 2001).

Some mechanisms have been identified to which abrupt climate changes could be attributed. For example, the meridional overturning circulation in the Atlantic Ocean appears to be sensitive to changes in freshwater fluxes into the North Atlantic such that above a certain threshold of freshwater input the meridional Atlantic circulation ceases. This process has been studied in a conceptual model by Stommel (1961) and in a coupled general circulation model by Schiller *et al.* (1997), for example. Abrupt transitions in the Atlantic thermohaline circulation have been used to explain the rapid swings in glacial climate (Stocker 2000; Ganopolski and Rahmstorf 2001) during the transition from glacial to interglacial climate (Mikolajewicz *et al.* 1997), or in various scenarios of future climate change (Manabe and Stouffer 1988; Rahmstorf and Ganopolski 1999; Knutti and Stocker 2002). Furthermore, rapid and presumably complete glaciations of the Earth, which probably happened some 800 to 700 million years ago – the so-called "snowball Earth" events – have been used to illustrate a positive feedback between the high albedo of ice caps and the atmospheric energy balance (see Hoffman and Schrag 2002). Likewise, a rapid onset of the last glacial was simulated by Calov *et al.* (2005), again, based on earlier theories of the atmosphere–cryosphere feedback by Budyko (1969) and Sellers (1969).

Ecosystems, when exposed to gradual changes in climate, nutrient loading, or habitat fragmentation, for example, may exhibit drastic switches to a new state. Examples include the dynamics of freshwater and marine ecosystems, forests, and arid lands (Scheffer *et al.* 2001). Jennerjahn *et al.* (2004) found that rapid shifts of tropical ecosystems occurred during the last ice age, and Hughen *et al.* (2004) showed the same for ecosystems in South America during the last deglaciation.

In the Holocene, the western Sahara is hypothesized to have expanded rapidly some 5500 years BP (deMenocal *et al.* 2000), and the apparent forcing, the global distribution of insolation, has changed steadily for the past several thousand years (see Figure 9.1). Petit-Maire and Guo (1996) reported large climatic fluctuations in the Sahara, with wet periods occurring during the early Holocene and around 5000 years BP, and two more arid episodes occurring at 6700–5500 years BP and at 4000–3600 years BP. The latter episode was severe, ruining ancient civilizations and socio-economic systems. In addition Cremaschi *et al.* (2006) described major drought spells at 5900–5760 years BP and at 5120 years BP in the central Sahara, followed by phases of enhanced precipitation and by the onset of extreme arid conditions at 1500 years BP. At high northern latitudes, tree macrofossils apparently vanished between 4000 and 3000 years BP (MacDonald *et al.* 2000; Figure 9.1, bottom). Most of these periods characterized by rapid land-cover change correspond to one of the six periods of Holocene rapid climate change identified by Mayewski *et al.* (2004).

In this paper, large-scale Holocene land-cover change and associated bio-geophysical feedbacks are addressed. Special emphasis is given to the underlying theory of rapid land-cover change in order to explore the possibility of predicting other so-called hot spots, i.e., regions on Earth that react very sensitively to changes in external forcing, in particular to anthropogenic land-cover change or emissions of greenhouse gases related to fossil-fuel use.

Rapid land-cover change in North Africa

Paleobotanic and paleoclimatic evidence indicates that during the early and mid-Holocene, some 11 500 to 6000 years BP, the Sahara was much greener than today (Prentice *et al.* 2000), and the Sahel reached at least as far north as 23°N in some parts of North Africa (Jolly *et al.* 1998). The greening was attributed to an increase in the North African summer monsoon as a response to changing orbital forcing (Kutzbach and Guetter 1986). The increase in summer monsoon triggered by changes in the Earth's orbit around the Sun and in the tilt of the Earth's axis, however, did not seem to be large enough to explain a large-scale greening (Joussaume *et al.* 1999). Claussen and Gayler (1997) found a strong feedback

Figure 9.1 (top) Changes in Northern Hemisphere insolation over the past 9000 years from Berger (1978), (middle) changes in terrigenous sediment off the coast of North Africa (deMenocal *et al.* 2000) indicating an abrupt expansion of the western part of the Sahara around 5500 years ago, and (bottom) changes in the number of Russian tree macrofossils recorded (MacDonald *et al.* 2000) indicating a southward retreat of Arctic trees between 5000 and 3000 years ago. Please note that the latter is presented in uncalibrated radiocarbon years before present, so that the time axis of the upper and middle figure might not match, if the latter data were calibrated.

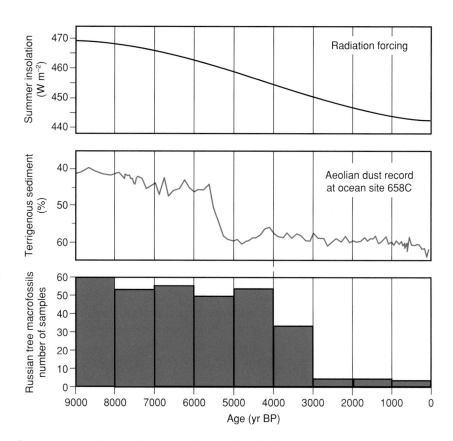

between vegetation and precipitation, mainly in the western part of the Sahara, which could amplify the increase in summer monsoon to foster a northward shift of Sahelian vegetation. Claussen and Gayler explained the positive feedback by an interaction between high albedo of Saharan sand deserts and tropical atmospheric circulation, as assumed by Otterman (1974) and described in a model by Charney (1975). Charney (1975) hypothesized that the high albedo over sub-tropical deserts causes a radiative cooling because the sum of incoming solar radiation, reflected solar radiation, and outgoing long-wave radiation is negative at the top of the atmosphere: more radiation leaves than enters the atmosphere above a sub-tropical desert. The local radiative cooling induces a subsidence of air masses, which compensates the cooling by adiabatic heating. The sinking motion suppresses convective precipitation. The reduction of precipitation is supposed to cause further vegetation degradation, thus enhancing the growth of desert-like conditions. Interestingly, Otterman (1974) and Charney (1975) focused not on the Sahara, but on the question of whether ongoing Sahelian drought could be caused by an increase in albedo due to overgrazing. Later on, however, it was found that changes in Sahelian albedo are too small to explain a strong desert-albedo effect (Xue and Shukla 1993), and that presumably mechanisms other than Charney's biogeophysical feedback dominate (see, e.g. Eltahir and Gong 1995).)

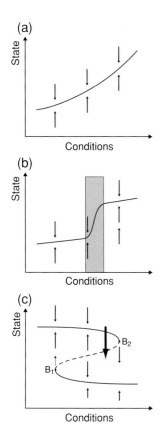

Figure 9.2 Possible changes of ecosystem equilibrium states as a function of variations in conditions, such as climate or nutrient loading. In case (a), the system changes gradually with external conditions. Any perturbation away from the equilibrium state will decay – as indicated by the arrows – towards the equilibrium line (solid line). Case (b) shows a system which is rather inert over certain ranges of conditions while responding more strongly when conditions exceed a critical level (shaded area). A particularly interesting case (c) arises when a system exhibits multiple equilibria for certain environmental conditions. In such a case abrupt shifts from one state to the other state are possible if conditions exceed, perhaps randomly, some threshold near points B_1 or B_2. In the presence of strong perturbations (indicated by the thick arrow) the system might jump from the upper stable equilibrium to the lower stable one. The dashed line depicts an unstable equilibrium. Any infinitesimally small perturbation would drive the system away from the unstable equilibrium towards a stable one. (This figure is taken with modifications from Scheffer *et al.* 2001.)

To study Holocene land-cover changes in the Sahara, Brovkin *et al.* (1998; and later on, others, cited below) analyzed the stability of the atmosphere–vegetation system in West Africa. The basic concept of such a stability analysis is outlined by Scheffer *et al.* (2001). They showed that, generally speaking, a system can reveal different stability characteristics. The system could change gradually, or even linearly, when external conditions change (see Figure 9.2a). In other cases, a system may be rather inert over certain ranges of conditions while responding more strongly, however, when conditions approach a critical level (Figure 9.2b). A particularly interesting case arises when a system exhibits alternative states, or multiple equilibria, for certain environmental conditions. In such a case, abrupt shifts from one state to the other state are possible if ubiquitous perturbations, either internally generated or externally imposed, exceed some threshold (Figure 9.2c). Regions in which the atmosphere–biosphere system, or more generally the climate system including atmosphere, biosphere, hydrosphere, pedosphere, and cryosphere, behaves like cases (b) and (c) are tentatively defined as "hot spots" in this paper.

How does this conceptual model apply to the dynamics of Saharan greening? Theoretical studies by Claussen (1994, 1997) revealed that the biogeophysical feedback in the western part of North Africa leads to multiple equilibrium solutions for present-day climate (see Figure 9.3a). For mid-Holocene climate, only one equilibrium solution for West African climate was found (Claussen and Gayler

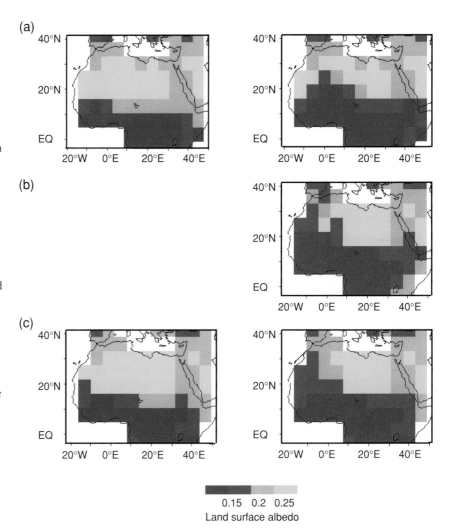

Figure 9.3 The equilibrium solutions obtained in the ECHAM-3–BIOME 1.0 model for (a) present-day climate according to Claussen (1997), (b) mid-Holocene climate (6000 years BP) after Claussen and Gayler (1997) and (c) for the climate of the Last Glacial Maximum some 21 000 years ago according to Kubatzki and Claussen (1998). The vegetation fraction is translated into land surface albedo. An albedo value above 0.25 corresponds to desert conditions. The left column shows the desert equilibrium, and the right column, the "green" equilibrium. Note that for the mid-Holocene situation (b), only a "green" solution was found. (This figure is taken with modifications from Brovkin *et al.* (1998). Copyright 1998 by the American Geophysical Union.)

1997) (Figure 9.3b). For glacial climate, however, the possibility of two solutions appears again (Kubatzki and Claussen 1998) (Figure 9.3b). Obviously, the atmosphere–biosphere system changes its stability depending on orbital forcing. While in present-day climate and in glacial climate the global insolation patterns are similar, Northern Hemisphere regions received up to 10 percent more insolation some 6000 years BP during boreal summer, and up to 15 percent less insolation during boreal winter (Berger 1978). Brovkin *et al.* (1998) hypothesized, therefore, that bifurcations and associated abrupt climate and vegetation changes could occur during the transition from middle Holocene to present-day climate and from the late glacial to early Holocene climate. To investigate this process and to explore the consequences of this hypothesis in detail, Brovkin *et al.* (1998) set up a mathematical version of the conceptual model, a minimal model of sub-tropical atmosphere–vegetation interaction.

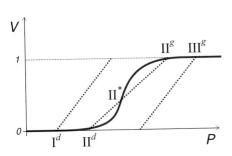

Figure 9.4 Stability diagram of a conceptual model of subtropical vegetation–precipitation feedback. The solid line indicates the changes in vegetation cover V^* as a function of precipitation P (Equation 1 in the text). The dashed lines represent the expected precipitation $P^*(V,E)$ for given vegetation cover without feedback between biosphere and atmosphere (Equation 2 in the text). The various dashed lines differ in assumed external forcing E, e.g. changes in insolation. The intersections of both curves are the equilibrium solution of the coupled system. Solutions I^d, II^d, II^g, and III^g are stable equilibria, and II^* is an unstable equilibrium. (This figure is taken from Brovkin *et al.* (1998). Copyright 1998 by the American Geophysical Union.)

The minimal model of Brovkin *et al.* is depicted in Figure 9.4. The solid line represents the dependence of equilibrium vegetation cover $V^*(P)$ on annual mean precipitation P, assumed to be given by

$$V^*(P) = \begin{cases} 0 & \text{if } P < P_{cr} \\ 1 - \dfrac{1}{1 + a(P - P_{cr})^2} & \text{otherwise} \end{cases} \qquad (1)$$

where a and P_{cr} are parameters obtained by processing ecosystem data of Olson *et al.* (1985) and precipitation data of Leemans and Cramer (1991) following the approach of continuous vegetation description by Brovkin *et al.* (1997). V^* is assumed to depend only on precipitation P, but not on external conditions E, such as insolation or atmospheric CO_2 concentration, for example. Hence it is assumed that $V^*(P)$ is nearly constant over time. The dashed lines in Figure 9.4 represent a linear model of expected precipitation $P^*(V,E)$ for given vegetation V and external conditions E. No feedback between vegetation and precipitation is assumed, and $P^*(V,E)$ is approximated by

$$P^*(V,E) = P_d(E) + b(E)V \qquad (2)$$

where $P_d(E)$ and $b(E)$ are parameters to be adjusted to data or to results of comprehensive models in which precipitation is computed under prescribed, constant vegetation cover for the region in question, i.e., West Africa in this case. The intersections of the curves $V^*(P)$ and $P^*(V,E)$ represent the equilibrium solutions of the system when vegetation dynamics and atmospheric dynamics are allowed to interact. In Figure 9.4, three possible examples are shown: two examples of a system exhibiting only one equilibrium solution, a desert solution I^d and a green solution III^g. The latter case corresponds to the situation in the mid-Holocene according to the simulations shown in Figure 9.3. The present-day and glacial situation is characterized by a system revealing three solutions (II^d, II^*, II^g). It can be shown that the solution II* is unstable to infinitesimally small perturbations. Further analysis of the model results depicted in Figure 9.3 shows that for present-day conditions II^d is more stable than II^g, i.e., the system is expected to reside more often near II^d than near II^g in the presence of finite-amplitude perturbations in rainfall. This could explain the existence of the Sahara today to be more likely than a strongly reduced, or greener, Sahara.

The transient vegetation/climate change in West Africa can be interpreted in terms of changes in the parameter E, which determines slope and origin of $P^*(V,E)$. As the Earth's orbit around the Sun varies, the strength of the African summer monsoon changes, which takes the atmosphere–vegetation system from a uni-stable situation (represented by case III with a green equilibrium solution III^g in Figure 9.4) to a bistable situation (represented by case II with two stable solutions II^d and II^g). When the bistable situation is reached, then it depends on the stability of the green solution II^g and the strength of perturbation, such as variations in precipitation or vegetation cover, whether a jump from the green to the desert state occurs.

Using a box climate model of sub-tropical monsoon dynamics, Brovkin *et al.* (1998) predicted that before 6000 years BP the atmosphere–vegetation system in North Africa revealed a stability as indicated by case III in Figure 9.4 with only one, green solution III^g. Because of the change in insolation, the dashed line $P^*(V,E)$ shifts to the left until around 6000 years BP and a situation is reached that is sketched as case II in Figure 9.4, with a desert solution II^d in addition to a green solution II^g. Around 3600 years BP, the desert solution II^d and the green solution II^g appeared to exhibit the same stability, i.e., the likelihood that the system resides near II^d or near II^g is the same. As the variability in precipitation, which triggers switches between equilibrium states, appears to be larger the greener the region, Brovkin *et al.* (1998) concluded that the jump from the green to the desert state should have happened between 6000 and 3600 years BP. It is also conceivable that a jump between states does not occur once, but that for some period of time the system jumps back and forth between the desert and the green state.

By using the Earth System Model of Intermediate Complexity, CLIMBER-2, Claussen *et al.* (1999) predicted a fast expansion of the Sahara some 5500 years BP (see Figure 9.5), which agrees well with the reconstruction of an abrupt change in terrestrial dust transport from western Africa into the North Atlantic (deMenocal *et al.* 2000; see Figure 9.1). Sensitivity tests (not shown here) revealed that CLIMBER-2 does not exhibit distinct green and desert-like equilibrium solutions for the same values of insolation as the more comprehensive atmosphere–biome model used by Claussen (1997), Claussen and Gayler (1997), and Kubatzki and Claussen (1998) (see Figure 9.3) does. Hence Claussen *et al.* (1999) concluded that the fast transition seen in CLIMBER-2 was not a bifurcation of the atmosphere–vegetation system, but a fast transition corresponding to case (b) in Figure 9.2. The authors attribute this effect to the coarse spatial resolution of CLIMBER-2, which does not differentiate between the western and eastern Sahara.

Paleobotanic and paleoclimatic evidence suggests that Holocene climate trends in the western and eastern Sahara differ. The abrupt increase in terrestrial dust flux

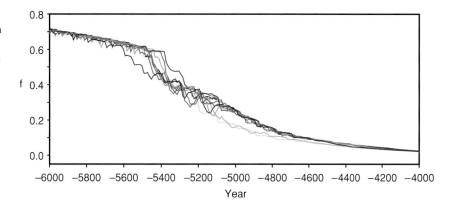

from the western part of Sahara as recorded in marine sediments off the coast of North Africa near Cape Blanc, Mauretania, by deMenocal *et al.* (2000) (see Figure 9.1) suggests a single abrupt aridification around 5500 years BP. Pachur (1999) found that the aridification in the eastern part of the Sahara increased more gradually with a period of larger variability between 7000 and 4500 years BP (Pachur 1999). Rapid swings between the arid and the wet state were also found in reconstructions from several hundred ^{14}C dates around the Tropic of Cancer by Petit-Maire and Guo (1996) and from tree rings in the central Sahara by Crematschi *et al.* (2006).

Strong variations in Saharan vegetation between 7000 and 5000 years BP were found (see Figure 9.6) by Renssen *et al.* (2003) in their model, ECBILT–CLIO–VECODE, a coupled atmosphere–ocean–vegetation model of intermediate complexity. In contrast to the CLIMBER-2, the model by Renssen *et al.* has a much stronger variability in rainfall. Renssen *et al.* (2003) showed that the change in variability in their model could be attributed to a bistability of the system during the period of strong variations. For that purpose they constructed a stability diagram, as in Figure 9.4, following Brovkin *et al.* (1998). They performed several experiments with prescribed vegetation cover to obtain a linear curve $P^*(V, tn)$ where *tn* stands for time slices 9000 years BP, 6000 years BP, and 0 years BP. The curve $V^*(P)$ was taken from the vegetation model with prescribed precipitation changes. For today, the stability diagram indicated a desert solution only, as case I in Figure 9.4. For 9000 years BP and 6000 years BP, two stable solutions appeared, as case II in Figure 9.4. Further experiments showed that for 9000 years BP the desert solution (represented as II^d in Figure 9.4) was not stable in the presence of simulated climate variations. The possibility of multiple solutions and rapid swings between the desert and green state in the model is qualitatively consistent with reconstructions by Petit-Maire and Guo (1996), Pachur (1999), and Cremaschi *et al.* (2006).

Zeng and Neelin (2000), Wang (2004), and Liu *et al.* (2006) explored the role of rainfall variability on the transition between green and desert-like conditions in North Africa. They found that strong rainfall variability could prevent the system from staying in one equilibrium. Instead, one would detect a strong variability in vegetation cover which, on average over decades and centuries, would decline

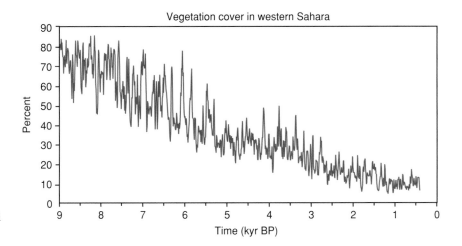

Figure 9.6 Simulated decrease in percentage of vegetation cover in the western Sahara/Sahel region (14°W to 3°E, 17°N to 28°N) as a function of time. (This figure is taken from Renssen *et al.* (2003). Copyright 2003 by the American Geophysical Union.)

smoothly. Only if variability occurs at sufficiently long time-scales (say, at decades instead of years), then abrupt transitions occur (this case is referred to as an "unstable collapse" in Liu *et al.* 2006). From their studies, one can conclude that bistability is neither a necessary nor sufficient condition for multiple transitions between green and desert-like conditions. These can occur in the presence of only one equilibrium solution, provided that the slope of curve $V^*(P)$ is sufficiently close to that of $(P^*(V,E))^{-1}$ and that there is slow (i.e., decadal or longer) rainfall variability (or "stable collapse" according to Liu *et al.* 2006).

Not all models reveal abrupt changes during the Holocene in northern Africa. The model by Wang *et al.* (2005) seems to follow insolation rather linearly, although the same vegetation model was used as implemented in CLIMBER-2 and the model by Renssen *et al.* (2003). Perhaps the atmospheric model, a sectorially averaged energy balance model, used by Wang *et al.* (2005) is not capable of properly simulating deviations from the zonal mean.

The idea of multiple equilibria in West Africa and vegetation/climate changes caused by bifurcations of the system was pursued further by Wang and Eltahir (2000), who used a zonally averaged atmospheric model for West Africa synchronously coupled to a dynamic vegetation model. They explained the decadal variability in the 20th century in terms of bifurcations of the West African atmosphere–vegetation system. Using Wang's and Eltahir's model, Irizarry-Oritz *et al.* (2003) obtained bistability of the atmosphere–vegetation system for West Africa for the mid-Holocene period, thereby corroborating the results by Renssen *et al.* (2003). Also Zeng and Neelin (2000), by using an atmospheric model of the tropics coupled to a dynamic vegetation model, explored the effect of multi-stability in the West African region. Hence multiple equilibrium solutions in West Africa are seen in atmosphere–vegetation models of different complexity and different structure. This suggests that the notion of multi-stability in the atmosphere–biosphere system and associated rapid land-cover changes in that region seems to be robust, although the details of the transition from a green Sahara to today's desert are not fully understood.

In the context of rapid land-cover changes in West Africa, the question arises whether early human land-use could have led to fast land-cover changes (e.g. Ruddiman (2003) in the context of mid-Holocene increase in atmospheric CO_2 concentration). Kubatzki (2000) tested this hypothesis by repeating the simulations of Claussen et al. (1999) and by perturbing vegetation cover in the simulations. Kubatzki (2000) reduced the vegetation of the green Sahara in the early to mid-Holocene periodically to a fractional coverage of 0.2. The resulting simulation revealed only small differences between perturbed and unperturbed cases, which suggests that early human land-cover change affected West African desert dynamics only marginally.

Rapid land-cover change in other regions of the world

After having discussed rapid land-cover change in North Africa, the question arises as to whether there is evidence of rapid land-cover changes in other regions of the world. After Fennoscandian and Laurentian ice sheets of the last Ice Age had disappeared, boreal forests colonized high northern latitudes and eventually reached farther north than today at least in Europe (Cheddadi et al. 1997) and Siberia (MacDonald et al. 2000). After the so-called Holocene optimum some 8000 to 6000 years BP (cf. Jansen et al. this volume), the boreal tree line shifted southward with perhaps some faster retreat between 4000 and 2000 years BP (uncalibrated ^{14}C ages) (MacDonald et al. 2000; see Figure 9.1). Other, more local reconstructions (e.g. Brovkin et al. 2002) do not reveal a fast change in boreal ecosystems, but a more steady change that parallels the change in insolation (Figure 9.7).

Levis et al. (1999) and Brovkin et al. (2003) analyzed several different atmosphere–vegetation models with respect to the dynamics of the boreal forest. In particular, Brovkin et al. (2003) applied a stability analysis to the dynamics of temperature-limited forests along the same lines as outlined above. In their analysis, they used the relative forest coverage F and growing degree days T (i.e., sum of daily temperatures above a certain threshold, 0°C in this case) as variables. Instead of prescribing the shape of curves $F^*(T)$ and $T^*(F)$, as done in stability analysis for sub-tropical vegetation (Equations 1, 2), they performed a number of sensitivity experiments with a comprehensive atmosphere–vegetation model (GENESISIBIS, Levis et al. 1999) and with models of intermediate complexity (MoBidiC, Crucifix et al. 2002; CLIMBER-2, Brovkin et al. 2002) to obtain the curves $T^*(F,E)$ and $F^*(T)$ for different external conditions E with present-day climate, indicated by $E = 0$, doubled atmospheric CO_2 concentrations, $E = C$, and insolation reduced by 2 percent, $E = S$ (see Figure 9.8). For all models under consideration and for all conditions of E, no multiple equilibrium solutions were found. Hence these studies suggest that a bifurcation of the atmosphere–biosphere system in the region of Siberia is not expected to occur. This statement is valid for a large region (as Siberia); however, local instabilities would still be possible, as only large-scale averages were considered in the studies above. The work by Levis et al. (1999) and

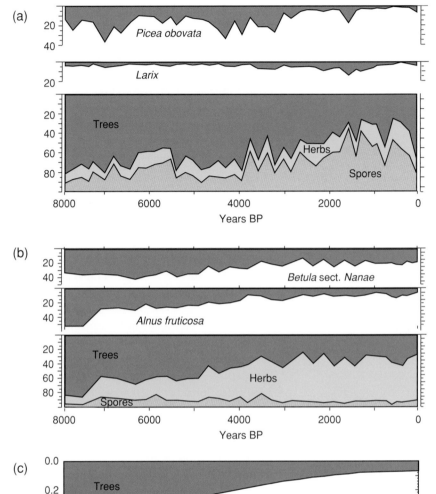

Figure 9.7 Pollen records from (a) the Lama lake (70°N, 90°E) and (b) the Levinson-Lessing Lake (74°N, 98°E). Green shaded areas represent percentages of pollen content of selected tree species and total pollen percentage of trees against herbs pollen and spores percentages. The bottom figure (c) depicts a model result by Brovkin *et al.* (2002). (This figure is taken from Brovkin *et al.* (2002). Copyright 2002 by the American Geophysical Union.)

Brovkin *et al.* (2003) not only excluded the possibility of bifurcations in the atmosphere–biosphere system over Siberia, even fast changes, such as those indicated in Figure 9.2b, are not expected when examining the phase diagrams of their models. An exception could be the MoBidiC model (Figure 9.8c) for which the slopes $dF^\star(T)/dT$ and $(dT^\star(F)/dF)^{-1}$ are rather close for small values of $F < 0.3$. Indeed, Crucifix *et al.* (2002) reported that in their model (MoBidiC) a change in boreal vegetation was most rapid between 4000 and 2000 years BP.

Figure 9.8 Stability diagram of boreal vegetation–temperature feedback obtained from sensitivity experiments with various climate system models. The solid blue lines indicate tree fraction F^* as a function of growing degree days T or, as labelled on the x-axis, GDD0 (i.e. the annual sum of daily mean temperatures above 0°C). The dashed red lines represent the expected growing degree days T^* for given forest cover F without feedback between biosphere and atmosphere. The curves $T^*(F,E)$ depend on the external forcing E. $E = 0$ indicates present-day insolation and atmospheric CO_2 concentration, $E^\wedge = S$, simulations with decreased insolation, and $E = C$, simulations with enhanced atmospheric CO_2 concentration. The intersections of curves $T^*(F,E)$ and $F^*(T)$ are the equilibrium solution of the coupled system, indicated by 1, 2, 3. Depicted are the results of sensitivity experiments with a comprehensive atmosphere–vegetation model (a) by Levis *et al.* (1999), and two models of intermediate complexity: (b) CLIMBER-2 (Brovkin *et al.* 2002) and (c) MoBidiC (Crucifix *et al.* 2002). The arrows in the top figure show the system dynamics in terms of decadal averages. (This figure is taken from Brovkin *et al.* (2003) with the permission of Springer Science and Business Media.)

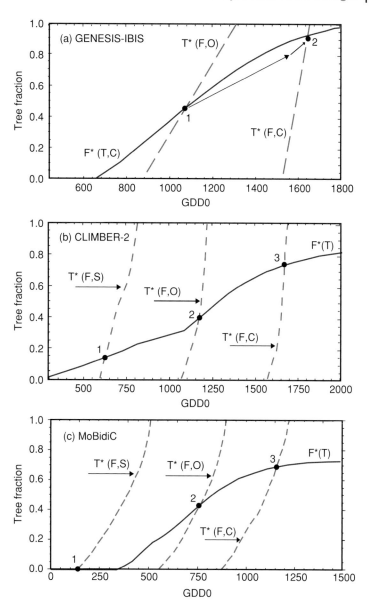

The analyses of Levis *et al.* (1999) and Brovkin *et al.* (2003) corroborate an earlier study by Claussen (1998) in which multiple equilibrium solutions were searched for globally by starting a coupled atmosphere–biome model from different initial conditions (all continents covered by forests, grassland, dark bare soil, and sand deserts, respectively). It appeared that only in northern Africa do two equilibria exist in the present-day climate: a "green Sahara", if the model was initialized with forests, grassland or dark soil, and an extended desert, if the model was initialized with continents covered by bright sand desert only.

Future land-cover change and surprises

The fastest land-cover change that currently occurs is related to (anthropogenic) land-use. Nonetheless, the question arises whether land-use or, more interestingly, anthropogenically induced climate change could as well lead to rapid land-cover change. Oyama and Nobre (2003) detected in their model (an atmospheric model asynchronously coupled to a potential vegetation model, as in Claussen 1997) multiple solutions for present-day climate in tropical South America, but only in this region, and not in northern Africa. In tropical South America, tropical forest was seen in the model when it was globally initialized with forest, but savanna emerged when the model was globally initialized with desert. In north-east Brazil, *caatinga*, Brazilian dry shrubland, was replaced by semi-desert vegetation in the model with deserts as initial conditions. This result is particularly interesting, given that at least one other climate model, the Hadley Centre climate model (Betts *et al.* 2004), yields a strong reduction in precipitation and subsequent dieback of Amazon forest if atmospheric CO_2 concentrations are strongly increased.

Also for the region of North Africa we might expect some land-cover change indirectly induced by humans. Wang and Eltahir (2002) concluded from their simulations that higher atmospheric CO_2 concentrations would make the atmosphere–biosphere system of West Africa more resilient to drought-inducing remote processes such as varying sea-surface temperatures. They expect that the regional climate in the Sahel, which tends to alternate between dry and wet spells, may experience longer or more frequent wet episodes and shorter or less frequent dry episodes in the future than in the past.

As mentioned above, West Africa is a region for which bistability of the atmosphere–vegetation system has been detected in various climate models for present-day climate. Brovkin *et al.* (1998) found in their conceptual model that under greenhouse-gas-induced climate warming a green Sahara (or better: a Sahara, greener than today) is likely to become more stable than a desert solution. Their statement somehow corroborated an earlier assertion by Petit-Maire (1990) that "greenhouse could green the Sahara" – in an analogy to the greening of North Africa during the so-called Holocene climate optimum. Subsequently, Claussen *et al.* (2003) explored the sub-tropical biogeophysical feedback in various scenarios of atmospheric CO_2 concentration increase, and they found a rapid increase of Sahara vegetation (see Figure 9.9). Despite this apparent similarity between mid-Holocene and greenhouse-gas induced greening, however, Claussen *et al.* (2003) concluded that the mid-Holocene climate would not be a direct analog for a potential future Saharan greening. Not only do the global patterns of climate change differ between the mid-Holocene model experiments and the greenhouse-gas sensitivity experiments, but also the relative role of mechanisms which lead to a reduction of the Sahara. Furthermore, the role of land-use might differ between the mid-Holocene and today. As mentioned above, human interference presumably had little influence on mid-Holocene desert dynamics. A future greening, however, could be suppressed by extensive land-use (Claussen *et al.* 2003).

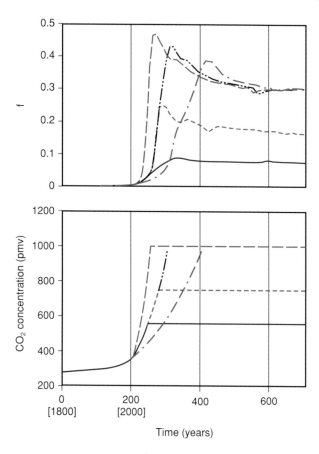

Figure 9.9 Saharan vegetation fraction (top) as a function of model years for different scenarios of changes in atmospheric CO₂ concentrations (bottom) (This figure is taken with modifications from Claussen *et al.* (2003) with the permission of Springer Science and Business Media.)

Whether or not the Sahara becomes greener because of a greenhouse-gas-induced warming presumably depends on the model set up. So far, only the study by Claussen *et al.* (2003) describes an interactive atmosphere–biosphere–ocean model experiment which yields such a greening. An overview of model experiments on greenhouse-gas-induced warming (Cubasch *et al.* 2001) shows an increase in most tropical areas and a decrease in most sub-tropical areas. There appears, however, to be little consistency among models with respect to simulated changes of precipitation in northern Africa.

Possible methods to detect "hot spots"

So far, only North Africa and, in one model only, north-west Brazil have been identified as regions in which bifurcations of the atmosphere–vegetation system could occur. Is this result robust or could some sensitive regions have been overlooked? Up to now, two methods have been used to explore the stability of the atmosphere–land-cover system. The first method relies on finding multiple solutions in simulations which are initialized with different land-cover conditions, everything else being unchanged. If a region exhibits multiple equilibrium solutions, then it is considered a "hot spot" in the sense outlined above. This method has been applied by Claussen (1998), Wang and Eltahir (2000), Zeng and Neelin (2000), and Oyama and Nobre (2003), for example. Wang (2004) pointed out that this detection method might not be reliable because of internal atmospheric variability which, if strong enough, eliminates multiple solutions, in line with simulations by Zeng and Neelin (2000). Hence, the results by Claussen (1998) might not be conclusive. On the other hand, Wang (2004) also argued that if two equilibrium solutions exist in a system, then the variability might be much stronger than in a system with only one equilibrium solution.

Perhaps then, the analysis of the amplitude of variability in the atmosphere–biosphere system would be an alternative method of searching for "hot spots". This argument is in line with a proposal by Kleinen *et al.* (2003) or in a modified, extended version by Held and Kleinen (2004). Kleinen *et al.* (2003) show that the power spectral properties change; the low frequency variability becomes larger as the system moves close to a bifurcation point.

In a third method used so far, a stability diagram as proposed in Brovkin *et al.* (1998) is constructed from a number of sensitivity experiments with comprehensive coupled models. Successful examples are given in Brovkin *et al.* (1998, 2003). This method could certainly be extended when a more general minimal model is applied to the global results of comprehensive, coupled models.

A further method would make use of exploring physical mechanisms. For example, a key indicator of Charney's (1975) theory of desert–albedo interaction is the negative anomaly of the radiation balance at the top of the atmosphere. As mentioned above, the annual mean radiation balance, i.e., the annual mean sum of incoming solar radiation, reflected solar radiation, and outgoing long-wave radiation, is positive in the tropics and negative at high latitudes. This gradient indirectly drives the general circulation of the atmosphere. Anomalies of the more or less zonally symmetric radiation balance are found over sub-tropical deserts, in particular the Sahara and Arabian deserts (Figure 9.10). Other, secondary, minima over land occur over central East Asia, Australia, and, only moderately, over Namibia. It is conceivable to assume that Charney's feedback could work over these areas (with the exception of the maritime region where mainly low clouds cause the local minima in the radiation balance).

Indeed, when applying the minimal model of Brovkin *et al.* (1998) to the Namib region, multiple solutions are possible within a range of plausible parameters. This supposition, however, has to be critically assessed by taking into account that the most important climatic factor is the upwelling of cold water along the Namibian coast.

Conclusions

If climate, either in its mean values or in its extremes, shifts abruptly and if these changes come unexpectedly, then human civilization is affected. There is ample

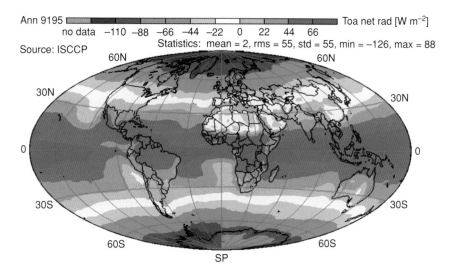

Figure 9.10 Annual averaged radiation balance at the top of the atmosphere for the period of 1991 to 1995. (This figure is taken from Raschke *et al.* (2005). Copyright by the Royal Meteorological Society.)

evidence that such climate-related crises have happened in the past (deMenocal 2001). In this paper, Holocene rapid climate change in relation to rapid land-cover change is the focus. It was shown that only for North Africa is there unambiguous evidence, in comparison with driving insolation changes, of an abrupt expansion of Saharan deserts. North African land-cover change is assumed to have occurred either as one single aridification event around 5500 years BP (deMenocal *et al.* 2000) or as multiple switches from arid to wet phases in the mid-Holocene around the Tropic of Cancer (Petit-Maire and Guo 1996), in the central Sahara (Cremaschi *et al.* 2006), and in the eastern part of the Sahara (Pachur 1996). These shifts presumably had a profound impact on Neolithic culture in this region. Neolithic groups who lived in the green Sahara, perhaps mostly around then permanent lakes, migrated from the aridified land between 4500 and 4000 years BP (Petit-Maire and Bryson 1998). Malville *et al.* (1998) suggest that the swing to arid conditions in southern Egypt forced an exodus and that the arrival of the well-organized nomadic groups in the Nile valley was a factor in the rise of civilization in Upper Egypt.

The abrupt Holocene climate changes in North Africa have qualitatively been recaptured by different climate system models of different physical complexity and different spatial resolution. A sectorially averaged model of intermediate complexity (Wang *et al.* 2005) reveals just a smooth transition. Similar, slightly more comprehensive models, but with explicit geographic resolution, however, yield either a fast transition around 5500 years BP (Claussen *et al.* 1999) or a period of enhanced variability (Irizarry-Ortiz *et al.* 2003; Renssen *et al.* 2003) between 7000 and 5000 years BP. Earlier numerical experiments with comprehensive atmosphere–biome models (Claussen 1994, 1997, 1998) showed multiple equilibrium solutions for North Africa for the pre-industrial or late Holocene climate, which indicates the possibility of abrupt shifts of the system. Hence the notion of rapid land-cover change in North Africa is supported by theoretical studies.

So far, there is no unambiguous sign for the potential of abrupt land-cover change in other regions of the Earth. The Amazon region and adjacent areas could be a "hot spot" in the present-day climate (Oyama and Nobre 2003) or in a world with strong greenhouse-gas-induced warming (Betts *et al.* 2004). The jury is still out whether the Arctic tree line shifted abruptly. Some data by MacDonald *et al.* (2000) seem to indicate a shift at a somewhat faster pace than climate forcing. Model studies hint at gradual shifts in the Arctic tree line.

In this study, only the interaction of vegetation with annual precipitation and/or temperature has been considered. There might be other climate thresholds involved such as the length of a dry season, for example, and climate and ecological thresholds might not always be the same (Jennerjahn *et al.* 2004; Maslin 2004). Furthermore, it is conceivable that climate-related thresholds and ecological thresholds are effective at different spatial scales. This issue has been raised by Scheffer *et al.* (2005), who argue that the synergy between small- and large-scale feedbacks could change the hysteresis of the system.

Biogeophysical feedbacks, i.e., processes which affect the near-surface energy, moisture, and momentum fluxes, have been mainly considered here. It is quite possible, however, that also biogeochemical feedbacks, i.e., processes which affect the chemical composition of the atmosphere, could directly or indirectly lead to rapid land-cover change. For example, in at least one model, a strong greenhouse-gas-induced climate warming enhances heterotrophic respiration such that the terrestrial ecosystems become a carbon source (Cox *et al.* 2000). The additional (global) warming together with a shift in atmospheric dynamics leads to a drastic warming over the Amazon region and a dieback of the Amazon forest within a few decades (Betts *et al.* 2004). Although this problem is still disputed, one has to consider the potential of biogeochemical feedbacks as drivers of rapid land-cover change in the future, and also in the past.

Generally, the question of how to detect "hot spots", i.e., regions on Earth which are sensitive to any changes in climate forcing, has to be raised. Several methods have been identified here that are in use, or could be applied to the problem of rapid land-cover change: (i) the initial-value method; (ii) the construction of a stability diagram from climate system models; (iii) analysis of trends in the spectrum of long-term time series; and (iv) analysis of key physical processes. Since only some 10 years have passed after large-scale instabilities in the atmosphere–vegetation system were detected in numerical models, it is expected that further progress in this matter will be made soon. After all, the land in the climate system through its changing surface is now being realized as an important interactive component in the system and it is rightly attracting more interest in the research community.

Acknowledgments

The author would like to thank Christian Reick, Max Planck Institute for Meteorology, Hamburg, Germany, Victor Brovkin, Potsdam Institute for Climate

Impact Research, Germany, Michel Crucifix, Institut d'Astronomie et Géophysique George Lemaître, Louvainla-Neuve, Belgium, Guiling Wang, University of Connecticut, Storrs, CT, USA, as well as two anonymous reviewers for constructive discussion. Thanks are also due to Barbara Zinecker and Norbert Noreiks, Max Planck Institute for Meteorology, for editorial and technical assistance.

References

Alley R.B., Marotzke J., Nordhaus W.D., *et al.* (2003) Abrupt climate change. *Science*, **299**, 2005–2010.

Berger A. (1978) Long-term variations of daily insolation and Quaternary climatic change. *Journal of Atmospheric and Oceanic Science*, **35**, 2362–2367.

Betts R.A., Cox P.M., Collins M., Harris P.P., Huntingford C. & Jones C.D. (2004) The role of ecosystem–atmosphere interactions in simulated Amazonian precipitation decrease and forest dieback under global climate warming. *Theoretical and Applied Climatology*, **78**(1–3), 157–175.

Brovkin V., Ganopolski A. & Svirezhev Yu. (1997) A continuous climate-vegetation classification for use in climate-biosphere studies. *Eco Model*, **101**, 251–261.

Brovkin V., Claussen M., Petoukhov V. & Ganopolski A. (1998) On the stability of the atmosphere-vegetation system in the Sahara/Sahel region. *Journal of Geophysical Research*, **103**(D24), 31613–31624.

Brovkin V., Bendtsen J., Claussen M., *et al.* (2002) Carbon cycle, vegetation and climate dynamics in the Holocene: Experiments with the CLIMBER-2 model. *Global Biogeochemical Cycles*, **16**(4), 1139, doi: 10.1029/2001GB001662.

Brovkin V., Levis S., Loutre M.F., *et al.* (2003) Stability analysis of the climate-vegetation system in the northern high latitudes. *Climatic Change*, **57**(1), 119–138.

Budyko M.I. (1969) The effect of solar radiation variations on the climate of the Earth. *Tellus*, **21**, 611–619.

Calov R., Ganopolski A., Petoukhov V., Claussen M. & Greve R. (2005) Transient simulation of the last glacial inception. Part I: Glacial inception as a bifurcation in the climate system. *Climate Dynamics*, **24**(6): 545–561, doi: 10.1007/s00382-005-0007-6.

Charney J.G. (1975) Dynamics of deserts and droughts in the Sahel. *Quarterly Journal of the Royal Meteorological Sociecty*, **101**, 193–202.

Cheddadi R., Yu G. & Guiot J., Harrison S.P. & Prentice I.C. (1997) The climate of Europe 6000 years ago. *Climate Dynamics*, **13**, 1–9.

Claussen M. (1994) On coupling global biome models with climate models. *Climate Ressearch*, **4**, 203–221.

Claussen M. (1997) Modelling biogeophysical feedback in the African and Indian Monsoon region. *Climate Dynamics*, **13**, 247–257.

Claussen M. (1998) On multiple solutions of the atmosphere–vegetation system in present-day climate. *Global Change Biology*, **4**, 549–559.

Claussen M. & Gayler V. (1997) The greening of Sahara during the mid-Holocene: results of an interactive atmosphere – biome model. *Global Ecology and Biogeography Letters*, **6**, 369–377.

Claussen M., Kubatzki C., Brovkin V., Ganopolski A., Hoelzmann P. & Pachur H.J. (1999) Simulation of an abrupt change in Saharan vegetation at the end of the mid-Holocene, *Geophysical Research Letters*, **24**(14), 2037–2040.

Claussen M., Brovkin V., Ganopolski A., Kubatzki C. & Petoukhov V. (2003) Climate change in northern Africa: the past is not the future. *Climatic Change*, **57**(1), 99–118.

Cox P.M., Betts R.A., Jones C.D., Spall S.A. & Totterdell I.J. (2000) Acceleration of global warming due to carbon-cycle feedbacks in a coupled climate model. *Nature*, **404**, 184–187.

Cremaschi M., Pelfini M. & Santilli M. (2006) *Cupressus dupreziana*: a dendroclimatic record for the middle–late Holocene in the central Sahara. *The Holocene*, **16**(2), 293–303, doi: 10.1191/0959683606hl926rr.

Crucifix M., Loutre M.F., Tulkens P., Fichefet T. & Berger A. (2002) Climate evolution during the Holocene: a study with an Earth system model of intermediate complexity. *Climate Dynamics*, **19**(1), 43–60.

Cubasch U., Meehl G.A., Boer G.J., *et al.* (2001) Projections of future climate change. In: *Climate Change 2001: The Scientific Basis. Contribution of Working Group I to the Third Assessment Report of the Intergovernmental Panel on Climate Change* (Eds J.T. Houghton, Y. Ding, D.J. Griggs, *et al.*), pp. 526–582. Cambridge University Press, Cambridge.

DeMenocal P.B. (2001) Cultural responses to climate change during the late Holocene. *Science*, **292**, 667–673.

DeMenocal P.B., Ortiz J., Guilderson T., *et al.* (2000) Abrupt onset and termination of the African Humid Period: rapid climate response to gradual insolation forcing. *Quaternary Science Review*, **19**, 347–361.

Eltahir E.A.B. & Gong C. (1995) Dynamics of wet and dry years in West Africa. *Journal of Climate*, **9**(5), 1030–1042.

Ganopolski A. & Rahmstorf S. (2001) Simulation of rapid glacial climate changes in a coupled climate model. *Nature*, **409**, 153–158.

Held H. & Kleinen T. (2004) Detection of climate system bifurcations by degenerate fingerprinting. *Geophysical Research Letters*, **31**, L23207.

Hoffman P.F. & Schrag D.P. (2002) The snowball Earth hypothesis: testing the limits of global change. *Terra Nova*, **14**(3), 129–155.

Hughen K.A., Eglinton T.I., Xu L. & Makou M. (2004) Abrupt tropical vegetation response to rapid climate changes. *Science*, **304**, 1955–1959.

Irizarry-Ortiz M.M., Wang G. & Elthair E.A.B. (2003) Role of the biosphere in the mid-Holocene climate of West Africa, *Journal of Geophysical Research*, **108**(D2), 4042.

Jansen E., Andersson C., Moros M., Nisancioglu K.H., Nyland B.F. & Telford R.J. (this volume) The early to mid-Holocene thermal optimum in the North Atlantic. In: *Global Warming and Natural Climate Variability: a Holocene Perspective* (Eds R.W. Battarbee & H.A. Binney), pp. 123–137. Blackwell, Oxford.

Jennerjahn T.C., Ittekkot V., Arz H.W., Behling H., Pätzold J. & Wefer G. (2004) Asynchronous terrestrial and marine signals of climate change during Heinrich events. *Science*, **306**, 2236–2239.

Jolly D., Prentice I.C., Bonnefille R., *et al.* (1998) Biome reconstruction from pollen and plant macrofossil data for Africa and Arabian Peninsula at 0 and 6k. *Journal of Biogeography*, **25**, 1007–1027.

Joussaume S., Taylor K.E., Braconnot P., *et al.* (1999) Monsoon changes for 6000 years ago: Results of 18 simulations from the Paleoclimate Modeling Intercomparison Project (PMIP). *Geophysical Research Letters*, **26**(7), 859–862.

Kleinen T., Held H. & Petschel-Held G. (2003) The potential role of spectral properties in detecting thresholds in the Earth system: application to the thermohaline circulation. *Ocean Dynamics*, **53**, 53–63, doi: 10.1007/s10236-002-0023-6.

Knutti R. & Stocker T.F. (2002) Limited predictability of the future thermohaline circulation close to an instability threshold. *Journal of Climate*, **15**, 179–186.

Kubatzki C. (2000) *Wechselwirkungen zwischen Klima und Landoberfläche im Holozän – Modellstudien*. Doktoral Thesis, Free University, Berlin.

Kubatzki C. & Claussen M. (1998) Simulation of the global biogeophysical interactions during the last glacial maximum. *Climate Dynamics*, **14**, 461–471.

Kutzbach J.E. & Guetter P.J. (1986) The influence of changing orbital parameters and surface boundary conditions on climate simulations for the past 18 000 years. *Journal of the Atmospheric Sciences*, **43**, 1726–1759.

Leemans R. & Cramer W. (1991) *The IIASA Database for Mean Monthly Values of Temperature, Precipitation, and Cloudiness on a Global Terrestrial Grid*. Research Report RR-91-18, International Institute for Applied Systems Analysis, Laxenburg.

Levis S., Foley J.A., Brovkin V. & Pollard D. (1999) On the stability of the high-latitude climate–vegetation system in a coupled atmosphere–biosphere model. *Global Ecology and Biogeography*, **8**, 489–500.

Liu Z., Wang Y., Gallimore R., Notaro M. & Prentice I.C. (2006) On the cause of abrupt vegetation collapse in North Africa during the Holocene: climate variability vs. vegetation feedback. *Geophysical Research Letters*, **33**, L22709.

MacDonald G.M., Velichko A.A., Kremenetski C.V., *et al.* (2000) Holocene tree-line history and climate change across Northern Eurasia. *Quaternary Research*, **53**, 302–311, doi: 10.1006/qres.1999.2123.

Malville J.M., Wendorf F., Mazar A.A. & Schild R. (1998) Megaliths and Neolithic astronomy in southern Egypt. *Nature*, **392**, 488–491.

Manabe S. & Stouffer R.J. (1988) Two stable equilibria of a coupled ocean–atmosphere model. *Journal of Climate*, **1**, 841–866.

Maslin M. (2004) Ecological versus climatic thresholds. *Science*, **306**, 2197–2198.

Mayewski P.A., Rohling E.E., Stager J.C., *et al.* (2004) Holocene climate variability. *Quaternary Research*, **62**, 243–255, doi: 10.1016/j.yqres.2004.07.001.

Mikolajewicz U., Crowley T.J., Schiller A. & Voss R. (1997) Modelling teleconnections between the North Atlantic and North Pacific during the Younger Dryas. *Nature*, **387**, 384–387.

Olson J.S., Watts J.A. & Allison L.J. (1985) *Major World Ecosystem Complexes Ranked by Carbon in Live Vegetation: a Database.* CDIAC Numerical Data Collection, NDP-017, Oak Ridge National Laboratory, Oak Ridge, 164 pp.

Otterman J. (1974) Baring high-albedo soils by overgrazing: a hypothesized desertification mechanism. *Science*, **186**, 531–533.

Oyama M.D. & Nobre C.A. (2003) A new climate-vegetation equilibrium state for Tropical South America. *Geophysical Research Letters*, **30**(23), 2199, doi: 10.1029/2003GL018600.

Pachur H.-J. (1999) Paläo-Environment und Drainagesysteme der Ostsahara im Spätpleistozän und Holozän. In: *Nordost-Afrika: Strukturen und Ressourcen; Ergebnisse aus dem Sonderforschungsbereich "Geowissenschaftliche Probleme in Ariden und Semiariden Gebieten"* (Eds E. Klitzsch & U. Thorweihe), pp. 366–455. Wiley–VCH, Weinheim.

Petit-Maire N. (1990) Will greenhouse green the Sahara? *Episodes*, **13**(2), 103–107.

Petit-Maire N. & Bryson R. (1998) Holocene climatic change and man in the Sahara. *International Conference "Tropical Climatology, Meteorology and Hydrology"*, Brussels, 22–24 May 1996, Proceedings edited by G. Demarée, J. Alexandre & M. De Dapper, pp. 51–67.

Petit-Maire N. & Guo Z. (1996) Mise en évidence de variations climatiques holocènes rapides, en phase dans les déserts actuels de Chine et du Nord de l'Afrique. *Sciences de la Terre et des Planètes*, **322**, 847–851.

Prentice I.C., Jolly D. & BIOME 6000 Members (2000) Mid-Holocene and glacial maximum vegetation geography of the northern continents and Africa. *Journal of Biogeography*, **27**, 507–519.

Rahmstorf S. & Ganopolski A. (1999) A. Long-term global warming scenarios computed with an efficient coupled climate model. *Climatic Change*, **43**, 353–367.

Raschke E., Ohmura A., Rossow W.B., *et al.* (2005) Cloud effects on the radiation budget based on ISCCP Data (1991 to 1995). *International Journal of Climatology*, **25**, 1103–1125, doi: 10.1002/joc.1147.

Renssen H., Brovkin V., Fichefet T. & Goosse H. (2003) Holocene climate instability during the termination of the African Humid Period, *Geophysical Research Letters*, **30**(4), 1184.

Rial J.A., Pielke Sr. R.A. Beniston M., *et al.* (2004) Nonlinearities, feedbacks and critical thresholds within the Earth's climate system. *Climatic Change*, **65**(1–2), 11–38.

Ruddiman W.F. (2003) The anthropocene greenhouse era began thousands of years ago. *Climatic Change*, **61**, 261–293.

Scheffer M., Carpenter S. Foley J.A., Folke C. & Walker B. (2001) Catastrophic shifts in ecosystems. *Nature*, **413**, 591–596.

Scheffer M., Homgreen M., Brovkin V. & Claussen M. (2005) Synergy between small- and large-scale feedbacks of vegetation on the water cycle. *Global Change Biology*, **11**, 1003–1012.

Schiller A., Mikolajewicz U. & Voss R. (1997) The stability of the North Atlantic thermohaline circulation in a coupled ocean–atmosphere general circulation model. *Climate Dynamics*, **13**, 325–347.

Sellers W.D. (1969) A global climatic model based on the energy balance of the Earth–atmosphere system. *Journal of Applied Meteorology*, **8**, 392–400.

Stocker T.F. (2000) Past and future reorganisations in the climate system, *Quaternary Science Reviews*, **19**, 301–319.

Stommel H. (1961) Thermohaline convection with two stable regimes of flow. *Tellus*, **13**, 224–230.

Wang G. (2004) A conceptual modeling study on biosphere–atmosphere interactions and its implications for physically based climate models. *Journal of Climate*, **17**(13), 2572–2583.

Wang G. & Eltahir E.A.B. (2000) Biosphere–atmosphere interactions over West Africa. 2. Multiple Equilibira. *Quarterly Journal of the Royal Meteorological Society*, **126**, 1261–1280.

Wang G. & Eltahir E.A.B. (2002) Response of the biosphere–atmosphere system in West Africa to CO_2 concentration changes. *Global Change Biology*, **8**, 1169–1182.

Wang Y., Mysak L.A., Wang Z. & Brovkin V. (2005) The greening of the McGill Paleoclimate Model: Part II. Simulation of Holocene millennial-scale natural climate changes. *Climate Dynamics*, **24**, 481–496.

Xue Y. & Shukla J. (1993) The influence of land surface properties on Sahel climate. Part I: Desertification. *Journal of Climate*, **6**, 2232–2245.

Zeng N. & Neelin J.D. (2000) The role of vegetation–climate interaction and interannual variability in shaping the African savanna. *Journal of Climate*, **13**, 2665–2670.

Zeng N., Neelin J.D., Lau K.-M. & Tucker C.J. (1999) Enhancement of interdecadal climate variability in the Sahel by vegetation interaction. *Science*, **286**, 1537–1540.

10 Holocene perspectives on future climate change

Raymond S. Bradley

Keywords

ENSO (or El Niño), monsoon (or Asian monsoon), climate forcing, drought

Introduction

Whatever anthropogenic climate changes occur in the future, they will be superimposed on, and interact with, underlying natural variability. Therefore, to anticipate future changes, we must understand how and why climates varied in the past. This requires well-dated records of forcing factors, as well as paleoclimate; both are available from a variety of natural archives. Relying on instrumental data to understand the spectrum of climate variability is completely inadequate; the record of large-scale (hemispheric or global) temperature extends back only ca. 150 years and the same can be said for most of the major modes of climate variability (Table 10.1). Even the longest instrumental records barely cover 300 years, and for most of the world such records are rarely longer than 120 years. This means that our understanding of climate system variability is largely limited to the interannual to decadal scale. To examine longer term (centennial to millennial-scale) variability requires much longer datasets. The Holocene provides a particularly relevant period for such an endeavor, as large-scale boundary conditions (continental ice extent, topography, sea-level) have remained very close to modern conditions for much of the Holocene, and low frequency (orbital) forcing is well understood for this period. Thus, Holocene paleoclimate data are able to resolve the full spectrum of climate variability and to place the limited instrumental records in a long-term perspective. This is particularly germane to the issue of anthropogenic climate change, as it provides a context for recent changes. Mann *et al.* (1999) argued that, for the Northern Hemisphere, the mean annual temperature of the past few decades of the 20th century was the highest for at least 1000 years, and this hypothesis has been supported by several later studies (National Research Council 2006). Moreover, Osborn and Briffa (2006) showed that both the magnitude and spatial

Table 10.1 Examples of some key climatic time series, based on instrumental records (Sources: http://www.cru.uea.ac.uk/cru/data/ and http://www.jisao.washington.edu/data.html)

Instrumental data	Starts	Indices	Starts
Central England temperature	1659	Southern Oscillation Index (Tahiti–Darwin pressure)	1866
Global and hemispheric temperatures	1850	Global SST ENSO index	1845
England and Wales precipitation	1766	Multi-variate ENSO	1950
		NAO	1821
		AMO	1860
		Northern Annular Mode/Arctic Oscillation	1899
		PDO	1900
		Southern Annular Mode/Antarctic Oscillation	1948
		Indian Monsoon	1844

extent of recent warming was unprecedented in the past millennium. Such studies provide a check on energy balance and general circulation models that seek to simulate temperature changes over multi-centennial periods in order to examine the role of natural versus anthropogenic forcing (e.g. Crowley 2000; Ammann *et al.* 2003; Goosse *et al.* 2005). All of these studies conclude that the rise in temperature during the 20th century (particularly the late 20th century) is unprecedented, and the spatial pattern of warming cannot be explained by natural forcing alone; greenhouse gases provide the necessary additional forcing to account for the recent changes in temperature. Thus, paleoclimate data contribute to both detection and attribution studies related to anthropogenic climate change.

Paleoclimate data are also important for the broader debate about the role of greenhouse gases in recent climate change. Those who seek to dismiss concerns over greenhouse gases frequently point to episodes in the past when temperatures were higher (even though these episodes were commonly just in a limited geographic region and/or in a particular season). For example, critics sometimes point to the early Holocene (a time when many Northern Hemisphere alpine glaciers disappeared and Arctic sea-ice was of limited extent) as "evidence" that human-induced climate change is a fiction and recent changes are merely a manifestation of underlying natural variability. Paleoclimatic research, however, clearly shows that orbital forcing (largely due to changes in precession) was responsible for the pattern of early Holocene warming. Furthermore, the early Holocene was a time of summer warmth mainly in the Northern Hemisphere; in that hemisphere, insolation receipts steadily declined during the Holocene. In Northern Hemisphere winters, and in the southern hemisphere summer and autumn (from December to June), insolation receipts in the early Holocene were below modern values, but increased through the Holocene. The rapid warming of the late 20th century is unrelated to this forcing; rather, the rate, seasonal pattern, and geographic extent of recent changes all point to greenhouse gases as the primary cause. Paleoclimate research can thus help to differentiate the cause and effect of climate changes, thereby elevating the discussion of strategies needed to deal with current environmental problems above the noisy distractions of those more interested in diverting attention from the issues at hand.

Although there has been an understandable emphasis on global (or hemispheric) temperature reconstructions, regional hydrologic variability is likely to be of most concern in the future, and so paleoclimatic reconstructions must pay particular attention to documenting past changes in precipitation and/or regional-scale water balance (as discussed by Verschuren and Charman, this volume). The Holocene record of climate variability is replete with examples of (largely unexplained) hydrologic instability. Multi-decadal- to multi-century-length droughts often started abruptly, were unprecedented (in the experience of societies at the time), and thus were highly disruptive to their agricultural, economic, and social foundations (AUP episodes: abrupt, unprecedented [in magnitude/duration], and persistent). One of the best documented examples is from Quintana Roo (Yucatan Peninsula), Mexico where there were a series of AUP droughts between ca. AD 800 and AD 1000. These were accompanied by severe economic and social disruption, leading to the "Classic Maya" collapse (Hodell *et al.* 1995, 2005). Future changes, that involve both natural and unprecedented anthropogenic forcing, may be just as disruptive if not more so. Furthermore, these changes will affect a world population that is expected to increase from ca. 6 billion people today to ca. 9–10 billion by 2070. In spite of technological advances, most of these people will continue to be subsistence or small-scale market agriculturalists, and will be just as vulnerable to climatic fluctuations as late prehistoric/early historic societies were. In an increasingly crowded world, moving on to greener pastures is not an option, making conflicts more likely (Campbell *et al.* 2007). Paleoclimatologists have a responsibility to ensure that the public, and politicians they elect, fully understand these issues so that they can better appreciate the consequences of inaction over controlling greenhouse gas emissions.

Holocene temperature changes

Long-term Holocene temperature changes largely reflect the dominant orbital forcing, with high northern latitudes receiving positive summer insolation anomalies of ca. 8–10 percent (outside the atmosphere) at the beginning of the Holocene (see Beer and van Geel, this volume). This led to warming that was especially pronounced in northern continental interiors, but there was also an increase in sea-surface temperatures in the North Atlantic, especially in the Norwegian Sea (Anderson *et al.* 2004) in the early Holocene. Orbital effects are clearly revealed, both in paleoclimate archives and in general circulation model simulations (see Crucifix, this volume). In the tropics, the changes in orbital forcing were manifested in large-scale circulation changes involving monsoon systems (as discussed further below) and these resulted in dramatic hydrologic changes with important consequences for regional hydrology and societies living in the regions (see Verschuren and Charman, this volume; Oldfield, this volume). In the Southern Hemisphere, the effect of an increase in summer insolation over the course of the Holocene was minimized by the strong buffering effect of the oceans and the limited land area in that hemisphere.

As summer temperatures declined in the Northern Hemisphere over the course of the Holocene, high mountain regions began to experience positive mass balance conditions; this led to an increase in ice extent in most glacierized areas, and the redevelopment of glaciers in many areas where ice had disappeared. The timing of this resurgence of glacial activity (neoglaciation) varies from region to region (perhaps related to winter balance conditions, and the height of mountain ranges, etc.) but in general glaciers across the Northern Hemisphere had a new lease of life by ca. 4500 years BP, and ice shelves re-formed along the northern coast of Ellesmere Island (Nunavut, Canada) (Bradley 1990). Many detailed studies of late Holocene glacier variations have been carried out (see Grove (2004), Birks, this volume, and Verschuren and Charman, this volume) and these show that the most recent advance of glaciers was the most extensive in almost all areas. Again, the exact timing of maximum ice advances varies from one region to another, but the period between AD 1250 and 1850 (ca. 150–750 years BP) brackets this interval, known as the Little Ice Age (LIA). Given that LIA ice extent in many mountainous areas was the most extensive since the Younger Dryas (that is, the first moraines encountered upslope from the Younger Dryas moraines commonly date from the LIA) it is clear that this latest neoglacial episode was the coldest of the Holocene. Since it was quite recent, we have abundant evidence of climatic conditions during this period, including documentary records, paintings, and even photographs from the mid-19th century. Many Dutch painters of the 17th century lived their entire lives during the height of LIA conditions and so, not surprisingly, their landscape paintings reflect the snow-covered hills, frozen lakes, and icy canals that were simply normal conditions in those days. By contrast, temperatures in the interval prior to the LIA were relatively warm in many areas, particularly in north-western Europe in High Medieval time (ca. AD 1100–1200) (Bradley *et al.* 2003a). Evidence for earlier warm intervals is also found (within the generally cool neoglacial period), but with limited geographic coverage it is not possible to assess the global nature of such episodes. Even for the most recent warm period, the global extent and magnitude of temperature anomalies remains unresolved (cf. Broecker 2001; Bradley *et al.* 2003b). One approach to a better understanding of past conditions is to use paleoclimate model simulations that reproduce known patterns of temperature anomalies as a tool to estimate larger scale (hemispheric or global) temperature changes (see Goosse *et al.*, this volume). Several multi-proxy temperature reconstructions extend back into Medieval time (e.g. Mann *et al.* 1999; Moberg *et al.* 2006; Osborn and Briffa 2006). All show a sharp increase in temperatures after the LIA, to levels at the end of the 20th century that were unprecedented in the previous millennium (National Research Council 2006). Consequently, in the brief interval of less than two centuries, the Northern Hemisphere (at least) has experienced the warmest and the coldest extremes of the late Holocene (Bradley 2000).

The temperature spectrum of Holocene temperature change is dominated at the low-frequency end by orbital forcing. At higher frequencies, there is evidence that solar forcing may have played a role, not necessarily by direct radiative forcing, but probably via stratospheric changes (related to the absorption of ultraviolet radiation) with subsequent effects on tropospheric circulation (Bradley 2003). Solar irradiance changes may also affect modes of climate variability (ENSO, NAO) with large-scale anomaly patterns as a result (discussed further, below). Consequently, solar effects seem to vary geographically, and since the forcing is small the response signal is easily lost within the ongoing variability of the climate system, a fact that has made detection of solar forcing a contentious issue. Moreover, explosive volcanic eruptions have also had radiative effects at various times, further confusing the picture. Both solar forcing and volcanic forcing appear to have played important roles at different times during the Holocene (see further discussion in Beer and van Geel, this volume). Thus the higher frequency part of the Holocene temperature spectrum is dominated by these two factors. Most recently, within the past ca. 50 years, these underlying natural forcings have become increasingly overwhelmed by anthropogenic forcing due to greenhouse gases. If these gases continue to increase as currently envisaged, model simulations indicate that orbitally modulated glaciations that might have been expected in the millennia ahead may not occur (as discussed further in Crucifix, this volume). To what extent anthropogenic land-use and land-cover changes might have affected global temperature prior to the past century (Ruddiman 2005) remains an intriguing open question, and an area of very active research involving both modeling and paleovegetation reconstructions (see Claussen, this volume; Birks, this volume; Oldfield, this volume).

Precipitation and hydrologic variability

Because of the importance of assessing the global effects of greenhouse gas increases, there has been much emphasis on reconstructing global or hemispheric mean temperature change, in order to place recent variations in a long-term perspective. Global warming now appears to be inevitable for the foreseeable future, but what is often overlooked are the associated changes in precipitation and hydrologic conditions that will certainly be associated with the rise in global temperature. Placing such changes in context is more difficult than reconstructing paleotemperatures because the correlation field for precipitation is much smaller (more geographically restricted) than for temperature. Consequently, precipitation and associated hydrologic variability must be considered on a regional scale.